Smart Antennas with MATLAB®

Frank B. Gross, Ph.D.

Second Edition

New York Chicago San Francisco
Athens London Madrid
Mexico City Milan New Delhi
Singapore Sydney Toronto

McGraw-Hill Education books are available at special quantity discounts to use as premiums and sales promotions or for use in corporate training programs. To contact a representative, please visit the Contact Us page at www.mhprofessional.com.

Smart Antennas with MATLAB®, Second Edition

Copyright © 2015, 2005 by McGraw-Hill Education. All rights reserved. Printed in the United States of America. Except as permitted under the United States Copyright Act of 1976, no part of this publication may be reproduced or distributed in any form or by any means, or stored in a data base or retrieval system, without the prior written permission of the publisher.

1 2 3 4 5 6 7 8 9 0   QVS/QVS   1 2 1 0 9 8 7 6 5

ISBN 978-0-07-182238-1
MHID 0-07-182238-0

This book is printed on acid-free paper.

Sponsoring Editor
Michael McCabe

Editing Supervisor
Stephen M. Smith

Production Supervisor
Lynn M. Messina

Project Manager
Namita Gahtori,
Cenveo® Publisher Services

Copy Editor
Ragini Pandey,
Cenveo Publisher Services

Proofreader
Surendra Nath Shivam,
Cenveo Publisher Services

Indexer
Cenveo Publisher Services

Art Director, Cover
Jeff Weeks

Composition
Cenveo Publisher Services

Originally published as *Smart Antennas for Wireless Communications: With MATLAB*, copyright © 2005.

Information contained in this work has been obtained by McGraw-Hill Education from sources believed to be reliable. However, neither McGraw-Hill Education nor its authors guarantee the accuracy or completeness of any information published herein, and neither McGraw-Hill Education nor its authors shall be responsible for any errors, omissions, or damages arising out of use of this information. This work is published with the understanding that McGraw-Hill Education and its authors are supplying information but are not attempting to render engineering or other professional services. If such services are required, the assistance of an appropriate professional should be sought.

About the Author
Frank B. Gross, Ph.D., is professor and department chair of Electrical Engineering at Georgia Southern University. His expertise lies in a variety of physics-based disciplines, including electromagnetics, antennas, propagation, metamaterials, and array processing.

Contents

Preface xiii
Acknowledgments xvii

1 **Introduction** 1
   1.1 What Is a Smart Antenna? 1
   1.2 Why Are Smart Antennas Emerging Now? ... 2
   1.3 What Are the Benefits of Smart Antennas? 3
   1.4 Smart Antennas Involve Many Disciplines ... 5
   1.5 Overview of the Book 6
   1.6 References 7

2 **Fundamentals of Electromagnetic Fields** 9
   2.1 Maxwell's Equations 9
   2.2 The Helmholtz Wave Equation 11
   2.3 Propagation in Rectangular Coordinates 12
   2.4 Propagation in Spherical Coordinates 14
   2.5 Electric Field Boundary Conditions 16
   2.6 Magnetic Field Boundary Conditions 19
   2.7 Planewave Reflection and Transmission
       Coefficients 21
       2.7.1 Normal Incidence 21
       2.7.2 Oblique Incidence 24
   2.8 Propagation over Flat Earth 28
   2.9 Knife-Edge Diffraction 30
   2.10 References 33
   2.11 Problems 33

3 **Antenna Fundamentals** 37
   3.1 Antenna Field Regions 37
   3.2 Power Density 39
   3.3 Radiation Intensity 42
   3.4 Basic Antenna Nomenclature 43
       3.4.1 Antenna Pattern 44
       3.4.2 Antenna Boresight 45
       3.4.3 Principal Plane Patterns 45
       3.4.4 Beamwidth 47
       3.4.5 Directivity 47
       3.4.6 Beam Solid Angle 48
       3.4.7 Gain 48
       3.4.8 Effective Aperture 49

	3.5	Friis Transmission Formula	50
	3.6	Magnetic Vector Potential and the Far Field	51
	3.7	Linear Antennas	52
		3.7.1 Infinitesimal Dipole	52
		3.7.2 Finite Length Dipole	54
	3.8	Loop Antennas	57
		3.8.1 Loop of Constant Phasor Current	57
	3.9	References	60
	3.10	Problems	60
4	**Array Fundamentals**		**63**
	4.1	Linear Arrays	63
		4.1.1 Two-Element Array	63
		4.1.2 Uniform N-Element Linear Array	65
		4.1.3 Uniform N-Element Linear Array Directivity	73
	4.2	Array Weighting	77
		4.2.1 Beamsteered and Weighted Arrays	86
	4.3	Circular Arrays	87
		4.3.1 Beamsteered Circular Arrays	88
	4.4	Rectangular Planar Arrays	89
	4.5	Fixed Beam Arrays	91
		4.5.1 Butler Matrices	91
	4.6	Fixed Sidelobe Canceling	93
	4.7	Retrodirective Arrays	96
		4.7.1 Passive Retrodirective Array	96
		4.7.2 Active Retrodirective Array	98
	4.8	References	99
	4.9	Problems	100
5	**Principles of Random Variables and Processes**		**103**
	5.1	Definition of Random Variables	103
	5.2	Probability Density Functions	104
	5.3	Expectation and Moments	105
	5.4	Common Probability Density Functions	107
		5.4.1 Gaussian Density	107
		5.4.2 Rayleigh Density	108
		5.4.3 Uniform Density	109
		5.4.4 Exponential Density	110
		5.4.5 Rician Density	111
		5.4.6 Laplace Density	112
	5.5	Stationarity and Ergodicity	113
	5.6	Autocorrelation and Power Spectral Density	114

5.7	Covariance Matrix	116
5.8	References	117
5.9	Problems	117

6 Propagation Channel Characteristics 119

6.1	Flat Earth Model	120
6.2	Multipath Propagation Mechanisms	122
6.3	Propagation Channel Basics	124
	6.3.1 Fading	125
	6.3.2 Fast Fading Modeling	126
	6.3.3 Channel Impulse Response	136
	6.3.4 Power Delay Profile	137
	6.3.5 Prediction of Power Delay Profiles	139
	6.3.6 Power Angular Profile	139
	6.3.7 Prediction of Angular Spread	142
	6.3.8 Power Delay–Angular Profile	145
	6.3.9 Channel Dispersion	146
	6.3.10 Slow-Fading Modeling	147
6.4	Improving Signal Quality	149
	6.4.1 Equalization	150
	6.4.2 Diversity	151
	6.4.3 Channel Coding	153
	6.4.4 MIMO	154
6.5	References	157
6.6	Problems	159

7 Angle-of-Arrival Estimation 163

7.1	Fundamentals of Matrix Algebra	163
	7.1.1 Vector Basics	164
	7.1.2 Matrix Basics	165
7.2	Array Correlation Matrix	169
7.3	AOA Estimation Methods	171
	7.3.1 Bartlett AOA Estimate	171
	7.3.2 Capon AOA Estimate	172
	7.3.3 Linear Prediction AOA Estimate	175
	7.3.4 Maximum Entropy AOA Estimate	176
	7.3.5 Pisarenko Harmonic Decomposition AOA Estimate	177
	7.3.6 Min-Norm AOA Estimate	178
	7.3.7 MUSIC AOA Estimate	179
	7.3.8 Root-MUSIC AOA Estimate	183
	7.3.9 ESPRIT AOA Estimate	189
7.4	References	193
7.5	Problems	194

Contents

8 Smart Antennas **197**
   8.1 Introduction 197
   8.2 The Historical Development
       of Smart Antennas 199
   8.3 Fixed Weight Beamforming Basics 201
       8.3.1 Maximum Signal-to-Interference
             Ratio 201
       8.3.2 Minimum Mean-Square Error 207
       8.3.3 Maximum Likelihood 211
       8.3.4 Minimum Variance 212
   8.4 Adaptive Beamforming 215
       8.4.1 Least Mean Squares 216
       8.4.2 Sample Matrix Inversion 218
       8.4.3 Recursive Least Squares 223
       8.4.4 Constant Modulus 227
       8.4.5 Least Squares Constant Modulus ... 229
       8.4.6 Conjugate Gradient Method 234
       8.4.7 Spreading Sequence Array
             Weights 238
       8.4.8 Description of the New
             SDMA Receiver 240
   8.5 References 248
   8.6 Problems 250

9 Direction Finding **255**
   by Robert Kellogg
   9.1 Loop Antennas 255
       9.1.1 Early Direction Finding
             with Loop Antennas 255
       9.1.2 Loop Antenna Fundamentals 256
       9.1.3 Vertical Loop Antennas 257
       9.1.4 Vertical Loop Matched
             Polarization 258
       9.1.5 Vertical Loop with Polarized
             Signal 258
       9.1.6 Cross-Loop Array and the
             Bellini-Tosi Radio Goniometer 260
       9.1.7 Loop Array Calibration 265
   9.2 Adcock Dipole Antenna Array 267
       9.2.1 Watson-Watt DF Algorithm 268
   9.3 Modern DF Applied to Adcock
       and Cross-Loop Arrays 270

	9.4	Geolocation	272
		9.4.1 Stansfield Algorithm	272
		9.4.2 Weighted Least-Square Solution	275
		9.4.3 Confidence Error Ellipse	276
		9.4.4 Mahalanobis Statistics	279
	9.5	References/Notes	281
	9.6	Problems	283

10 Vector Sensors — 285

by Jeffrey Connor

- 10.1 Introduction — 285
- 10.2 Vector Sensor Antenna Array Response — 289
  - 10.2.1 Single Vector Sensor Steering Vector Derivation — 289
  - 10.2.2 Vector Sensor Array Signal Model and Steering Vector — 293
- 10.3 Vector Sensor Direction Finding — 294
  - 10.3.1 Cross-Product Direction Finding — 294
  - 10.3.2 Super Resolution Direction Finding — 297
- 10.4 Vector Sensor Beamforming — 300
- 10.5 Vector Sensor Cramer-Rao Lower Bound — 304
- 10.6 Acknowledgments — 308
- 10.7 References — 308
- 10.8 Problems — 309

11 Smart Antenna Design — 313

by Jeffrey Connor

- 11.1 Introduction — 313
- 11.2 Global Optimization Algorithms — 315
  - 11.2.1 Description of Algorithms — 318
- 11.3 Optimizing Smart Antenna Arrays — 334
  - 11.3.1 Thinning Array Elements — 335
  - 11.3.2 Optimizing Array Element Positions — 340
- 11.4 Adaptive Nulling — 344
- 11.5 NEC in Smart Antenna Design — 348
  - 11.5.1 NEC2 Resources — 348
  - 11.5.2 Setting Up the NEC2 Simulation — 349
  - 11.5.3 Integrating NEC2 with MATLAB — 353
  - 11.5.4 Example: Simple Half-Wavelength Dipole Antenna — 354
  - 11.5.5 Monopole Array Example — 356

11.6 Evolutionary Antenna Design 358
11.7 Current and Future Trends 363
       11.7.1 Reconfigurable Antennas
               and Arrays 363
       11.7.2 Open-Source Computational
               Electromagnetics Software 364
11.8 References 365
11.9 Problems 367

Index 371

Preface

We live in an age of "smart" technologies. We have smart cards, smartphones, smart environments, smart thermometers, smart buildings, smart sensors, smart whiteboards, and now smart antennas. Smart antennas previously were called *adaptive arrays* and were reported on extensively in the literature. Many published papers that date back to the 1950s demonstrate the application of adaptive algorithms to enhance array performance. With evolving technological advances, we now have smart antennas that include arrays but often don't resemble arrays at all. Many of these modern antennas have been reported in my book *Frontiers in Antennas: Next Generation Design & Engineering* (McGraw-Hill, 2011). My definition of a smart antenna is: "A self-optimizing intelligently interactive antenna." Sometimes these smart antennas are referred to as *cognitive antennas*. Overall, these antennas range from simple beamsteering arrays to self-healing antennas. In other words, they can be relatively dumb to extremely smart.

Smart antennas can allow the user to search for specific signals even in a background of noise and interference. In the case of self-healing antennas, they can modify their performance if and when damaged so as to minimize the performance loss. Some single smart antennas can measure all six vector components of the incident electric and magnetic fields in order to determine the angle of arrival (AOA). Whether simple or complex, a smart antenna always uses an algorithm to optimize performance under changing conditions.

The title for this new edition of *Smart Antennas* has dropped "for Wireless Communications" because I think it is too restrictive for this ever-evolving science. Smart antennas are applied to many disciplines far beyond wireless communications. That being said, this revision has added three new chapters: Chap. 9, Direction Finding; Chap. 10, Vector Sensors; and Chap. 11, Smart Antenna Design.

I have an extensive background in signal processing, radar, communications, and electromagnetics. Having worked in industry and in academia, I have had difficulty finding a smart antenna text that can be equally shared by both communities. In addition, only a few books in print even address the subject of smart antennas. Because smart antennas involve an amalgamation of many different disciplines, a background in each related area must be possessed in order to appreciate this topic as a whole. Thus, one overriding goal of this text is to present the fundamentals of several different science and engineering disciplines. The intent is to show how all of these disciplines converge in the study of smart antennas. In order to understand smart antenna behavior, one must be versed in various topics such as electromagnetics, antennas, array processing, propagation, channel characterization, random processes, spectral estimation, and adaptive methods. Thus, the book lays a background in each of these disciplines before tackling smart antennas in particular.

This text is organized into 11 chapters. Chapter 1 gives the background, motivation, justification, and benefits for the study of this important topic. Chapter 2 provides a summary of electromagnetics, reflection, diffraction, and propagation. These concepts help in understanding the path gain factor, coverage diagrams, and fading. This material will be used later in order to understand the nature of multipath propagation and the phase relationship between different array elements. Chapter 3 deals with general antenna theory including antenna metrics such as beamwidth, gain, principal plane patterns, and effective apertures. The Friis transmission formula is discussed because it aids in the understanding of spherical spreading and reception issues. Finally, the specific behavior of dipoles and loops is addressed. The goal is to help the reader gain a basic understanding of how individual antenna elements affect the behavior of arrays. Chapter 4 addresses the subject of antenna arrays. Array phenomenology is addressed in order to help in understanding the relationship of array shape to beam patterns. Linear, circular, and planar arrays are discussed. Array weighting or "shading" is explored in order to help the reader understand how array weights influence the radiation pattern. Specific arrays are discussed, such as fixed beam arrays, beamsteered arrays, Butler matrices, and retrodirective arrays. This treatment of array behavior is invaluable in understanding smart antenna behavior and limitations. Chapter 5 lays a foundation in random variables and processes. This is necessary because multipath signals and noise are characterized by random behavior. Also, channel delays and angles of arrival tend to be random variables. Thus, a minimum background in random processes must be established in order to understand the nature of arriving signals and how to process array inputs. Many smart antenna applications require the computation of the array correlation matrix. The topics of ergodicity and stationarity are discussed in order to help the reader understand the nature of the correlation matrix, prediction of the matrix, and what information can be extracted in order to compute the optimum array weights. It is assumed that students taking a smart antenna class are already familiar with random processes. However, this chapter is provided in order to help undergird concepts that are addressed in later chapters. Chapter 6 addresses propagation channel characteristics.

Such critical issues as fading, delay spread, angular spread, dispersion, and equalization are discussed. In addition, MIMO (multiple-input multiple-output) is briefly defined and addressed. If one understands the nature of multipath fading, one can better design a smart antenna that minimizes the deleterious effects. Chapter 7 discusses numerous different spectral estimation methods. The topics range from Bartlett beamforming to Pisarenko harmonic decomposition to eigenstructure methods such as MUSIC and ESPRIT. This chapter helps in understanding the many useful properties of the array correlation matrix as well as demonstrating that AOA can be predicted with greater accuracy than the array resolution can allow. In addition, many of the techniques discussed in Chap. 7 help aid in understanding adaptive methods. Chapter 8 shows the historical development of smart antennas and how weights can be computed by minimizing cost functions. The minimum mean-squared-error method lends itself to understanding iterative methods such as least mean squares. Several iterative methods are developed and discussed, and their performance is contrasted with numerous examples. Current popular methods such as the constant modulus algorithm, sample matrix inversion, and conjugate gradient method are explored. A waveform diversity concept is discussed wherein a different waveform is applied to each array element in order to determine angles of arrival. This method has application to both MIMO communications and MIMO radar. Chapter 9 addresses direction finding and angle-of-arrival (AOA) estimation. In addition, this chapter addresses the practicals on how to determine AOA accuracies given that noise is present in the system. Chapter 10 addresses vector antennas. Vector antennas are a relatively new antenna that is capable of measuring electric and magnetic field vectors in three dimensions. If we know precisely the incident electric and magnetic field vectors, we can determine the AOA with a single antenna. Chapter 11 addresses reconfigurable antennas in which the antenna has multiple parts that are capable of being switched on or off in order to allow the antenna to physically modify its geometry and thus modify its behavior. This can allow a damaged antenna to heal itself or it can allow a single antenna to modify its beam pattern. One method of optimizing reconfigurable antennas is to use a genetic algorithm.

Numerous MATLAB examples are given in the text and most homework questions require the use of MATLAB. It is felt that if the student can program smart antenna algorithms in MATLAB, a further depth of understanding can be achieved. All MATLAB codes written and used for the completion of the book are available for use by the reader. It is intended that these codes serve as templates for further work.

Examples and Solutions on the Web

The MATLAB codes are available at www.mhprofessional.com/gross. The codes are organized into three categories: examples, figures, and problems. There are codes produced in conjunction with example problems in most chapters. The codes are denoted by sa_ex#_#.m. For example, Example 8.4 will have an associated MATLAB code labeled sa_ex8_4.m. There are codes used to produce most of the figures in the book. These codes are denoted by sa_fig#_#.m. Thus, if one is interested in replicating a book figure or modifying a book figure, the code associated with that figure can be downloaded and used. For example, Fig. 2.1 has an associated code called sa_fig2_1.m. There are also codes used to produce most of the homework solutions. These codes are denoted by sa_prob#_#.m. For example, homework Prob. 5.2 will have a solution partly generated with a MATLAB code labeled sa_prob5_2.m. Students will have download access to the codes associated with figures and examples. Instructors will have access to the codes associated with figures, examples, and problem solutions.

For Instructors

This book can be used as a one-semester graduate or advanced undergraduate text. The extent to which beginning chapters are treated depends on the background of the students. It is normally assumed that the prerequisites for the course are an undergraduate course in communications, an undergraduate course in advanced electromagnetics, and a basic course in random processes. The students may be able to take this course, without a prior course in random processes, if they have a background in random processes learned in an undergraduate course in communications. It is assumed that all students have had an advanced engineering math course that covers matrix algebra, including the calculation of eigenvalues and eigenvectors.

I hope that this book will open new doors of understanding for the uninitiated and serve as good resource material for the practitioner. Enjoy.

Frank B. Gross, Ph.D.

Acknowledgments

I wish to acknowledge the tremendous help and support of many individuals. A special thanks goes to the very supportive staff at McGraw-Hill. The senior editor, Michael McCabe, has fully supported the second edition of *Smart Antennas*. I am thankful for the initial expert review and helpful comments from Professor Fernando Teixeira of the Ohio State University. His comments and insights have helped to strengthen the book. I am appreciative of my former EEL 5930 Smart Antennas class. The first book was road-tested on them and they happily survived the experience. A deep debt of gratitude is owed to Jeff Connor and Robert Kellogg for adding the new chapter material on direction finding, vector antennas, and reconfigurable antennas.

Smart Antennas with MATLAB®

CHAPTER 1
Introduction

This text has been written in response to the recent extreme interest in the rapidly growing field of smart antennas. Although some of the principles of smart antennas have been around for over 50 years, new wireless applications demanding smart antenna technology are growing exponentially. In addition, the latest algorithms that control smart antennas have matured to the point of being extremely effective in dynamic and dispersive multipath environments. Thus, smart antennas are a critical adjunct for increasing the performance of a myriad of wireless applications. This new technology has a major role in all forms of wireless systems ranging from mobile cellular to *personal communications services* (PCS) to radar. This text does not address specific applications as much as it introduces the reader to the basic principles that underlie smart antennas. A solid foundation is necessary in order to understand the full applicability and benefit of this rapidly growing technology.

1.1 What Is a Smart Antenna?

The term "smart antenna" generally refers to any antenna array, terminated in a sophisticated signal processor, which can adjust or adapt its own beam pattern in order to emphasize signals of interest and to minimize interfering signals.

Smart antennas generally encompass both switched beam and beamformed adaptive systems. Switched beam systems have several available fixed beam patterns. A decision is made as to which beam to access, at any given point in time, based on the requirements of the system. Beamformed adaptive systems allow the antenna to steer the beam to any direction of interest while simultaneously nulling interfering signals. The smart antenna concept is opposed to the fixed beam "dumb antenna," which does not attempt to adapt its radiation pattern to an ever-changing electromagnetic environment. In the past, smart antennas have alternatively been labeled adaptive arrays or digital beamforming arrays. This new terminology reflects our penchant for "smart" technologies and more accurately identifies an adaptive array that is controlled by sophisticated signal processing. Figure 1.1 contrasts two antenna arrays. The first is a traditional, fixed beam array where the mainlobe can

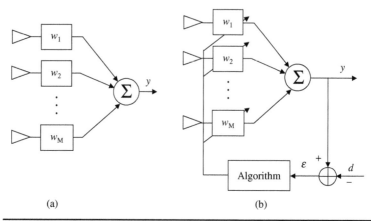

FIGURE 1.1 (a) Traditional array, (b) smart antenna.

be steered, by defining the fixed array weights $\bar{w}$. However, this configuration is neither smart nor adaptive.

The second array in this figure is a smart antenna designed to adapt to a changing signal environment in order to optimize a given algorithm. An optimizing criterion, or cost function, is normally defined based on the requirements at hand. In this example, the cost function is defined as the *magnitude of the error squared*, $|e|^2$, between the desired signal d and the array output y. The array weights $\bar{w}$ are adjusted until the output matches the desired signal and the cost function is minimized. This results in an optimum radiation pattern.

1.2 Why Are Smart Antennas Emerging Now?

The rapid growth in demand for smart antennas is fueled by two major reasons. First, the technology for high-speed *analog-to-digital converters* (ADC) and high-speed digital signal processing is burgeoning at an alarming rate. Even though the concept of smart antennas has been around since the late 1950s [1–3], the technology required in order to make the necessary rapid and computationally intense calculations has only emerged recently. Early smart antennas, or adaptive arrays, were limited in their capabilities because adaptive algorithms were usually implemented in analog hardware. With the growth of ADC and *digital signal processing* (DSP); what was once performed in hardware can now be performed digitally and quickly [4]. ADCs, which have resolutions that range from 8 to 24 bits, and sampling rates approaching 20 *gigasamples per second* (GSa/s), are now a reality [5]. In time, superconducting data converters will be able to sample data at rates up to 100 GSa/s [6]. This makes the direct digitization of most *radio frequency* (RF) signals possible in many wireless applications. At the very least, ADC can be applied to IF frequencies in higher RF frequency applications. This allows most of the signal processing to be defined in software near the front end of the receiver. In addition, DSP can be implemented with high-speed parallel processing using *field programmable gate arrays* (FPGA). Current commercially available FPGAs have speeds of up to 256 BMACS.[1] Thus, the benefits of smart antenna integration will only flourish, as the exponential growth in the enabling digital technology continues.

Second, the global demand for all forms of wireless communication and sensing continues to grow at a rapid rate. Smart antennas are the practical realization of the subject of adaptive array signal processing and have a wide range of interesting applications. These applications include, but are not limited to, the following: mobile wireless communications [7], software-defined radio [8, 9], *wireless local area networks* (WLAN) [10], *wireless local loops* (WLL) [11], mobile Internet, *wireless metropolitan area networks* (WMAN) [12], satellite-based personal communications services, radar [13], ubiquitous radar [14], many forms of remote sensing, *mobile ad hoc networks* (MANET) [15], high data rate communications [16], satellite communications [17], *multiple-in-multiple-out* (MIMO) systems [18], and waveform diversity systems [19].

The rapid growth in telecommunications alone is sufficient to justify the incorporation of smart antennas to enable higher system capacities and data rates. In 2018, $1.94 trillion are projected to be spent globally on telecommunications.

1.3 What Are the Benefits of Smart Antennas?

Smart antennas have numerous important benefits in wireless applications as well as in sensors such as radar. In the realm of mobile wireless applications, smart antennas can provide higher system capacities by directing narrow beams toward the users of interest, while nulling other users not of interest. This allows for higher signal-to-interference ratios, lower power levels, and permits greater frequency reuse within the same cell. This concept is called *space division multiple access* (SDMA). In the United States, most base stations sectorize each cell into three 120° swaths as seen in Fig. 1.2*a*. This allows the system capacity to potentially triple within a single cell because users in each of the three sectors can share the same spectral resources. Most base stations can be modified to include smart antennas within each sector. Thus the 120° sectors can be further subdivided as shown in Fig. 1.2*b*. This further subdivision enables the use

---

[1]BMACS: Billion multiply accumulates per second.

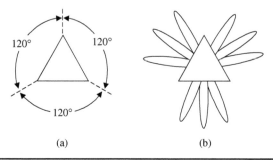

FIGURE 1.2 (a) Sectorized, (b) smart antennas.

of lower power levels, and provides for even higher system capacities and greater bandwidths.

Another benefit of smart antennas is that the deleterious effects of multipath can be mitigated. As is discussed in Chap. 8, a constant modulus algorithm, which controls the smart antenna, can be implemented in order to null multipath signals. This will dramatically reduce fading in the received signal. Higher data rates can be realized because smart antennas can simultaneously reduce both co-channel interference and multipath fading. Multipath reduction not only benefits mobile communications but also applies to many applications of radar systems.

Smart antennas can be used to enhance *direction-finding* (DF) techniques by more accurately finding *angles of arrival* (AOA) [20]. A vast array of spectral estimation techniques can be incorporated, which are able to isolate the AOA with an angular precision that exceeds the resolution of the array. This topic is discussed in detail in Chap. 7. The accurate estimation of AOA is especially beneficial in radar systems for imaging objects or accurately tracking moving objects. Smart antenna DF capabilities also enhance geo-location services enabling a wireless system to better determine the location of a particular mobile user. Additionally, smart antennas can direct the array main beam toward signals of interest even when no reference signal or training sequence is available. This capability is called *blind adaptive beamforming*.

Smart antennas also play a role in MIMO communications systems [18] and in waveform diverse MIMO radar systems [21, 22]. Because diverse waveforms are transmitted from each element in the transmit array and are combined at the receive array, smart antennas play a role in modifying radiation patterns in order to best capitalize on the presence of multipath. With MIMO radar, the smart antenna can exploit the independence between the various signals at each array element in order to use target scintillation for improved performance, increase array resolution, and mitigate clutter [19].

Many smart antenna benefits are discussed in detail in Chaps. 7 and 8. In summary, let us list some of the numerous potential benefits of smart antennas.

- Improved system capacities
- Higher permissible signal bandwidths
- Space division multiple access (SDMA)
- Higher signal-to-interference ratios
- Increased frequency reuse
- Sidelobe canceling or null steering
- Multipath mitigation
- Constant modulus restoration to phase modulated signals
- Blind adaptation
- Improved angle-of-arrival estimation and direction finding
- Instantaneous tracking of moving sources
- Reduced speckle in radar imaging
- Clutter suppression
- Increased degrees of freedom
- Improved array resolution
- MIMO compatibility in both communications and radar

1.4 Smart Antennas Involve Many Disciplines

The general subject of smart antennas is the necessary union between such related topics as electromagnetics, antennas, propagation, communications, random processes, adaptive theory, spectral estimation, and array signal processing. Figure 1.3 demonstrates the important relationship between each discipline. Many previous attempts have been made to explain smart antennas from the background of a single discipline; however, this myopic approach appeals only to small segments of the engineering community and does not yield a full appreciation for this valuable subject. No single engineering discipline can be the sole province of this rapidly growing field. The subject of smart antennas transcends specific applications and thus merits a more global treatment. In order to fundamentally understand smart antennas, one must be versed in many varied and related topics. One could argue that some of the disciplines displayed in Fig. 1.3 can be merged to create a smaller list. However, the specialist, in each of these specific disciplines, brings a unique contribution to the general field of smart antennas. Thus, this book is an attempt to preview all of these disciplines and to relate each of them to the subject as a whole.

6 Chapter One

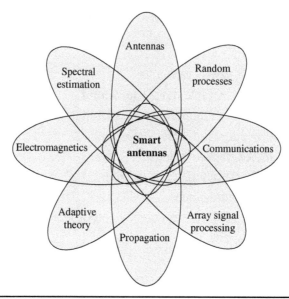

FIGURE 1.3 Venn diagram relating various disciplines to smart antennas.

1.5 Overview of the Book

As has been mentioned, this text has been written in order to provide the reader with a broad and fundamental understanding of the subject of smart antennas. A foundation of basic principles is laid in each of the supporting disciplines that relate to smart antennas as a whole. These various disciplines are outlined chapter by chapter in this book. Electromagnetics, antennas, and arrays are discussed in Chaps. 2 to 4. This foundation is critical in better understanding the physics supporting smart antenna behavior. Random processes along with specific probability distributions are discussed in Chap. 5. This aids in the understanding of noise, channels, delay spread, angular spread, and channel profiles. Propagation channel characterization is discussed in Chap. 6. This helps the reader understand the basics of multipath and fading. This will prove invaluable in understanding the limitations of smart antenna performance. AOA estimation is discussed in Chap. 7. This chapter explores the various techniques used to estimate AOA and lays a foundation for understanding eigenstructure methods. Smart antennas are discussed at length in Chap. 8. The history of smart antenna development is discussed. Numerous adaptive algorithms are explained and explored. The overall topic of direction finding (DF) and geolocation is discussed in Chap. 9. This is a valuable topic under the umbrella of smart antennas in which a user can locate the source of RF signals. Many smart antenna methods are employed in DF and geolocation. Even single antennas can be used to locate the angle of arrival of a source. Chapter 10 addresses vector sensor processing, where single antennas can be used to determine angles of arrival for signals over all possible angles of arrival. Vector sensors typically employ three nested loops and three nested dipoles, and use polarization information to determine signal direction. Lastly, Chap. 11 covers smart antenna design using a full array of optimization algorithms such as particle swarm and genetic.

The intent is that the interested reader can use the principles outlined in this book in order to become very knowledgeable in the fundamentals of smart antennas. In addition, numerous MATLAB examples are given. It is believed that a concept can be more fully understood if it can be modeled in software and solutions can be visualized. Numerous MATLAB script files are provided at www.mhprofessional.com/gross so that the student can understand how to program these various algorithms and observe their performance. It is ultimately hoped, with this fundamental understanding, that the student can use this information as a springboard for more advanced work in this fascinating field. The chapters are summarized as follows:

Chapter 1: Introduction
Chapter 2: Fundamentals of Electromagnetic Fields
Chapter 3: Antenna Fundamentals
Chapter 4: Array Fundamentals
Chapter 5: Principles of Random Variables and Processes
Chapter 6: Propagation Channel Characteristics
Chapter 7: Angle-of-Arrival Estimation
Chapter 8: Smart Antennas
Chapter 9: Direction Finding
Chapter 10: Vector Sensors
Chapter 11: Smart Antenna Design

1.6 References

1. Van Atta, L. "Electromagnetic Reflection," U.S. Patent 2908002, Oct. 6, 1959.
2. Howells, P. "Intermediate Frequency Sidelobe Canceller," U.S. Patent 3202990, Aug. 24, 1965.
3. Howells, P. "Explorations in Fixed and Adaptive Resolution at GE and SURC," *IEEE Transactions on Antenna and Propagation*, Special Issue on Adaptive Antennas, Vol. AP-24, No. 5, pp. 575–584, Sept. 1976.
4. Walden, R. H. "Performance Trends for Analog-to-Digital Converters," *IEEE Commn. Mag.*, pp. 96–101, Feb. 1999.
5. Litva, J., and T. Kwok-Yeung Lo, *Digital Beamforming in Wireless Communications*, Artech House, Boston, MA, 1996.
6. Brock, D. K., O. A. Mukhanov, and J. Rosa, "Superconductor Digital RF Development for Software Radio," *IEEE Commun. Mag.*, pp. 174, Feb. 2001.
7. Liberti, J., and T. Rappaport, *Smart Antennas for Wireless Communications: IS-95 and Third Generation CDMA Applications*, Prentice Hall, New York, 1999.
8. Reed, J., *Software Radio: A Modern Approach to Radio Engineering*, Prentice Hall, New York, 2002.

9. Mitola, J., "Software Radios," *IEEE Commun. Mag.*, May 1995.
10. Doufexi, A., S. Armour, A. Nix, P. Karlsson, and D. Bull, "Range and Through put Enhancement of Wireless Local Area Networks Using Smart Sectorised Antennas," *IEEE Transactions on Wireless Communications*, Vol. 3, No. 5, pp. 1437–1443, Sept. 2004.
11. Weisman, C., *The Essential Guide to RF and Wireless*, 2d ed., Prentice Hall, New York 2002.
12. Stallings, W., *Local and Metropolitan Area Networks*, 6th ed., Prentice Hall, New York, 2000.
13. Skolnik, M., *Introduction to Radar Systems*, 3d ed., McGraw-Hill, New York, 2001.
14. Skolnik, M., "Attributes of the Ubiquitous Phased Array Radar," *IEEE Phased Array Systems and Technology Symposium*, Boston, MA, Oct. 14–17, 2003.
15. Lal, D., T. Joshi, and D. Agrawal, "Localized Transmission Scheduling for Spatial Multiplexing Using Smart Antennas in Wireless Adhoc Networks," *13th IEEE Workshop on Local and Metropolitan Area Networks*, pp. 175–180, April 2004.
16. Wang Y., and H. Scheving, "Adaptive Arrays for High Rate Data Communications," 48th *IEEE Vehicular Technology Conference*, Vol. 2, pp. 1029–1033, May 1998.
17. Jeng, S., and H. Lin, "Smart Antenna System and Its Application in Low-Earth-Orbit Satellite Communication Systems," *IEE Proceedings on Microwaves, Antennas, and Propagation*, Vol. 146, No. 2, pp. 125–130, April 1999.
18. Durgin, G., *Space-Time Wireless Channels*, Prentice Hall, New York, 2003.
19. Ertan, S., H. Griffiths, M. Wicks, et al., "Bistatic Radar Denial by Spatial Waveform Diversity," *IEE RADAR 2002*, Edinburgh, pp. 17–21, Oct. 15–17, 2002.
20. Talwar, S., M. Viberg, and A. Paulraj, "Blind Estimation of Multiple Co-Channel Digital Signals Using an Atnenna Array," *IEEE Signal Processing Letters*, Vol. 1, No. 2, Feb. 1994.
21. Rabideau, D., and P. Parker, "Ubiquitous MIMO Multifunction Digital Array Radar," *IEEE Signals, Systems, and Computers*, 37th Asilomar Conference, Vol. 1, pp. 1057–1064, Nov. 9–12, 2003.
22. Fishler, E., A. Haimovich, R. Blum, et al., "MIMO Radar: An Idea Whose Time Has Come," *Proceedings of the IEEE Radar Conference*, pp. 71–78, April 26–29, 2004.

CHAPTER 2
Fundamentals of Electromagnetic Fields

The foundation for all wireless communications is based on understanding the radiation and reception of wireless antennas, as well as the propagation of electromagnetic fields between these antennas. Regardless of the form of wireless communications used or the particular modulation scheme chosen, wireless communication is based on the laws of physics. Radiation, propagation, and reception can be explained through the use of Maxwell's four foundational equations.

2.1 Maxwell's Equations

It was the genius of James Clerk Maxwell[1] to combine the previous work of Michael Faraday,[2] Andre Marie Ampere,[3] and Carl Fredrick Gauss[4] into one unified electromagnetic theory. (Some very useful references describing electromagnetics basics are Sadiku [1], Hayt [2], and Ulaby [3].) Maxwell's equations are given as follows:

$$\text{Faraday's law} \quad \nabla \times \overline{E} = -\frac{\partial \overline{B}}{\partial t} \quad (2.1)$$

---

[1] James Clerk Maxwell (1831–1879): A Scottish born physicist who published his treatise on electricity and magnetism in 1873.
[2] Michael Faraday (1791–1867): An English born chemist and experimenter who connected time-varying magnetic fields with induced currents.
[3] Andre Marie Ampere (1775–1836): A French born physicist who found that current in one wire exerts force on another wire.
[4] Carl Fredrick Gauss (1777–1855): A German born mathematical genius who helped establish a worldwide network of terrestrial magnetism observation points.

Chapter Two

$$\text{Ampere's law} \quad \nabla \times \bar{H} = -\frac{\partial \bar{D}}{\partial t} + \bar{J} \quad (2.2)$$

$$\text{Gauss's law} \begin{cases} \nabla \cdot \bar{D} = \rho & (2.3) \\ \nabla \cdot \bar{B} = 0 & (2.4) \end{cases}$$

where $\bar{E}$ = electric field intensity vector (V/m)
$\bar{D}$ = electric flux density vector (C/m²)
$\bar{H}$ = magnetic field intensity vector (A/m)
$\bar{B}$ = magnetic flux density vector (W/m²)
$\bar{J}$ = volume current density vector (A/m²)
ρ = volume charge density (C/m³)

The electric flux density and the electric field intensity are related through the permittivity of the medium as given by

$$\bar{D} = \varepsilon \bar{E} \quad (2.5)$$

The magnetic flux density and the magnetic field intensity are related through the permeability of the medium as given by

$$\bar{B} = \mu \bar{H} \quad (2.6)$$

where $\varepsilon = \varepsilon_r \varepsilon_0$ = permittivity of the medium (F/m)
ε_0 = permittivity of free space = 8.85×10^{-12} F/m
$\mu = \mu_r \mu_0$ = permeability of the medium (H/m)
μ_0 = permeability of free space = $4\pi \times 10^{-7}$ H/m

With no sources present and expressing the fields as the phasors $\bar{E}_s$ and $\bar{H}_s$, Maxwell's equations can then be written in phasor form as

$$\nabla \times \bar{E}_s = -j\omega \mu \bar{H}_s \quad (2.7)$$

$$\nabla \times \bar{H}_s = (\sigma + j\omega \varepsilon)\bar{E}_s \quad (2.8)$$

$$\nabla \cdot \bar{E}_s = 0 \quad (2.9)$$

$$\nabla \cdot \bar{H}_s = 0 \quad (2.10)$$

The phasor form of Maxwell's equations assumes that the fields are expressed in complex form as sinusoids or can be expanded in sinusoids, that is, $\bar{E} = \text{Re}\{\bar{E}_s e^{j\omega t}\}; \bar{H} = \text{Re}\{\bar{H}_s e^{j\omega t}\}$. Thus solutions stemming from the use of Eqs. (2.7) to (2.10) must be sinusoidal solutions. One such solution is the Helmholtz wave equation.

2.2 The Helmholtz Wave Equation

We can solve for the propagation of waves in free space by taking the curl of both sides of Eq. (2.7) and eliminating $\bar{H}_s$ by using Eq. (2.8). The result can be rewritten as

$$\nabla \times \nabla \times \bar{E}_s = -j\omega\mu(\sigma + j\omega\varepsilon)\bar{E}_s \quad (2.11)$$

We also can invoke a well-known vector identity where $\nabla \times \nabla \times \bar{E}_s = \nabla(\nabla \cdot \bar{E}_s) - \nabla^2 \bar{E}_s$. Because we are in free space where no sources exist, the divergence of $\bar{E}_s$ equals zero as given by Eq. (2.9). Equation (2.11) can thus be rewritten as

$$\nabla^2 \bar{E}_s - \gamma^2 \bar{E}_s = 0 \quad (2.12)$$

where

$$\gamma^2 = j\omega\mu(\sigma + j\omega\varepsilon) \quad (2.13)$$

Equation (2.12) is called the *vector Helmholtz[5] wave equation* and γ is known as the *propagation constant*. Because γ is obviously a complex quantity, it can be more simply expressed as

$$\gamma = \alpha + j\beta \quad (2.14)$$

where α is attenuation constant (Np/m) and β is phase constant (rad/m).

Through a simple manipulation of the real part of γ^2 and the magnitude of γ^2, one can derive separate equations for α and β as given by

$$\alpha = \omega\sqrt{\frac{\mu\varepsilon}{2}\left[\sqrt{1+\left(\frac{\sigma}{\omega\varepsilon}\right)^2} - 1\right]} \quad (2.15)$$

$$\beta = \omega\sqrt{\frac{\mu\varepsilon}{2}\left[\sqrt{1+\left(\frac{\sigma}{\omega\varepsilon}\right)^2} + 1\right]} \quad (2.16)$$

It can be seen that the attenuation constant in Eq. (2.15) and the phase constant in Eq. (2.16) are functions of the radian frequency ω, constitutive parameters μ and ε, and also of the conductivity of the medium σ. The term $\sigma/\omega\varepsilon$ is typically referred to as the *loss tangent*.

When a medium has a loss tangent < .01, the material is said to be a good insulator. Indoor building materials, such as brick or concrete have loss tangent values near .1 at 3 GHz. When the loss tangent is > 100, the material is said to be a good conductor. Figure 2.1 shows a plot of α/β versus the loss tangent.

---

[5]Hermann Helmholtz (1821–1894): A German born physician who served in the Prussian army fighting Napolean. He was a self-taught mathematician.

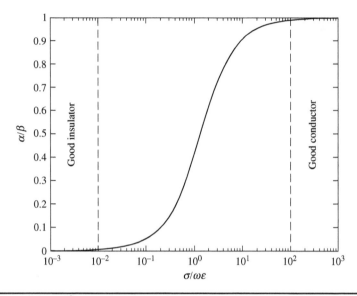

Figure 2.1 α/β versus loss tangent.

We can also solve for the propagation of magnetic fields in free space by taking the curl of both sides of Eq. (2.8) and substituting Eq. (2.7) to get the Helmholtz equation for $\bar{H}_s$ as given by

$$\nabla^2 \bar{H}_s - \gamma^2 \bar{H}_s = 0 \qquad (2.17)$$

The propagation constant is identical to that given in Eq. (2.14).

2.3 Propagation in Rectangular Coordinates

The vector Helmholtz equation in Eq. (2.12) can be solved in any orthogonal coordinate system by substituting the appropriate del (∇) operator for that coordinate system. Let us first assume a solution in rectangular coordinates. Figure 2.2 shows a rectangular coordinate system relative to the earth's surface.

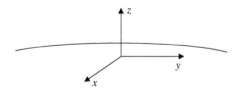

Figure 2.2 Rectangular coordinate system relative to the earth.

Fundamentals of Electromagnetic Fields

It is assumed that the z axis is perpendicular to the surface whereas the x and y coordinates are parallel. Let us also assume that the electric field is polarized in the z-direction and is only propagating in the x-direction. Thus Eq. (2.12) can be furthered simplified.

$$\frac{d^2 E_{xs}}{dx^2} - \gamma^2 E_{xs} = 0 \tag{2.18}$$

The solution is of the form

$$E_{zs}(x) = E_0 e^{-\gamma x} + E_1 e^{\gamma x} \tag{2.19}$$

Assuming that the field propagates only in the positive x-direction and is finite at infinity, E_1 must be equal to 0 giving

$$E_{zs}(x) = E_0 e^{-\gamma x} \tag{2.20}$$

We can revert the phasor of Eq. (2.20) back to the time domain by reintroducing $e^{j\omega t}$. Thus

$$\bar{E}(x,t) = \text{Re}\{E_0 e^{-\gamma x} e^{j\omega t} \hat{z}\} = \text{Re}\{E_0 e^{-\alpha x} e^{j(\omega t - \beta x)} \hat{z}\}$$

or

$$\bar{E}(x,t) = E_0 e^{-\alpha x} \cos(\omega t - \beta x) \hat{z} \tag{2.21}$$

Figure 2.3 shows an example plot of the normalized propagating E-field at a fixed point in time.

The attenuation constant in Eq. (2.21) is representative of a medium with an ideal homogeneous conductivity. In a more realistic radio wave propagation model, the attenuation is further affected by the

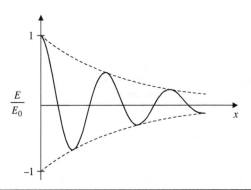

FIGURE 2.3 Propagating E-field.

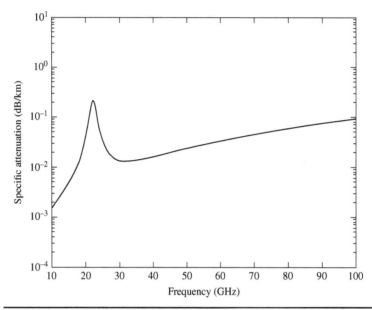

Figure 2.4 Attenuation by water vapor at sea level.

presence of atmospheric gases, clouds, fog, rain, and water vapor. This is especially true for propagation frequencies above 10 GHz. Thus a more sophisticated model is necessary to more accurately represent the full signal attenuation. Three good references explaining atmospheric attenuation, due to other factors, are Collin [4], Ulaby et al. [5], and Elachi [6]. Figure 2.4 shows the attenuation by the molecular resonance of uncondensed water vapor. A resonance condition can be seen to occur at about 22 GHz. The equation for plotting Fig. 2.4 was taken from Frey [7].

2.4 Propagation in Spherical Coordinates

We may also calculate the propagation of electric fields from an isotropic point source in spherical coordinates. The traditional wave equation approach is developed through vector and scalar potentials as shown in Collin [4] or Balanis [8]. However, a thumbnail derivation, albeit less rigorous, can be taken directly from Eq. (2.12) in spherical coordinates.[6] This derivation assumes an isotropic point source. Figure 2.5 shows a spherical coordinate system over the earth.

---

[6]In the far field $\bar{E} = -j\omega\bar{A}$; thus the Helmholtz wave equation is of the same form for either $\bar{E}$ or $\bar{A}$.

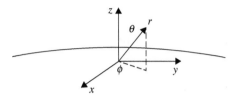

FIGURE 2.5 Spherical coordinate system relative to the earth.

We will assume that the source is isotropic such that the electric field solution is not a function of (θ,ϕ). (It should be noted that we have made no assumption that the isotropic source is an infinitesimal dipole. This allows a simplification of the solution.) Assuming that the electric field is polarized in the θ-direction and is only a function of r, we can write Eq. (2.12) in spherical coordinates as

$$\frac{d}{dr}\left(r^2 \frac{dE_{\theta s}}{dr}\right) - \gamma^2 r^2 E_{\theta s} = 0 \qquad (2.22)$$

For finite fields, the solution can be seen to be of the form

$$E_{\theta s}(r) = \frac{E_0 e^{-\gamma r}}{r} \qquad (2.23)$$

As before, we can express the phasor of Eq. (2.23) in the time domain to get

$$\bar{E}(r,t) = \frac{E_0 e^{-\alpha r}}{r}\cos(\omega t - \beta r)\hat{\theta} \qquad (2.24)$$

The difference between Eq. (2.23) in spherical coordinates and Eq. (2.20) in rectangular coordinates is because that there is a point source producing the propagating wave, thus giving rise to the $1/r$ dependence. This factor is termed *spherical spreading* implying that because the radiation emanates from a point source, the field spreads out as if over the surface of a sphere whose radius is r. Because all finite length antennas are used to generate radio waves, all propagating electric farfields undergo a spherical spreading loss as well as the attenuation loss due to factors discussed earlier. The solution depicted in Eqs. (2.23) and (2.24) is identical in form to the more classically derived solutions. The term E_0 can be viewed as being frequency dependent for finite length sources.

2.5 Electric Field Boundary Conditions

All electric and magnetic field behavior is influenced and disrupted by boundaries. Boundaries interrupt the normal flow of propagating fields and change the field strengths of static fields. All material discontinuities give rise to reflected, transmitted, refracted, diffracted, and scattered fields. These perturbed fields give rise to multipath conditions to exist within a channel. As the number of material discontinuities increases, the number of multipath signals increases. Boundary conditions on electric fields must be established in order to determine the nature of the reflection, transmission, or refraction between dielectric media. Scattering or diffraction conditions are accounted for by different mechanisms. These are discussed in Sec. 2.8.

Two of Maxwell's equations, in integral form, can be used to establish electric field boundary conditions. These are the conservation of energy as given by

$$\oint \bar{E} \cdot \overline{d\ell} = 0 \qquad (2.25)$$

and the conservation of flux as given by

$$\oint \bar{D} \cdot d\bar{S} = Q_{enc} \qquad (2.26)$$

Equation (2.25) can be applied to find tangential boundary conditions (E_t) and Eq. (2.26) can be applied to find normal boundary conditions (D_n). Let us recast the electric field strength and the electric flux density as having tangential and normal components relative to the boundary.

$$\bar{E} = \bar{E}_t + \bar{E}_n \qquad (2.27)$$

$$\bar{D} = \bar{D}_t + \bar{D}_n \qquad (2.28)$$

Figure 2.6 shows the boundary between two media and the corresponding tangential and normal electric fields on each side of the boundary.

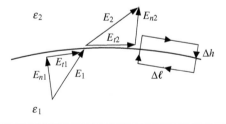

Figure 2.6 Dielectric boundary with E-fields.

Fundamentals of Electromagnetic Fields

Applying Eq. (2.25) to the loop shown in Fig. 2.6, and allowing the loop dimensions to become very small relative to the radii of curvature of the boundary, we obtain the following simplification of the line integral:

$$E_{t2}\Delta\ell - E_{n2}\frac{\Delta h}{2} - E_{n1}\frac{\Delta h}{2} - E_{t1}\Delta\ell = 0 \tag{2.29}$$

Allowing the loop height $\Delta h \to 0$, Eq. (2.29) becomes

$$E_{t1} = E_{t2} \tag{2.30}$$

Thus, tangential E is continuous across the boundary between two dielectrics.

Figure 2.7 shows the boundary between two media and the corresponding tangential and normal electric flux densities on each side of the boundary. The boundary surface has a surface charge density ρ_s.

Applying Eq. (2.26) to the cylindrical closed surface shown in Fig. 2.7, and allowing the cylinder dimensions to become very small relative to the radii of curvature of the boundary, we obtain the following simplification of the surface integral:

$$D_{n2}\Delta s - D_{n1}\Delta s = \rho_s \Delta s$$

or

$$D_{n2} - D_{n1} = \rho_s \tag{2.31}$$

Thus, normal D is discontinuous across a material boundary by the surface charge density at that point.

We can apply the two boundary conditions given in Eqs. (2.30) and (2.31) to determine the refraction properties of two dissimilar dielectric materials. Let us assume that the surface charge at the boundary between the two materials is zero ($\rho_s = 0$). Let us also construct a surface normal $\hat{n}$, pointing into region 2, as shown in Fig. 2.8. Then $\bar{E}_1$ and $\bar{D}_1$ are inclined at an angle θ_1 with respect to the surface normal.

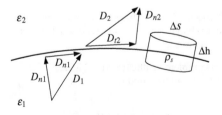

FIGURE 2.7 Dielectric boundary with electric flux density.

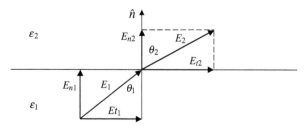

FIGURE 2.8 D and E at a dielectric boundary.

Also, $\bar{E}_2$ and $\bar{D}_2$ are inclined at an angle θ_2 with respect to the surface normal.

Applying the boundary condition in Eq. (2.30), we get

$$E_1 \sin\theta_1 = E_{t1} = E_{t2} = E_2 \sin\theta_2$$

or

$$E_1 \sin\theta_1 = E_2 \sin\theta_2 \tag{2.32}$$

In the same way, we can apply a similar procedure to satisfy the boundary conditions of Eq. (2.31) to yield

$$\varepsilon_1 E_1 \cos\theta_1 = D_{n1} = D_{n2} = \varepsilon_2 E_2 \cos\theta_2$$

or

$$\varepsilon_1 E_1 \cos\theta_1 = \varepsilon_2 E_2 \cos\theta_2 \tag{2.33}$$

Dividing Eq. (2.32) by Eq. (2.33), we can perform simple algebra to obtain a relationship between the two angles of the corresponding E-fields.

$$\frac{\tan\theta_1}{\tan\theta_2} = \frac{\varepsilon_{r1}}{\varepsilon_{r2}} \tag{2.34}$$

Example 2.1 Two semi-infinite dielectrics share a boundary in the $z = 0$ plane. There is no surface charge on the boundary. For $z \leq 0$, $\varepsilon_{r1} = 4$. For $z \geq 0$, $\varepsilon_{r2} = 8$. If $\theta_1 = 30°$, what is the angle θ_2?

Solution Let this problem be illustrated in Fig. 2.9.
Using Eq. (2.34), it can be found that

$$\theta_2 = \tan^{-1}\left(\frac{\varepsilon_{r2}}{\varepsilon_{r1}} \tan\theta_1\right) = 49.1°$$

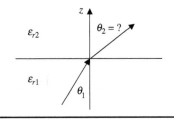

FIGURE 2.9 For Example 2.1.

Example 2.2 Two semi-infinite dielectrics share a boundary in the $z = 0$ plane. There is no surface charge on the boundary. For $z \leq 0$, $\varepsilon_{r1} = 4$. For $z \geq 0$, $\varepsilon_{r2} = 8$. Given that $\bar{E}_1 = 2\hat{x} + 4\hat{y} + 6\hat{z}$, what is the electric field in region 2?

Solution We can use Eq. (2.34) by finding the angle θ_1 from the equation for $\bar{E}_1$. However, it would be simpler to apply the boundary conditions of Eqs. (2.30) and (2.31). Thus using Eq. (2.30)

$$\bar{E}_{t1} = 2\hat{x} + 4\hat{y} = \bar{E}_{t2}$$

Also, $\bar{D}_{n2} = \varepsilon_{r2}\varepsilon_0 \bar{E}_{n2} = 8\varepsilon_0 E_{n2}\hat{z}$ and $\bar{D}_{n1} = \varepsilon_{r1}\varepsilon_0 \bar{E}_{n1} = 4\varepsilon_0(6\hat{z})$. Thus using Eq. (2.31), $E_{n2} = 3$ yielding

$$\bar{E}_2 = 2\hat{x} + 4\hat{y} + 3\hat{z}$$

2.6 Magnetic Field Boundary Conditions

The magnetic field boundary conditions are duals of the boundary conditions for electric fields. The remaining two Maxwell's equations, in integral form, can be used to establish these magnetic boundary conditions. These are Ampere's circuital law as given by

$$\oint \bar{H} \cdot \overline{d\ell} = I \qquad (2.35)$$

and the conservation of magnetic flux as given by

$$\oint \bar{B} \cdot \overline{dS} = 0 \qquad (2.36)$$

Equation (2.35) can be applied to find tangential boundary conditions (H_t) and Eq. (2.36) can be applied to find normal boundary conditions (B_n). Let us recast the magnetic field intensity and the magnetic flux density as having tangential and normal components relative to the magnetic boundary.

$$\bar{H} = \bar{H}_t + \bar{H}_n \qquad (2.37)$$

$$\bar{B} = \bar{B}_t + \bar{B}_n \qquad (2.38)$$

Figure 2.10 shows the boundary between two media and the corresponding tangential and normal magnetic fields on each side of

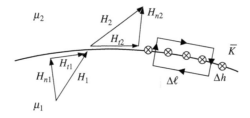

FIGURE 2.10 Magnetic boundary with E-fields.

the boundary. In addition, a surface current density $\bar{K}$ flows along the boundary.

Applying Eq. (2.37) to the loop shown in Fig. 2.10, and allowing the loop dimensions to become very small relative to the radii of curvature of the boundary, we obtain the following simplification of the line integral:

$$H_{t2}\Delta\ell - H_{n2}\frac{\Delta h}{2} - H_{n1}\frac{\Delta h}{2} - H_{t1}\Delta\ell = K\Delta\ell \tag{2.39}$$

Allowing the loop height $\Delta h \to 0$, Eq. (2.39) becomes

$$H_{t1} - H_{t2} = K \tag{2.40}$$

Thus, tangential H is discontinuous across the boundary between two magnetic materials.

Figure 2.11 shows the boundary between two media and the corresponding tangential and normal magnetic flux densities on each side of the boundary. The boundary has no corresponding magnetic surface charge because magnetic monopoles do not exist.

Applying Eq. (2.36) to the cylindrical closed surface shown in Fig. 2.11, and allowing the cylinder dimensions to become very small relative to the radii of curvature of the boundary, we obtain the following simplification of the surface integral:

$$B_{n2}\Delta s - B_{n1}\Delta s = 0$$

or

$$B_{n2} = B_{n1} \tag{2.41}$$

Thus, normal B is continuous across a magnetic material boundary.

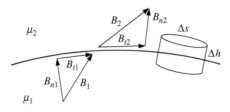

FIGURE 2.11 Dielectric boundary with electric flux density.

We can apply the two boundary conditions given in Eq. (2.41) to determine the magnetic refractive properties of two dissimilar magnetic materials. Let us assume that the surface current density at the boundary between the two materials is zero ($K = 0$). Performing a similar operation as in Sec. 2.5, we can use simple algebra to obtain a relationship between the two angles of the corresponding flux lines such that

$$\frac{\tan \theta_1}{\tan \theta_2} = \frac{\mu_{r1}}{\mu_{r2}} \qquad (2.42)$$

Examples 2.1 and 2.2 can be applied to magnetic fields with the same corresponding values for the relative permeabilities achieving the same results. This is not surprising since electric and magnetic fields and electric and magnetic media are duals of each other.

2.7 Planewave Reflection and Transmission Coefficients

Multipath signals are the consequence of the transmitted signal reflecting, transmitting, and diffracting from various structures along the route to the receiver. In this section we deal exclusively with the reflection and transmission of planewaves. One aspect of calculating each multipath term is being able to predict the reflection and transmission through various materials. The boundary conditions of Eqs. (2.30) and (2.40) can be invoked to allow us to determine the reflection and transmission coefficients. The simplest case is to predict the reflection and transmission across a planar boundary at normal incidence. The details of the derivation can be found in Sadiku [1].

2.7.1 Normal Incidence

Figure 2.12 shows a plane wave normally incident upon a planar material boundary.

E_{is} and H_{is} symbolize the incident fields, in phasor form, propagating in the positive z-direction. E_{rs} and H_{rs} symbolize the reflected fields propagating in the minus z-direction. E_{ts} and H_{ts} symbolize the transmitted fields propagating in the positive z-direction. The exact expressions for the E and H fields are given by the following expressions:

Incident fields:

$$\bar{E}_{is}(z) = E_{i0} e^{-\gamma_1 z} \hat{x} \qquad (2.43)$$

$$\bar{H}_{is}(z) = \frac{E_{i0}}{\eta_1} e^{-\gamma_1 z} \hat{y} \qquad (2.44)$$

Reflected fields:

$$\bar{E}_{rs}(z) = E_{r0} e^{\gamma_1 z} \hat{x} \qquad (2.45)$$

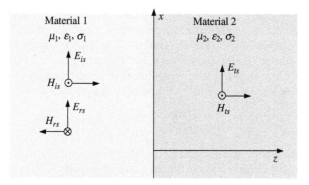

FIGURE 2.12 Plane wave normally incident on a material boundary at z = 0.

$$\bar{H}_{rs}(z) = -\frac{E_{r0}}{\eta_1} e^{\gamma_1 z} \hat{y} \tag{2.46}$$

Transmitted fields:

$$\bar{E}_{ts}(z) = E_{t0} e^{-\gamma_2 z} \hat{x} \tag{2.47}$$

$$\bar{H}_{ts}(z) = \frac{E_{t0}}{\eta_2} e^{-\gamma_2 z} \hat{y} \tag{2.48}$$

where the intrinsic impedances are given by

$$\eta_1 = \sqrt{\frac{\frac{\mu_1}{\varepsilon_1}}{1 - j\frac{\sigma_1}{\omega\varepsilon_1}}} = \text{intrinsic impedance of medium 1}$$

$$\eta_2 = \sqrt{\frac{\frac{\mu_2}{\varepsilon_2}}{1 - j\frac{\sigma_2}{\omega\varepsilon_2}}} = \text{intrinsic impedance of medium 2}$$

Obviously the intrinsic impedances are dependent upon the loss tangent as well as the propagation constants in both media.

Assuming that there is no surface current at the boundary and utilizing the tangential boundary conditions of Eqs. (2.30) and (2.40), one can derive the reflection and transmission coefficients, respectively, as

$$R = \frac{\eta_2 - \eta_1}{\eta_2 + \eta_1} = |R| e^{j\theta_R} \tag{2.49}$$

$$T = \frac{2\eta_2}{\eta_2 + \eta_1} = |T| e^{j\theta_T} \tag{2.50}$$

Fundamentals of Electromagnetic Fields

Knowing the reflection and transmission coefficients R and T, one can determine the total electric field in regions 1 and 2. The total electric field in region 1 is given by

$$\bar{E}_{1s} = \bar{E}_{is} + \bar{E}_{rs} = E_{i0}\left[e^{-\gamma_1 z} + Re^{\gamma_1 z}\right]\hat{x} \tag{2.51}$$

whereas the total electric field in region 2 is given by

$$\bar{E}_{2s} = TE_0 e^{-\gamma_2 z}\hat{x} \tag{2.52}$$

When a nonzero reflection coefficient exists and region 1 is lossless, a standing wave is established. This standing wave gives rise to an interference pattern, which is a function of distance from the boundary. This interference is a trivial example of fading that occurs in many wireless applications. It is instructive to derive this standing wave envelope.

The total field in region 1 can be reexpressed using the polar form of the reflection coefficient. If we assume that region 1 is lossless (i.e., $\sigma_1 = 0$), then

$$\bar{E}_{1s} = E_{i0}[e^{-j\beta_1 z} + |R|e^{j(\beta_1 z + \theta_R)}]\hat{x} \tag{2.53}$$

Combining the real and imaginary parts in Eq. (2.53), we get

$$\bar{E}_{1s} = E_{i0}[\cos(\beta_1 z) + |R|\cos(\beta_1 z + \theta_R) \\ + (|R|\sin(\beta_1 z + \theta_R) - \sin(\beta_1 z))e^{j(\pi/2)}]\hat{x} \tag{2.54}$$

We may now convert the phasor of Eq. (2.54) into instantaneous time form

$$E_1(z,t) = E_{i0}[(\cos(\beta_1 z) + |R|\cos(\beta_1 z + \theta_R))\cos\omega t \\ - (|R|\sin(\beta_1 z + \theta_R) - \sin(\beta_1 z))\sin\omega t] \tag{2.55}$$

Because Eq. (2.55) contains two components that are in phase quadrature, we may easily find the magnitude to be given as

$$|E_1(z)| \\ = E_{i0}\sqrt{(\cos(\beta_1 z) + |R|\cos(\beta_1 z + \theta_R))^2 + (\sin(\beta_1 z) - |R|\sin(\beta_1 z + \theta_R))^2} \\ = E_{i0}\sqrt{1 + |R|^2 + 2|R|\cos(2\beta_1 z + \theta_R)} \tag{2.56}$$

Equation (2.56) has extrema when the cosine term is either +1 or −1. Thus the maximum and minimum values are given as

$$|E_1|_{\max} = E_{i0}\sqrt{1 + |R|^2 + 2|R|} = E_{i0}(1 + |R|) \tag{2.57}$$

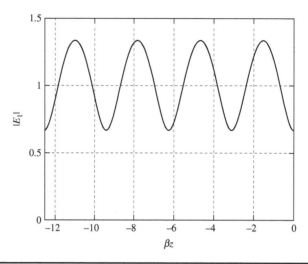

FIGURE 2.13 Standing wave pattern for normal incidence ($\eta_2 = \eta_1/2$).

$$|E_1|_{\min} = E_{i0}\sqrt{1 + |R|^2 - 2|R|} = E_{i0}(1 - |R|) \quad (2.58)$$

The standing wave ratio s is defined as the ratio of $|E_1|_{\max}/|E_2|_{\min}$.

Example 2.3 A boundary exists between two regions where region 1 is free space and region 2 has the parameters $\mu_2 = \mu_0$, $\varepsilon_2 = 4\varepsilon_0$, and $\sigma_2 = 0$. If $E_{i0} = 1$, use MATLAB to plot the standing wave pattern over the range $-4\pi < \beta_1 z < 0$.

Solution Solving for the reflection coefficient

$$R = \frac{\sqrt{\frac{\mu_0}{4\varepsilon_0}} - \sqrt{\frac{\mu_0}{\varepsilon_0}}}{\sqrt{\frac{\mu_0}{4\varepsilon_0}} + \sqrt{\frac{\mu_0}{\varepsilon_0}}} = -\frac{1}{3} = \frac{1}{3}e^{j\pi}$$

Using Eq. (2.56) and MATLAB, the standing wave pattern appears as shown in Fig. 2.13.

Normal incidence is a special case of the more interesting oblique incidence. Oblique incidence is discussed in the next section.

2.7.2 Oblique Incidence

The oblique incidence case is considerably more complicated than the normal incidence case and an intensive derivation of the reflection and transmission coefficients can be seen in Sadiku [1]. The oblique incidence reflection and transmission coefficients are called the *Fresnel coefficients*. Only the highlights are given in this discussion.

Figure 2.14 depicts an incident field upon a boundary. It is assumed that both media are lossless. The electric field is parallel to the plane of incidence. The plane of incidence is that plane containing

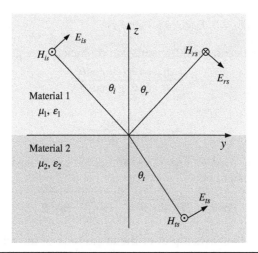

FIGURE 2.14 Parallel polarization reflection and transmission.

the surface normal and the direction of propagation. The angles θ_i, θ_r, and θ_t are the angles of incidence, reflection, and transmission, respectively, with respect to the surface normal ($\pm x$ axis). By a careful application of the boundary conditions, given in Eqs. (2.30) and (2.40), one can determine two laws.

The first is Snell's law of reflection, which states that the angle of reflection equals the angle of incidence. (This property is also termed *specular* reflection.)

$$\theta_r = \theta_i \tag{2.59}$$

The second result is the consequence of the conservation of phase and is also called *Snell's law of refraction*.

$$\beta_1 \sin \theta_i = \beta_2 \sin \theta_t \tag{2.60}$$

Parallel Polarization

The incident field, for the parallel polarization case, is indicated in Fig. 2.14. The coordinate system is rotated from the coordinate system in Fig. 2.12 so as to indicate reflection from a horizontal surface. This is often the case for elevated antennas over a flat earth. This is a parallel-polarized field because the E-field is in the y-z plane, which is the plane of incidence.

The incident, reflected, and transmitted electric fields are given by

$$\bar{E}_{is} = E_0(\cos\theta_i \hat{y} + \sin\theta_i \hat{z})e^{-j\beta_1(y\sin\theta_i - z\cos\theta_i)} \tag{2.61}$$

$$\bar{E}_{rs} = R_{\parallel} E_0(\cos\theta_i \hat{y} - \sin\theta_i \hat{z})e^{-j\beta_1(y\sin\theta_i + z\cos\theta_i)} \tag{2.62}$$

$$\bar{E}_{ts} = T_{\|} E_0 (\cos\theta_t \hat{y} + \sin\theta_t \hat{z}) e^{-j\beta_2 (y\sin\theta_t - z\cos\theta_t)} \tag{2.63}$$

where the reflection and transmission coefficients are given as

$$R_{\|} = \frac{\eta_2 \cos\theta_t - \eta_1 \cos\theta_i}{\eta_2 \cos\theta_t + \eta_1 \cos\theta_i} \tag{2.64}$$

and

$$T_{\|} = \frac{2\eta_2 \cos\theta_i}{\eta_2 \cos\theta_t + \eta_1 \cos\theta_i} \tag{2.65}$$

The term $\cos\theta_t$ in Eqs. (2.64) and (2.65) can be easily calculated using Eq. (2.60) to be

$$\cos\theta_t = \sqrt{1 - \sin^2\theta_t} = \sqrt{1 - \frac{\mu_1 \varepsilon_1}{\mu_2 \varepsilon_2} \sin^2\theta_i} \tag{2.66}$$

Figure 2.15 shows a plot of the magnitude of the reflection and transmission coefficients in the case that both media are nonmagnetic and lossless. The permittivities are given respectively as $\varepsilon_1 = \varepsilon_0$ and $\varepsilon_2 = 2\varepsilon_0, 8\varepsilon_0, 32\varepsilon_0$.

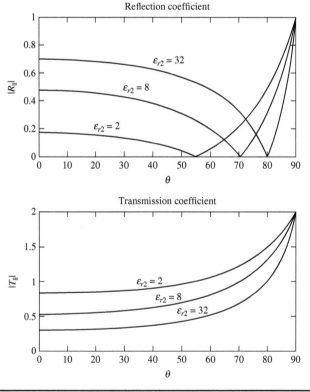

Figure 2.15 Reflection and transmission coefficient magnitude for parallel polarization.

Fundamentals of Electromagnetic Fields

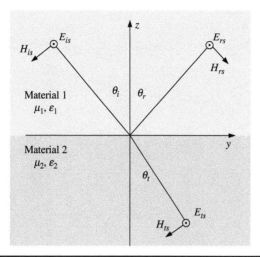

FIGURE 2.16 Perpendicular polarization reflection and transmission.

Perpendicular Polarization

The incident field, for the perpendicular polarization case, is indicated in Fig. 2.16.

The incident, reflected, and transmitted electric fields are given by

$$\bar{E}_{is} = E_0 \hat{x} e^{-j\beta_1(y\sin\theta_i - z\cos\theta_i)} \tag{2.67}$$

$$\bar{E}_{rs} = R_\perp E_0 \hat{x} e^{-j\beta_1(y\sin\theta_i + z\cos\theta_i)} \tag{2.68}$$

$$\bar{E}_{ts} = T_\perp E_0 \hat{x} e^{-j\beta_2(y\sin\theta_t - z\cos\theta_t)} \tag{2.69}$$

where the reflection and transmission coefficients are given as

$$R_\perp = \frac{\eta_2 \cos\theta_i - \eta_1 \cos\theta_t}{\eta_2 \cos\theta_i + \eta_1 \cos\theta_t} \tag{2.70}$$

and

$$T_\perp = \frac{2\eta_2 \cos\theta_i}{\eta_2 \cos\theta_i + \eta_1 \cos\theta_t} \tag{2.71}$$

Figure 2.17 shows a plot of the magnitude of the reflection and transmission coefficients in the case that both media are nonmagnetic and lossless. The permittivities are given as $\varepsilon_1 = \varepsilon_0$ and $\varepsilon_2 = 2\varepsilon_0, 8\varepsilon_0, 32\varepsilon_0$, respectively.

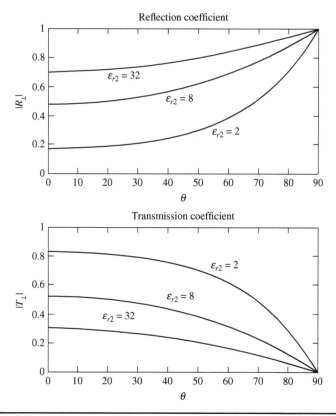

FIGURE 2.17 Reflection and transmission coefficient magnitude for perpendicular polarization.

2.8 Propagation over Flat Earth

Having discussed the planewave reflection coefficients for parallel and perpendicular polarization, we are now in a position to analyze the propagation of planewaves over flat earth. Even though the earth has curvature, and this curvature dramatically affects long-distance propagation, we will limit this discussion to short distances and assume that the earth is flat. This allows us to make some propagation generalizations. This topic is a critical start in understanding general multipath propagation problems because the flat earth model allows us to include a second indirect path. This second path will produce interference effects at the receiver.

Let us consider isotropic transmit and receive antennas as shown in Fig. 2.18. The transmitting antenna is at height h_1, and the receiving antenna is at height h_2, and the two antennas are separated by a

Fundamentals of Electromagnetic Fields

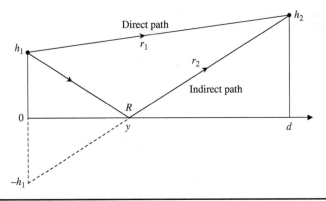

FIGURE 2.18 Flat earth model with two isotropic antennas.

horizontal distance d. The reflection coefficient from the surface of the earth is R and can be approximated either by Eq. (2.64) or (2.70). It should be noted that R is complex for most earth reflections.

The received signal at the receiver is composed of the direct path propagation as well as the reflected signal at the point y. Thus, the composite signal, due to the direct and indirect paths, is proportional to the following equation:

$$\frac{e^{-jkr_1}}{r_1} + R\frac{e^{-jkr_2}}{r_2} \qquad (2.72)$$

where k is the wavenumber as given by the dispersion relation

$$k^2 = k_x^2 + k_y^2 + k_z^2 = \beta^2 \qquad (2.73)$$

with $k = \omega\sqrt{\mu\varepsilon} = 2\pi/\lambda$.

The reflection coefficient R is normally complex and can alternatively be expressed as $R = |R|e^{-j\psi}$. Through some simple algebra, it can be shown that

$$r_1 = \sqrt{d^2 + (h_2 - h_1)^2} \qquad (2.74)$$

$$r_2 = \sqrt{d^2 + (h_2 + h_1)^2} \qquad (2.75)$$

Factoring the direct path term out of Eq. (2.72), we get

$$\frac{e^{-jkr_1}}{r_1}\left[1 + R\frac{r_1}{r_2}e^{-jk(r_2-r_1)}\right] \qquad (2.76)$$

The magnitude of the second term in Eq. (2.76) is called the *path gain factor F*. F is also similar to the *height gain* defined in Bertoni [9]. Thus

$$F = \left| 1 + R \frac{r_1}{r_2} e^{-jk(r_2-r_1)} \right| \qquad (2.77)$$

This factor is analogous to the array factor for a two-element array separated by a distance of $2h_1$. In addition, the point of reflection at y is the solution to a simple algebra problem and is given as $y = dh_1/(h_1 + h_2)$. Using the definition of the path gain factor, we can rewrite Eq. (2.76) as

$$\frac{e^{-jkr_1}}{r_1} F \qquad (2.78)$$

If we make the assumption that the antenna heights $h_1, h_2 \ll r_1, r_2$, we can use a binomial expansion on Eqs. (2.74) and (2.75) to simplify the distances.

$$r_1 = \sqrt{d^2 + (h_2 - h_1)^2} \approx d + \frac{(h_2 - h_1)^2}{2d} \qquad (2.79)$$

$$r_2 = \sqrt{d^2 + (h_2 + h_1)^2} \approx d + \frac{(h_2 + h_1)^2}{2d} \qquad (2.80)$$

Consequently, the difference in path lengths is given as

$$r_2 - r_1 = \frac{2h_1 h_2}{2} \qquad (2.81)$$

We additionally also can assume that $r_1/r_2 \approx 1$ for long distances. Under these conditions, we have a shallow grazing angle at the point of reflection. Thus, $R \approx -1$. By substituting R and Eq. (2.81) into the path gain factor, we now can get

$$F = 2 \left| \sin \frac{kh_1 h_2}{d} \right| = 2 \left| \sin \frac{2\pi h_1}{d} \frac{h_2}{\lambda} \right| \qquad (2.82)$$

Figure 2.19 shows a typical plot of F for a range of values of h_2/λ where $h_1 = 5$ m, $d = 200$ m. It is clear that the received signal can vary between zero and twice the direct path signal strength due to constructive and destructive interference created by the indirect path. The path gain factor can also be used to create a coverage diagram, the details of which can be found in [4].

2.9 Knife-Edge Diffraction

In addition to received fields being disrupted by reflection from the ground, additional propagation paths can be created by diffraction from hills, buildings, and other objects. These structures may not be positioned at angles so as to allow specular reflection; however, they might provide

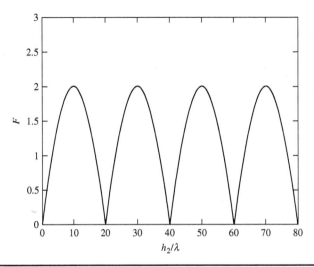

Figure 2.19 Path gain factor.

for a diffraction term in the total received field. Figure 2.20 shows a hill of height h located between the transmit and receive antennas. This hill can be modeled as a half-plane or knife edge. It is assumed that there is no specular reflection from the hilltop. It is also assumed that the hill blocks any possible reflections from the ground, which arrive at the receiving antenna. Thus, the received field is only composed of the direct path and the diffraction path terms. h_c is the clearance height from the knife-edge to the direct path. $h_c < 0$ corresponds to the knife-edge being

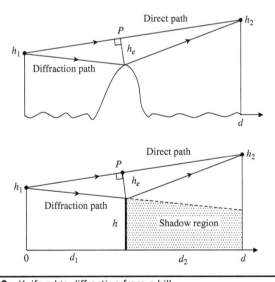

Figure 2.20 Knife-edge diffraction from a hill.

below the line of sight and therefore two propagation paths exist as shown in the figure. When $h_c > 0$ the knife-edge obstructs the direct path. Thus only the diffraction term is received. d_1 and d_2 are the respective horizontal distances to the knife-edge plane ($d = d_1 + d_2$). The diffracted field can allow for a received signal even when line of sight is not possible. If the receive antenna has no direct path to the transmit antenna, the receive antenna is said to be in the shadow region. If the receiver is in the line of sight of the transmitter, it is said to be in the lit region. The derivation of the solution for the diffracted field can be found in Collin [4] or in Jordan and Balmain [10]. It can be shown that the path gain factor due to diffraction is given by

$$F_d = \frac{1}{\sqrt{2}}\left|\int_{-H_c}^{\infty} e^{-j\pi u^2/2}\, du\right| \quad (2.83)$$

where

$$H_c \approx h_c \sqrt{\frac{2d}{\lambda d_1 d_2}}$$

Thus, we can replace the path gain factor F in the flat-earth model with the path gain factor for diffraction. We can therefore rewrite Eq. (2.78) using F_d to become

$$\frac{e^{-jkr}}{r} F_d \quad (2.84)$$

Figure 2.21 shows the path gain factor F_d for a range of H_c values. It can be seen that when the knife-edge is below the line of sight ($h_c < 0$), there is an interference pattern as the direct path and the

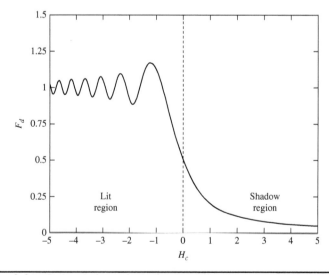

FIGURE 2.21 Path gain factor for knife-edge diffraction.

diffracted path phase in and out. However, when the knife-edge is above the line of sight ($h_c > 0$), no direct path exists and the field quickly diminishes deeper into the shadow region. In the case where $h_c = 0$, the path gain factor is .5 showing that the received field is 6 dB down from the direct path alone.

2.10 References

1. Sadiku, M. N. O., *Elements of Electromagnetics*, 3d ed., Oxford University Press, Oxford, 2001.
2. Hayt, W. H., Jr., and J. A. Buck, *Engineering Electromagnetics*, 6th ed., McGraw-Hill, New York, 2001.
3. Ulaby, F. T., *Fundamentals of Applied Electromagnetics*, Media ed., Prentice Hall, New York, 2004.
4. Collin, R. E., *Antennas and Radiowave Propagation*, McGraw-Hill, New York, 1985.
5. Ulaby, F. T., R. K. Moore, and A. K. Fung, *Microwave Remote Sensing Fundamentals and Radiometry*, Vol. I, Artech House, Boston, MA, 1981.
6. Elachi, C., *Introduction to the Physics and Techniques of Remote Sensing*, Wiley Interscience, New York, 1987.
7. Frey, T. L., Jr., "The Effects of the Atmosphere and Weather on the Performance of a mm-Wave Communication Link," *Applied Microwave & Wireless*, Vol. 11, No. 2, pp. 76–80, Feb. 1999.
8. Balanis, C., *Antenna Theory Analysis and Design*, 2d ed., Wiley, New York, 1997.
9. Bertoni, H., *Radio Propagation for Modern Wiereless Systems*, Prentice Hall, New York, 2000.
10. Jordan, E., and K. Balmain, *Electromagnetic Waves and Radiating Systems*, 2d ed., Prentice Hall, New York, 1968.

2.11 Problems

2.1 For a lossy medium ($\sigma \neq 0$) with the following constitutive parameters, $\mu = 4\mu_0$, $\varepsilon = 2\varepsilon_0$, $\sigma/\omega\varepsilon = 1$, $f = 1$ MHz, find α and β.

2.2 For a lossy material such that $\mu = 6\mu_0$, $\varepsilon = \varepsilon_0$. If the attenuation constant is 1 Np/m at 10 MHz, find
  (a) The phase constant β
  (b) The loss tangent
  (c) The conductivity σ
  (d) The intrinsic impedance

2.3 Use MATLAB to plot the ratio α/β for the range of $.01 < \sigma/\omega\varepsilon < 100$ with the horizontal scale being log base 10.

2.4 Using Eq. (2.21), if $\mu = \mu_0$, $\varepsilon = 4\varepsilon_0$, $\sigma/\omega\varepsilon = 1$, $f = 100$ MHz, how far must the wave travel in the z-direction before the amplitude is attenuated by 30 percent?

2.5 Two semi-infinite dielectrics share a boundary in the $z = 0$ plane as depicted in Fig. 2P.1. There is no surface charge on the boundary. For $z \leq 0$, $\varepsilon_{r1} = 2$. For $z \geq 0$, $\varepsilon_{r2} = 6$. If $\theta_1 = 45°$, what is the angle θ_2?

2.6 Two semi-infinite dielectrics share a boundary in the $z = 0$ plane. There is no surface charge on the boundary. For $z \leq 0$, $\varepsilon_{r1} = 2$. For $z \geq 0$, $\varepsilon_{r2} = 4$. Given that $\bar{E}_1 = 4\hat{x} + 2\hat{y} + 3\hat{z}$ in region 1, find the electric field in region 2($\bar{E}_2$)?

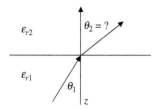

FIGURE 2P.1   For Prob. 2.5.

2.7 Two semi-infinite magnetic regions share a boundary in the $z = 0$ plane. There is no surface current density K on the boundary. For $z \le 0$, $\mu_{r1} = 4$. For $z \ge 0$, $\mu_{r2} = 2$. Given that $\bar{H}_1 = 4\hat{x} + 2\hat{y} + 3\hat{z}$ in region 1, what is the magnetic flux density in region 2 ($\bar{B}_2$)?

2.8 A plane wave normally incident on a material boundary as indicated in Fig. 2P.2. Both regions are nonmagnetic. $\varepsilon_{r1} = 2$, $\varepsilon_{r2} = 4$. If the loss tangent in region 1 is 1.732 and the loss tangent in region 2 is 2.8284, find the following:
  (a) The intrinsic impedance in region 1
  (b) The intrinsic impedance in region 2
  (c) The reflection coefficient R
  (d) The transmission coefficient T

2.9 Modify Prob. 2.8 such that the loss tangent in region 1 is zero. If $E_{i0} = 1$, use MATLAB to plot the standing wave pattern for the range $-2\pi < \beta z < 0$.

2.10 In the oblique incidence case with parallel polarization, as depicted in Fig. 2.14, both regions are nonmagnetic and $\varepsilon_{r1} = 1, \varepsilon_{r2} = 6$. If the loss tangent in region 1 is 1.732 and the loss tangent in region 2 is 2.8284, $\theta_i = 45°$. Find the following:
  (a) The intrinsic impedance in region 1
  (b) The intrinsic impedance in region 2
  (c) The reflection coefficient $R_{||}$
  (d) The transmission coefficient $T_{||}$

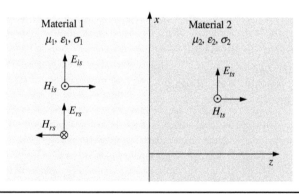

FIGURE 2P.2   Plane wave at normal incidence on a boundary.

2.11 Repeat Prob. 2.10 but for the perpendicular polarization case as depicted in Fig. 2.16.

2.12 Region 1 is free space and region 2 is lossless and nonmagnetic with $\varepsilon_{r2} = 2, 8, 64$. Use MATLAB and superimpose plots for all three dielectric constants as indicated below for a range of angles $0 < \theta < 90°$:
   (a) $R_{\parallel}$
   (b) $T_{\parallel}$
   (c) $R_{\perp}$
   (d) $T_{\perp}$

2.13 For two antennas over flat earth, derive the equation for the specular reflection point y in terms of d, h_1, and h_2.

2.14 Using Fig. 2.18 with $h_1 = 20$ m, $h_2 = 200$ m, and $d = 1$ km, $R = -1$, use MATLAB to plot the exact path gain factor [Eq. (2.77)] for 200 MHz $< f <$ 400 MHz. Assume free space.

2.15 Repeat Prob. 2.14 but allow $f = 300$ MHz and plot F for the range 500 m $< d < 1000$ m. Assume free space.

2.16 Repeat Prob. 2.14 but allow the incident field to be perpendicular to the plane of incidence such that E is horizontal to the ground. Allow the ground to be nonmagnetic and have a dielectric constant $\varepsilon_r = 4$.
   (a) What is the Fresnel reflection coefficient?
   (b) Plot the path gain factor F for 200 MHz $< f <$ 400 MHz.

2.17 Use MATLAB and plot the path gain factor due to diffraction (F_d) found in Eq. (2.82) for the range of values $-8 < H_c < 8$.

2.18 For the hill between two antennas as shown in Fig. 2P.3 where $h_1 = 150$ m, $h_2 = 200$ m, $d_1 = 300$ m, $d_2 = 700$ m, $d = 1$ km, $h = 250$ m, $f = 300$ MHz.
   (a) What is the direct path length r_1?
   (b) What is the indirect path length r_2?
   (c) What is the clearance height h_c (be sure your sign is right)?
   (d) What is the integration height H_c?
   (e) Using Eq. (2.84), what is the magnitude of the received signal if $E_0 = 1$ V/m?

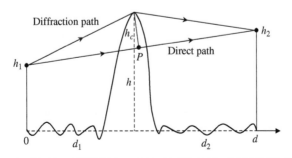

FIGURE 2P.3 Hill between two antennas.

CHAPTER 3
Antenna Fundamentals

The design and analysis of smart antennas assumes a working knowledge of many different but related disciplines. A smart antenna designer must rely on such disciples as (1) random processes, (2) electromagnetics, (3) propagation, (4) spectral estimation methods, (5) adaptive techniques, and (6) antenna fundamentals. Especially, though, smart antenna design is heavily dependent on a basic knowledge of antenna theory. It is critical to match the individual antenna behavior with the overall system requirements. Thus, this chapter covers the relevant antenna topics such as the near and far fields that surround antennas, power densities, radiation intensities, directivities, beamwidths, antenna reception, and fundamental antenna designs including dipoles and loops. It is not necessarily required that one have an extensive antenna background, in order to understand smart antennas, but it would be wise to review some of the literature referenced in this chapter. The foundation for much of the material in this chapter is taken from the books of Balanis [1], Kraus and Marhefka [2], and Stutzman and Thiele [3].

3.1 Antenna Field Regions

Antennas produce complex electromagnetic (EM) fields both near to and far from the antennas. Not all of the EM fields generated actually radiate into space. Some of the fields remain in the vicinity of the antenna and are viewed as reactive near fields; much the same way as an inductor or capacitor is a reactive storage element in lumped element circuits. Other fields do radiate and can be detected at great distances. An excellent treatment of the antenna field regions is given in [1] and [2]. Figure 3.1 shows a simple dipole antenna with four antenna regions.

The boundaries defined in the regions above are not arbitrarily assigned but are the consequence of solving the exact fields surrounding a finite length antenna.

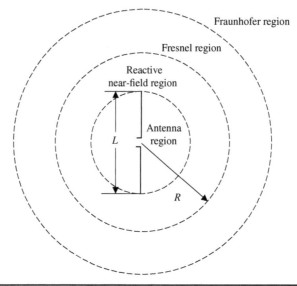

FIGURE 3.1 Antenna field regions.

The four regions and their boundaries are defined as follows:

Antenna region: The region that circumscribes the physical antenna boundaries is called the *antenna region,* as defined by

$$R \le \frac{L}{2}$$

Reactive near-field region: The region that contains the reactive energy surrounding the antenna is called the *reactive near-field region.* It represents energy stored in the vicinity of the antenna which does not radiate and thus is seen in the imaginary part of the antenna terminal impedance. This region is defined by

$$R \le 0.62\sqrt{\frac{L^3}{\lambda}}$$

Fresnel region (radiating near field): The region that lies between the reactive near field and the Fraunhofer far field is the *Fresnel region* or *radiating near-field region.* The antenna field radiates in this region, but the radiation pattern changes with distance from the phase center because the radiated field components diminish at different rates. This region is defined by

$$0.62\sqrt{\frac{L^3}{\lambda}} \le R \le \frac{2L^2}{\lambda}$$

Fraunhofer region (far field): The region that lies beyond the near field and where the radiation pattern is unchanging with distance is defined as the *Fraunhofer region*. This is the principal region of operation for most elemental antennas. This region is defined by

$$R \geq \frac{2L^2}{\lambda}$$

For practical purposes, this text generally assumes antenna radiation in the Fresnel or Fraunhofer regions. If array element coupling is to be considered, the reactive near-field region must be accounted for in all calculations.

3.2 Power Density

All radiated antenna fields carry power away from the antenna that can be intercepted by distant receiving antennas. It is this power that is of use in communication systems. As a trivial example, let us assume that the propagating phasor fields, generated by a point source isotropic antenna, are given further and are expressed in spherical coordinates.

$$\bar{E}_{\theta s} = \frac{E_0}{r} e^{-jkr} \hat{\theta} \text{ V/m} \tag{3.1}$$

$$\bar{H}_{\phi s} = \frac{E_0}{\eta r} e^{-jkr} \hat{\phi} \text{ A/m} \tag{3.2}$$

where η is the intrinsic impedance of the medium.

If the intrinsic medium is lossless, the time-varying instantaneous fields can be easily derived from Eqs. (3.1) and (3.2) to be

$$\bar{E}(r,t) = \text{Re}\left\{\frac{E_0}{r} e^{j(wt-kr)} \hat{\theta}\right\} = \frac{E_0}{r} \cos(wt - kr)\hat{\theta} \tag{3.3}$$

$$\bar{H}(r,t) = \text{Re}\left\{\frac{E_0}{\eta r} e^{j(wt-kr)} \hat{\phi}\right\} = \frac{E_0}{\eta r} \cos(wt - kr)\hat{\phi} \tag{3.4}$$

The electric field intensity in Eq. (3.3) is seen to radiate in the positive r direction and is polarized in the positive $\hat{\theta}$ direction. The magnetic field intensity in Eq. (3.4) is seen to radiate in the positive r direction and is polarized in the positive $\hat{\phi}$ direction. Figure 3.2 shows the field vectors in spherical coordinates. These far fields are mutually perpendicular and are tangent to the sphere whose radius is r.

The Poynting vector, named after J. H. Poynting,[1] is the cross product of the electric and magnetic field intensities and is given as

$$\bar{P} = \bar{E} \times \bar{H} \text{ W/m}^2 \tag{3.5}$$

---

[1]John Henry Poynting (1852–1914): A student of Maxwell who derived an equation to show energy flow from EM fields.

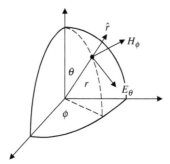

FIGURE 3.2 EM field radiating from a point source.

The cross product is in the right-handed sense and gives the direction of propagation of the power density. The Poynting vector is a measure of the instantaneous power density flow away from the source. By substituting Eqs. (3.3) and (3.4) into Eq. (3.5) and using a simple trigonometric identity, we get

$$\bar{P}(r,t) = \frac{E_0^2}{2\eta r^2}[1 + \cos(2\omega t - 2kr)]\hat{r} \qquad (3.6)$$

The first term in Eq. (3.6) represents the time average power density radiating away from the antenna whereas the second term represents an instantaneous ebb and flow. By taking the time average of Eq. (3.6), we can define the average power density.

$$\bar{W}(r) = \frac{1}{T}\int_0^T \bar{P}(r,t)dt$$
$$= \frac{E_0^2}{2\eta r^2}\hat{r} \ \text{W/m}^2 \qquad (3.7)$$

The calculation of the time average power density is equivalent to performing a calculation in phasor space.

$$\bar{W}(r,\theta,\phi) = \frac{1}{2}\text{Re}(\bar{E}_s \times \bar{H}_s^*) = \frac{1}{2\eta}|E_s|^2\hat{r} \qquad (3.8)$$

Equation (3.8) represents the average power density flow away from the isotropic antenna and thus is not a function of θ or ϕ. For practical antennas, the power density is always a function of r and at least one angular coordinate.

In general, the power density can be represented as a power flow through a sphere of radius r as shown in Fig. 3.3.

The total power radiated by an antenna is found by the closed surface integral of the power density over the sphere bounding the antenna. This is equivalent to applying the divergence theorem to the power density. The total power is thus given by

Antenna Fundamentals   41

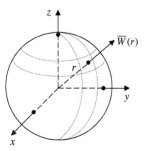

FIGURE 3.3 Power density from an isotropic point source.

$$P_{tot} = \oint \int \overline{W} \cdot d\overline{s} = \int_0^{2\pi} \int_0^{\pi} W_r(r,\theta,\phi) r^2 \sin\theta d\theta d\phi$$

$$= \int_0^{2\pi} \int_0^{\pi} W_r(r,\theta,\phi) r^2 d\Omega \ \text{W} \quad (3.9)$$

where $d\Omega = \sin\theta d\theta d\phi$ = element of solid angle or the differential solid angle.

In the isotropic case where the power density is not a function of θ or ϕ, Eq. (3.9) simplifies to become

$$P_{tot} = \int_0^{2\pi} \int_0^{\pi} W_r(r) r^2 \sin\theta d\theta d\phi = 4\pi r^2 W_r(r) \quad (3.10)$$

or conversely

$$W_r(r) = \frac{P_{tot}}{4\pi r^2} \quad (3.11)$$

Thus, for isotropic antennas, the power density is found by uniformly spreading the total power radiated over the surface of a sphere of radius r. Thus the density diminishes inversely with r^2. It is also interesting to observe that the power density is only a function of the real power (P_{tot}) delivered to the antenna terminals. The reactive power does not contribute to the radiated fields.

Example 3.1 Find the total power radiated by an isotropic antenna whose electric field intensity is given as

$$\overline{E}_s = \frac{2}{r} e^{-jkr} \hat{\theta} \ \text{V/m}$$

Solution Equation (3.7) shows the power density to be

$$\overline{W}(r) = \frac{5.3 \times 10^{-3}}{r^2} \hat{r} \ \text{W/m}^2$$

Substituting this result into Eq. (3.9) yields the total power to be

$$P_{tot} = 66.7 \ \text{mW}$$

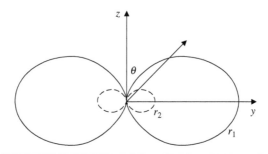

FIGURE 3.4 Pattern plots for two ranges r_1 and r_2.

Example 3.2 Plot the power density versus angle for the far electric field at two different distances r_1 and r_2. Where $r_1 = 100$ m and $r_2 = 200$ m. Given that

$$\overline{E}_s = \frac{100 \sin\theta}{r} e^{-jkr} \hat{\theta}$$

$$\overline{H}_s = \frac{100 \sin\theta}{\eta r} e^{-jkr} \hat{\phi}$$

Solution By using Eq. (3.8), the power density magnitude can easily be found to be

$$W(r,\theta) = \frac{13.3 \sin^2\theta}{r^2}$$

Using MATLAB code sa_ex3_2.m and the polar plot command, we get a pattern plot for both distances as shown in Fig. 3.4.

3.3 Radiation Intensity

The radiation intensity can be viewed as a distance normalized power density. The power density in Eq. (3.8) is inversely proportional to the distance squared and thus diminishes rapidly moving away from the antenna. This is useful in indicating power levels but is not useful in indicating distant antenna patterns. The radiation intensity removes the $1/r^2$ dependence, thus making far-field pattern plots distance independent. The radiation intensity is thus defined as

$$U(\theta,\phi) = r^2 |\overline{W}(r,\theta,\phi)| = r^2 W_r(r,\theta,\phi) \tag{3.12}$$

It is easy to show Eq. (3.12) can alternatively be expressed by

$$U(\theta,\phi) = \frac{r^2}{2\eta} |\overline{E}_s(r,\theta,\phi)|^2$$

$$= \frac{\eta r^2}{2} |\overline{H}_s(r,\theta,\phi)|^2 \tag{3.13}$$

This definition also simplifies the calculation of the total power radiated by the antenna. Equation (3.9) can be repeated substituting the radiation intensity.

Antenna Fundamentals

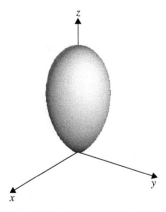

FIGURE 3.5 Antenna 3D pattern.

$$P_{tot} = \int_0^{2\pi}\int_0^{\pi} W_r(r,\theta,\phi)r^2 \sin\theta\, d\theta\, d\phi$$

$$= \int_0^{2\pi}\int_0^{\pi} U(\theta,\phi)d\Omega \text{ W} \qquad (3.14)$$

The general radiation intensity indicates the radiation pattern of the antenna in three dimensions. All anisotropic antennas have a nonuniform radiation intensity and therefore a nonuniform radiation pattern. Figure 3.5 shows an example of a three-dimensional (3D) pattern as displayed in spherical coordinates. This antenna pattern or *beam* pattern is an indication of the directions in which the signal is radiated. In the case of Fig. 3.5, the maximum radiation is in the $\theta = 0$ direction or along the z axis.

Example 3.3 In the Fraunhofer region (far field), a small dipole has an electric field intensity given by

$$\bar{E}_s(r,\theta,\phi) = \frac{E_0 \sin\theta}{r} e^{-jkr}\hat{\theta}$$

What are the radiation intensity and the total power radiated by this antenna?

Solution Using Eq. (3.13), the radiation intensity is given as

$$U(\theta,\phi) = \frac{E_0^2 \sin^2\theta}{2\eta}$$

Using Eq. (3.14), the total power radiated is given as

$$P_{tot} = 0.011 E_0^2 \text{ W}$$

3.4 Basic Antenna Nomenclature

With an understanding of the derivation of the radiation intensity $(U(\theta,\phi))$, we can now define some metrics that help define an antenna's performance.

3.4.1 Antenna Pattern

An antenna pattern is either a function or a plot describing the directional properties of an antenna. The pattern can be based on the function describing the electric or magnetic fields. In that case, the pattern is called a *field pattern*. The pattern can also be based on the radiation intensity function defined in the previous section. In that case, the pattern is called a *power pattern*. The antenna pattern may not come from a functional description but also may be the result of antenna measurements. In this case the measured pattern can be expressed as a *field pattern* or as a *power pattern*.

Figure 3.6a and b shows a typical two-dimensional (2D) field pattern plot displayed in both rectangular coordinates and in polar coordinates. The mainlobe and sidelobes of the pattern are indicated. The mainlobe is that portion of the pattern that has maximum intended radiation. The sidelobes are generally unintended radiation directions.

Figure 3.6a and b can be viewed as demonstrating a 2D slice of what is typically a 3D pattern. Figure 3.7 shows the same pattern displayed in three dimensions. The 3D perspective is useful for illustration purposes, but often antenna designers display 2D plots in the principal planes of 3D patterns.

Example 3.4 Use MATLAB to produce a 3D radiation pattern for a radiation intensity given by $U(\theta) = \cos^2\theta$.

Solution Radiation patterns are normally calculated in spherical coordinates. That being understood, the patterns have corresponding x, y, and z coordinate values. We can simply use the coordinate transformation to transform points on the radiation intensity to (x, y, z) coordinates. Thus

$$x = U(\theta,\phi)\sin\theta\cos\phi \quad y = U(\theta,\phi)\sin\theta\sin\phi \quad z = U(\theta,\phi)\cos\theta$$

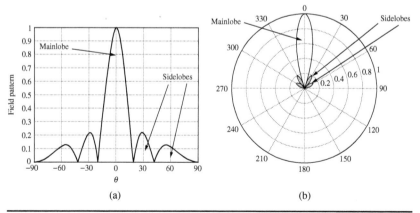

FIGURE 3.6 (a) Field pattern plot in rectangular coordinates, (b) field pattern plot in polar coordinates.

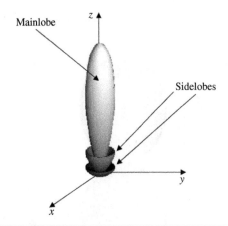

FIGURE 3.7 Field pattern plot in three dimensions.

MATLAB has a feature called *ezmesh* that allows us to easily plot a surface in three dimensions. The corresponding MATLAB commands are

```
fx = inline('cos(theta)^2*sin(theta)*cos(phi)')
fy = inline('cos(theta)^2*sin(theta)*sin(phi)')
fz = inline('cos(theta)^2*cos(theta)')
figure
ezmesh(fx,fy,fz,[ 0  2*pi  0  pi ],100)
colormap([ 0 0 0 ])
axis equal
set(gca,'xdir','reverse','ydir','reverse')
```

The plot from MATLAB code `sa_ex3_4.m` is shown in Fig. 3.8.

3.4.2 Antenna Boresight

The antenna boresight is the intended physical aiming direction of an antenna or the direction of maximum gain. This is also the central axis of the antenna's mainlobe. In other words, it is the normally intended direction for maximum radiation. The boresight in Fig. 3.7 is central axis of the mainlobe, which corresponds to the z axis, where $\theta = 0$.

3.4.3 Principal Plane Patterns

The field patterns or the power patterns are usually taken as 2D slices of the 3D antenna pattern. These slices can be defined in several different ways. One option is to plot the E- and H-plane patterns when

$x = \cos(\theta)^2 \sin(\theta) \cos(\phi)$ $y = \cos(\theta)^2 \sin(\theta) \sin(\phi)$ $z = \cos(\theta)^2 \cos(\theta)$

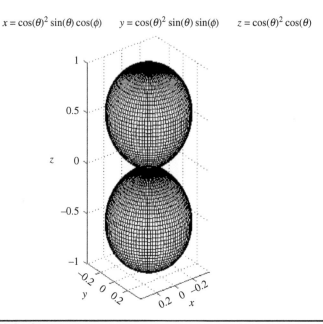

FIGURE 3.8 3D radiation pattern.

the radiation is linearly polarized. In this case, the E-plane pattern is the plot containing the E-field vector and the direction of maximum radiation. In the other case, the H-plane pattern is the plot containing the H-field vector and the direction of maximum radiation. It is most convenient to orient the antenna in spherical coordinates such that the E and H planes correspond to the θ and ϕ constant planes. These planes are called the *azimuth* and *elevation* planes, respectively. Figure 3.9 shows the spherical coordinate system with polarized fields and the azimuth and elevation planes parallel to these field vectors.

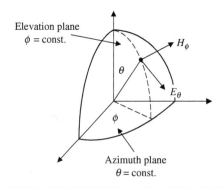

FIGURE 3.9 Principal planes in spherical coordinates.

Antenna Fundamentals   47

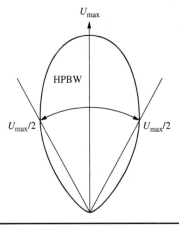

FIGURE 3.10 Half-power beamwidth.

3.4.4   Beamwidth

The beamwidth is measured from the 3-dB points of a radiation pattern. Figure 3.10 shows a 2D slice of Fig. 3.5. The beamwidth is the angle between the 3-dB points. As this is a power pattern, the 3-dB points are also the half power points.

In the case of field patterns instead of power patterns, the 3-dB points would be when the normalized pattern amplitude = $\frac{1}{\sqrt{2}}$ = 0.707.

3.4.5   Directivity

The directivity is a measure of how directive an individual antenna is relative to an isotropic antenna radiating the same total power. In other words, the directivity is the ratio of the power density of an anisotropic antenna relative to an isotropic antenna radiating the same total power. Thus the directivity is given as

$$D(\theta,\phi) = \frac{W(\theta,\phi)}{\frac{P_{tot}}{4\pi r^2}} = \frac{4\pi U(\theta,\phi)}{P_{tot}} \qquad (3.15)$$

The directivity can be more explicitly expressed by substituting Eq. (3.14) into Eq. (3.15) to get

$$D(\theta,\phi) = \frac{4\pi U(\theta,\phi)}{\int_0^{2\pi}\int_0^{\pi} U(\theta,\phi)\sin\theta\, d\theta\, d\phi} \qquad (3.16)$$

The maximum directivity is a constant and is simply the maximum of Eq. (3.16). The maximum directivity is normally denoted by D_0. Thus, the maximum directivity is found by a slight modification of Eq. (3.16) to be

$$D_0 = \frac{4\pi U_{max}}{\int_0^{2\pi}\int_0^{\pi} U(\theta,\phi)\sin\theta\, d\theta\, d\phi} \qquad (3.17)$$

The directivity of an isotropic source is always equal to 1 because isotropic sources radiate equally in all directions and therefore are not inherently directive.

Example 3.5 Find the directivity and the maximum directivity of the small dipole radiation pattern given in Example 3.3.

Solution Using Eq. (3.16), we find

$$D(\theta,\phi) = \frac{4\pi \sin^2\theta}{\int_0^{2\pi}\int_0^{\pi} \sin^3\theta \, d\theta \, d\phi} = 1.5\sin^2\theta$$

As can intuitively be seen in the Example 3.5, the directivity is not affected by the radiation intensity amplitude but is only dependent on the functional form of $U(\theta, \phi)$. The scalar amplitude terms divide out. The maximum directivity is a constant and is simply given by Eq. (3.17) to be $D_0 = 1.5$.

Plotting the directivity of an antenna is more useful than plotting the radiation intensity because the amplitude indicates the performance relative to an isotropic radiator irrespective of the distance. This is helpful in indicating not just the antenna pattern but also indicating a form of the antenna gain.

3.4.6 Beam Solid Angle

The beam solid angle (Ω_A) is that angle through which all of the antenna power radiates if its radiation intensity were equal to its maximum radiation intensity (U_{max}). The beam solid angle can be seen by expressing Eq. (3.17) as

$$D_0 = \frac{4\pi}{\int_0^{2\pi}\int_0^{\pi} \frac{U(\theta,\phi)}{U_{max}} \sin\theta \, d\theta \, d\phi} = \frac{4\pi}{\Omega_A} \quad (3.18)$$

where

$$\Omega_A = \text{Beam solid angle} = \int_0^{2\pi}\int_0^{\pi} \frac{U(\theta,\phi)}{U_{max}} \sin\theta \, d\theta \, d\phi \quad (3.19)$$

The beam solid angle is given in *steradians* where one steradian is defined as the solid angle of a sphere subtending an area on the surface of the sphere equal to r^2. Thus there are 4π steradians in a sphere. The beam solid angle is the spatial version of the equivalent noise bandwidth in communications. An explanation of the noise equivalent bandwidth is given by Haykin [4].

3.4.7 Gain

The directivity of an antenna is an indication of the directionality of an antenna. It is the ability of an antenna to direct energy in preferred directions. The directivity assumes that there are no antenna losses through conduction losses, dielectric losses, and transmission line mismatches. The antenna gain is a modification of the directivity so

Antenna Fundamentals

as to include the effects of antenna inefficiencies. The gain is more reflective of an actual antenna's performance. The antenna gain expression is given by

$$G(\theta,\phi) = eD(\theta,\phi) \quad (3.20)$$

where e is the total antenna efficiency including the effects of losses and mismatches. The pattern produced by the gain is identical to the pattern produced by the directivity except for the efficiency scale factor e.

3.4.8 Effective Aperture

Just as an antenna can radiate power in various preferred directions, it can also receive power from the same preferred directions. This principle is called *reciprocity*. Figure 3.11 shows transmit and receive antennas. The transmit antenna is transmitting with power P_1 (watts) and radiates a power density W_1 (watts/m²).

The receive antenna intercepts a portion of the incident power density W_1, thereby delivering power P_2 to the load. The receive antenna can be viewed as an effective aperture of area A_{e2} that captures a portion of the available power density. Thus, using Eqs. (3.15) and (3.20), we can write the received power as

$$P_2 = A_{e2}W_1 = \frac{A_{e2}P_1 e_1 D_1(\theta_1,\phi_1)}{4\pi r_1^2} W \quad (3.21)$$

where r_1, θ_1, ϕ_1 are the local spherical coordinates for antenna 1.

If the antennas in Fig. 3.10 are reversed such that the receive antenna transmits and the transmit antenna receives, it can be shown that

$$P_1 = A_{e1}W_2 = \frac{A_{e1}P_2 e_2 D_2(\theta_2,\phi_2)}{4\pi r_2^2} W \quad (3.22)$$

where r_2, θ_2, ϕ_2 are the local spherical coordinates for antenna 2.

The derivation is beyond the scope of this text, but it can be shown from Eqs. (3.21) and (3.22) [1, 2] that the effective aperture is related to the directivity of an antenna by

$$A_e(\theta,\phi) = \frac{\lambda^2}{4\pi}eD(\theta,\phi) = \frac{\lambda^2}{4\pi}G(\theta,\phi) \quad (3.23)$$

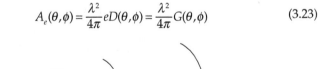

FIGURE 3.11 Transmit and receive antennas.

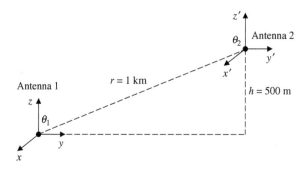

FIGURE 3.12 Two antennas separated by a distance r.

3.5 Friis Transmission Formula

Harald Friis[2] devised a formula relating the transmit and received powers between two distant antennas. We will assume that the transmit and receive antennas are polarization matched. (That is, the polarization of the receive antenna is perfectly matched to the polarization created by the transmit antenna.) By substituting Eq. (3.23) into Eq. (3.21) one can derive the following relationship:

$$\frac{P_2}{P_1} = \left(\frac{\lambda}{4\pi r}\right)^2 G_1(\theta_1,\phi_1) G_2(\theta_2,\phi_2) \qquad (3.24)$$

If polarization must be taken into account, we can multiply Eq. (3.24) by the polarization loss factor (PLF) $= |\hat{\rho}_1 \cdot \hat{\rho}_2|^2$, where $\hat{\rho}_1$ and $\hat{\rho}_2$ are the polarizations of antennas 1 and 2, respectively. An intensive treatment of polarization and the polarization loss factor can be found in [2].

Example 3.6 Calculate the receive power P_2 at antenna 2 if the transmit power $P_1 = 1$ kW. The transmitter gain is $G_1(\theta_1,\phi_1) = \sin^2(\theta_1)$ and the receiver gain is $G_2(\theta_2,\phi_2) = \sin^2(\theta_2)$. The operating frequency is 2 GHz. Use the Fig. 3.12 to solve the problem. The y-z and y'-z' planes are coplanar.

Solution The geometry indicates $\theta_1 = 60°$ and $\theta_2 = 120°$. Using Eq. (3.24), we have

$$P_2 = P_1 \left(\frac{\lambda}{4\pi r}\right)^2 G_1(\theta_1,\phi_1) G_2(\theta_2,\phi_2)$$

$$= P_1 \left(\frac{\lambda}{4\pi r}\right)^2 \sin^2(\theta_1) \sin^2(\theta_2)$$

$$= 10^3 \left(\frac{15 \text{ cm}}{4\pi 10^3}\right)^2 \sin^2(60) \sin^2(120)$$

$$= 80.1 \text{ nW}$$

---

[2]Harald T. Friis (1893–1976): In 1946 he developed a transmission formula.

3.6 Magnetic Vector Potential and the Far Field

All antennas radiate electric and magnetic fields by virtue of charges accelerating on the antenna. These accelerating charges are normally in the form of an ac current. Though it is possible to calculate distant radiated fields directly from the antenna currents, in most cases it is mathematically unwieldy. Therefore an intermediate step is used where we instead calculate the magnetic vector potential. From the vector potential we can then find the distant radiated fields.

The point form of Gauss' law states that $\nabla \cdot \bar{B} = 0$. Because the divergence of the curl of any vector is always identically equal to zero ($\nabla \cdot \nabla \times \bar{A} = 0$), we then can define the $\bar{B}$ field in terms of the magnetic vector potential $\bar{A}$.

$$\bar{B} = \nabla \times \bar{A} \qquad (3.25)$$

Equation (3.25) solves Gauss' law.

Because $\bar{B}$ and $\bar{H}$ are related by a constant, we can also write

$$\bar{H} = \frac{1}{\mu} \nabla \times \bar{A} \qquad (3.26)$$

The electric field, in a source free region, can be derived from the magnetic field by

$$\bar{E} = \frac{1}{j\omega\varepsilon} \nabla \times \bar{H} \qquad (3.27)$$

Thus, if we know the vector potential, we can subsequently calculate $\bar{E}$ and $\bar{H}$ fields.

Figure 3.13 shows an arbitrary current source $\bar{I}$ creating a distant vector potential $\bar{A}$.

The vector potential is related to the current source by

$$\bar{A} = \frac{\mu}{4\pi} \int \bar{I}(r') \frac{e^{-jkR}}{R} dl' \qquad (3.28)$$

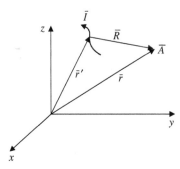

FIGURE 3.13 Current source and distant vector potential.

where $\bar{I}(\bar{r}') = I_x(\bar{r}')\hat{x} + I_y(\bar{r}')\hat{y} + I_z(\bar{r}')\hat{z}$ = current in three dimensions
$\bar{r}'$ = position vector in source coordinates
$\bar{r}$ = position vector in field coordinates
$\bar{R}$ = distance vector = $\bar{r} - \bar{r}'$
$R = |\bar{R}|$
dl' = differential length at current source

The vector potential can be found from any line current source using Eq. (3.28). The results can be substituted into Eqs. (3.26) and (3.27) to find the distant fields. Two of the easier antenna problems to solve are the linear antenna and the loop antenna.

3.7 Linear Antennas

Fundamental to understanding antenna radiation is to understand the behavior of a straight wire or linear antenna. Not only is the math greatly simplified by a straight wire segment, but the linear antenna solution gives insight into the behavior of many more complicated structures that can often be viewed as a collection of straight wire segments.

3.7.1 Infinitesimal Dipole

The infinitesimal dipole is a short wire segment antenna, where the length $L \ll \lambda$. It is aligned along the z axis symmetrically placed about the x-y plane as shown in Fig. 3.14.

The phasor current is given by $\bar{I} = I_0 \hat{z}$. The position and distance vectors are given by $\bar{r} = r\hat{r} = x\hat{x} + y\hat{y} + z\hat{z}$, $\bar{r}' = z'\hat{z}$, and $\bar{R} = x\hat{x} + y\hat{y} + (z - z')\hat{z}$. Thus, the vector potential is given by

$$\bar{A} = \frac{\mu_0}{4\pi} \int_{-L/2}^{L/2} I_0 \hat{z} \frac{e^{-jk\sqrt{x^2+y^2+(z-z')^2}}}{\sqrt{x^2 + y^2 + (z - z')^2}} dz' \tag{3.29}$$

Because we are assuming an infinitesimal dipole ($r \gg z'$), $R \approx r$. Thus, the integral can easily be solved yielding

$$\bar{A} = \frac{\mu_0}{4\pi} \int_{-L/2}^{L/2} I_0 \hat{z} \frac{e^{-jkr}}{r} dz' = \frac{\mu_0 I_0 L}{4\pi r} e^{-jkr} \hat{z} = A_z \hat{z} \tag{3.30}$$

Because most antenna fields are more conveniently expressed in spherical coordinates, we may apply a vector transformation to Eq. (3.30) or

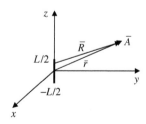

FIGURE 3.14 Infinitesimal dipole.

Antenna Fundamentals

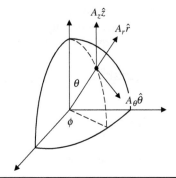

Figure 3.15 Vector potential for an infinitesimal dipole.

we may graphically determine $\bar{A}$ in spherical coordinates. Figure 3.15 shows the vector potential in both rectangular and spherical coordinates.

The A_r and A_θ components are vector projections of A_z onto the corresponding axes. It can thus be shown that

$$A_r = A_z \cos\theta = \frac{\mu_0 I_0 L e^{-jkr}}{4\pi r} \cos\theta \tag{3.31}$$

$$A_\theta = -A_z \sin\theta = -\frac{\mu_0 I_0 L e^{-jkr}}{4\pi r} \sin\theta \tag{3.32}$$

Because $\bar{H}$ is related to the curl of $\bar{A}$ in Eq. (3.26), we can take the curl in spherical coordinates to get

$$H_\phi = \frac{jk I_0 L \sin\theta}{4\pi r}\left[1+\frac{1}{jkr}\right]e^{-jkr} \tag{3.33}$$

where $H_r = 0$ and $H_\theta = 0$.

The electric field can be found from Eq. (3.33) by substituting into Eq. (3.27) giving

$$E_r = \frac{\eta I_0 L \cos\theta}{2\pi r^2}\left[1+\frac{1}{jkr}\right]e^{-jkr} \tag{3.34}$$

$$E_\theta = \frac{jk\eta I_0 L \sin\theta}{4\pi r}\left[1+\frac{1}{jkr}-\frac{1}{(kr)^2}\right]e^{-jkr} \tag{3.35}$$

where $E_\phi = 0$ and η = intrinsic impedance of the medium.

In the far field, the higher-order terms involving $1/r^2$ and $1/r^3$ become negligibly small simplifying Eqs. (3.34) and (3.35) to become

$$E_\theta = \frac{jk\eta I_0 L \sin\theta}{4\pi r} e^{-jkr} \tag{3.36}$$

$$H_\phi = \frac{jk I_0 L \sin\theta}{4\pi r} e^{-jkr} \tag{3.37}$$

It should be noted that in the far field $\frac{E_\theta}{H_\phi} = \eta$.

Power Density and Radiation Intensity

We can calculate the far-field power density and the radiation intensity by substituting Eq. (3.36) into Eqs. (3.8) and (3.12) to get

$$W_r(\theta,\phi) = \frac{1}{2\eta}\left|\frac{k\eta I_0 L \sin\theta}{4\pi r}\right|^2 = \frac{\eta}{8}\left|\frac{I_0 L}{\lambda}\right|^2 \frac{\sin^2\theta}{r^2} \quad (3.38)$$

$$U(\theta) = \frac{\eta}{8}\left|\frac{I_0 L}{\lambda}\right|^2 \sin^2\theta \quad (3.39)$$

where I_0 = complex phasor current = $|I_0|e^{j\varsigma}$
λ = wavelength

Figure 3.16 shows a plot of the normalized radiation intensity given in Eq. (3.39). The radiation intensity is superimposed on a rectangular coordinate system. Because the antenna is aligned along the z axis, the maximum radiation is broadside to the infinitesimal dipole.

Directivity

The directivity, as defined in Eq. (3.16), can be applied to the infinitesimal dipole radiation intensity. The constant terms divide out in the numerator and denominator yielding

$$D(\theta) = \frac{4\pi \sin^2\theta}{\int_0^{2\pi}\int_0^{\pi} \sin^3\theta\, d\theta\, d\phi} = 1.5\sin^2\theta \quad (3.40)$$

This solution is exactly the same as derived in Example 3.3.

3.7.2 Finite Length Dipole

The same procedure can also be applied to a finite length dipole to determine the far fields and the radiation pattern. However, because a finite length dipole can be viewed as the concatenation of numerous infinitesimal dipoles, we can use the principle of superposition to

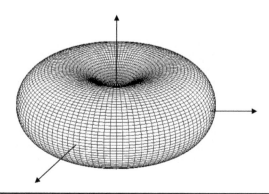

FIGURE 3.16 3D plot of infinitesimal dipole radiation.

find the fields. Superposing numerous infinitesimal dipoles of length dz' results in an integral given as [1]

$$E_\theta = \frac{j\eta k e^{-jkr}}{4\pi r} \sin\theta \int_{-L/2}^{L/2} I(z') e^{jkz'\cos\theta} dz' \qquad (3.41)$$

Because dipoles are center fed and the currents must terminate at the ends, a good approximation for the dipole current is given as sinusoidal (King [5]). It is well known that a twin lead transmission line with an open-circuit termination generates sinusoidal standing waves along the conductors. If the ends of the leads are bent, so as to form a dipole, the currents can still be approximated as piece-wise sinusoidal. Figure 3.17a shows a twin lead transmission line with sinusoidal currents. Figure 3.17b shows a twin lead transmission line with sinusoidal currents, which is terminated into a dipole. The dipole currents can be viewed as an extension of the existing transmission line currents.

Because sinusoidal currents are a good approximation to the currents on a linear antenna, we can devise an analytic expression for the current in Eq. (3.41) [1, 3] as

$$I(z') = \begin{cases} I_0 \sin\left[k(\tfrac{L}{2}-z')\right] & 0 \le z' \le L/2 \\ I_0 \sin\left[k(\tfrac{L}{2}-z')\right] & -L/2 \le z' \le 0 \end{cases} \qquad (3.42)$$

By substituting Eq. (3.42) into Eq. (3.41), we can solve for the approximate electric far field as

$$E_\theta = \frac{j\eta I_0 e^{-jkr}}{2\pi r} \left[\frac{\cos\left(\tfrac{kL}{2}\cos\theta\right) - \cos\left(\tfrac{kL}{2}\right)}{\sin\theta}\right] \qquad (3.43)$$

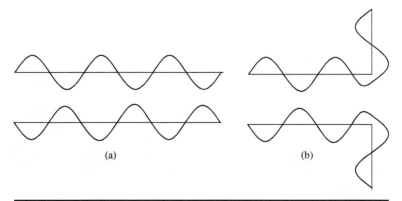

FIGURE 3.17 Standing waves on a transmission line and on a dipole.

Chapter Three

The magnetic field can easily be found to be

$$H_\phi = \frac{jI_0 e^{-jkr}}{2\pi r}\left[\frac{\cos\left(\frac{kL}{2}\cos\theta\right)-\cos\left(\frac{kL}{2}\right)}{\sin\theta}\right] \quad (3.44)$$

Power Density and Radiation Intensity

We can again calculate the far-field power density and the radiation intensity to be

$$W_r(\theta,\phi) = \frac{1}{2\eta}|E_\theta|^2 = \frac{\eta}{8}\left|\frac{I_0}{\pi r}\right|^2 \left[\frac{\cos\left(\frac{kL}{2}\cos\theta\right)-\cos\left(\frac{kL}{2}\right)}{\sin\theta}\right]^2 \quad (3.45)$$

$$U(\theta) = \frac{\eta}{8}\left|\frac{I_0}{\pi}\right|^2 \left[\frac{\cos\left(\frac{kL}{2}\cos\theta\right)-\cos\left(\frac{kL}{2}\right)}{\sin\theta}\right]^2 \quad (3.46)$$

Figure 3.18 shows plots of the normalized radiation intensity given in Eq. (3.46) in three dimensions for $\frac{L}{\lambda} = 0.5, 1$, and 1.5. It can be seen that as the dipole length increases, the mainlobe narrows. However, in the $L/\lambda = 1.5$ case, the mainlobe no longer is perpendicular to the dipole axis. Usually dipoles are designed to be of length $L = \lambda/2$.

Directivity

The directivity of the finite length dipole is given by substituting Eq. (3.46) into Eq. (3.16).

$$D(\theta) = \frac{4\pi\left[\dfrac{\cos\left(\pi\frac{L}{\lambda}\cos\theta\right)-\cos\left(\pi\frac{L}{\lambda}\right)}{\sin\theta}\right]^2}{\displaystyle\int_0^{2\pi}\int_0^{\pi}\left[\dfrac{\cos\left(\pi\frac{L}{\lambda}\cos\theta\right)-\cos\left(\pi\frac{L}{\lambda}\right)}{\sin\theta}\right]^2 \sin\theta\, d\theta\, d\phi} \quad (3.47)$$

where L/λ is the length in wavelengths.

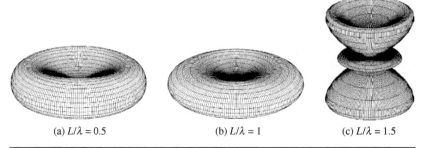

(a) $L/\lambda = 0.5$ (b) $L/\lambda = 1$ (c) $L/\lambda = 1.5$

Figure 3.18 Finite dipole radiation intensity.

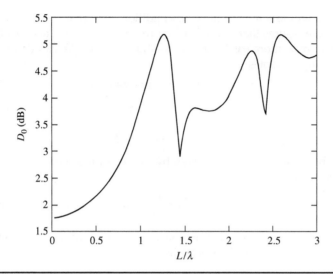

Figure 3.19 Maximum directivity of finite length dipole.

The maximum directivity is given when the numerator is at its maximum value. We can plot the finite dipole maximum directivity (D_0) versus the length L in wavelengths to yield Fig. 3.19.

3.8 Loop Antennas

In addition to solving for the fields radiated from linear antennas, it is also instructive to calculate fields radiated by loop antennas. Both linear and loop antennas form the backbone for numerous antenna problems and both forms of antennas shed light on general antenna behavior.

3.8.1 Loop of Constant Phasor Current

Figure 3.20 shows a loop antenna of radius a, centered on the z axis and residing in the x-y plane. The loop current I_0 flows in the $\hat{\phi}'$ direction.

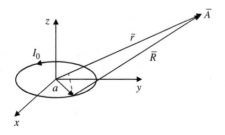

Figure 3.20 Small loop antenna.

We will assume the far-field condition such that ($a \ll r$). We can therefore assume that $\bar{r}$ and $\bar{R}$ are approximately parallel to each other. We can now more easily determine the vector $\bar{R}$. We can define the following variables:

$$\bar{r} = r\sin\theta\cos\phi\hat{x} + \sin\theta\sin\phi\hat{y} + \cos\theta\hat{z} \quad (3.48)$$

$$\bar{r}' = a\cos\phi'\hat{x} + a\sin\phi'\hat{y} \quad (3.49)$$

$$R \approx r - \bar{r}'\cdot\hat{r} = r - a\sin\theta\cos(\phi - \phi') \quad (3.50)$$

The vector potential is given by the integral of Eq. (3.27) and is found to be

$$\bar{A} = \frac{\mu_0}{4\pi}\int_0^{2\pi} I_0\hat{\phi}'\frac{e^{-jkR}}{R}ad\phi' \quad (3.51)$$

where $\hat{\phi}'$ is the current direction in source coordinates. It can be converted to field coordinates by the relation

$$\hat{\phi}' = \cos(\phi - \phi')\hat{\phi} + \sin(\phi - \phi')\hat{\rho} \quad (3.52)$$

We can safely approximate the denominator term in Eq. (3.51) as $R \approx r$ without appreciably changing the solution. The exponent term, however, must preserve the phase information, and therefore we will use R exactly as defined by Eq. (3.50). Substituting Eqs. (3.52) and (3.50) into Eq. (3.51), we get

$$\bar{A} = \frac{a\mu_0}{4\pi}\int_0^{2\pi} I_0(\cos(\phi-\phi')\hat{\phi} + \sin(\phi-\phi')\hat{\rho})\frac{e^{-jk(r-a\sin\theta\cos(\phi-\phi'))}}{r}d\phi' \quad (3.53)$$

Because the loop is symmetric, the same solution will result regardless of the ϕ value chosen. For simplification purposes, let us choose $\phi = 0$. Also, because the $\hat{\rho}$ term has odd symmetry, the $\hat{\rho}$ term integrates to zero. Thus, Eq. (3.53) thus simplifies to

$$\bar{A} = \frac{a\mu_0 I_0\hat{\phi}}{4\pi}\frac{e^{-jkr}}{r}\int_0^{2\pi}\cos\phi' e^{jka\sin\theta\cos\phi'}d\phi' \quad (3.54)$$

Equation (3.54) is solvable in closed form and the integral solution can be found in Gradshteyn and Ryzhik [6]. The vector potential solution is given as

$$\bar{A} = \frac{ja\mu_0 I_0\hat{\phi}}{2}\frac{e^{-jkr}}{r}J_1(ka\sin\theta) \quad (3.55)$$

where J_1 is the Bessel function of the first kind and order 1.

The electric and magnetic field intensities can be found from Eqs. (3.26) and (3.27) to be

$$E_\phi = \frac{ak\eta I_0}{2}\frac{e^{-jkr}}{r}J_1(ka\sin\theta) \quad (3.56)$$

where $E_r \approx 0$ and $E_\theta \approx 0$.

Antenna Fundamentals

$$H_\theta = -\frac{E_\phi}{\eta} = -\frac{akI_0}{2}\frac{e^{-jkr}}{r}J_1(ka\sin\theta) \qquad (3.57)$$

where H_r and $H_\phi \approx 0$.

Power Density and Radiation Intensity

The power density and radiation intensity can be easily found to be

$$W_r(\theta,\phi) = \frac{1}{2\eta}|E_\phi|^2 = \frac{\eta}{8}\left(\frac{2\pi a}{\lambda}\right)^2 \frac{|I_0|^2}{r^2}J_1^2(ka\sin\theta) \qquad (3.58)$$

$$U(\theta,\phi) = \frac{\eta}{8}\left(\frac{2\pi a}{\lambda}\right)^2 |I_0|^2 J_1^2(ka\sin\theta) \qquad (3.59)$$

The variable ka can also be written as $\frac{2\pi a}{\lambda} = \frac{C}{\lambda}$, where C is the circumference of the loop. Figure 3.21 shows plots of the normalized radiation intensity given in Eq. (3.59) for $\frac{C}{\lambda} = 0.5, 1.25,$ and 3. The radiation intensity is superimposed on a rectangular coordinate system.

Directivity

The directivity of the constant phasor current loop is again given by Eq. (3.16). We can substitute Eq. (3.59) into Eq. (3.16) to yield

$$D(\theta) = \frac{4\pi J_1^2(ka\sin\theta)}{\int_0^{2\pi}\int_0^{\pi} J_1^2(ka\sin\theta)\sin\theta\, d\theta\, d\phi} = \frac{2J_1^2(\frac{C}{\lambda}\sin\theta)}{\int_0^{\pi} J_1^2(\frac{C}{\lambda}\sin\theta)\sin\theta\, d\theta} \qquad (3.60)$$

where $C = 2\pi a =$ loop circumference.

The maximum directivity (D_0) is given when the numerator is maximum. The numerator is maximum when the Bessel function is maximum. For $C/\lambda > 1.84$ the Bessel function maximum is always 0.582. For $C/\lambda > 1.84$ the maximum is given as $2J_1^2(C/\lambda)$. We can plot

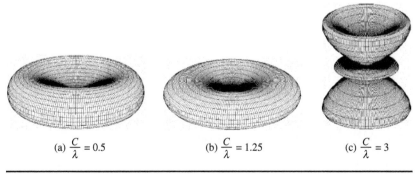

(a) $\frac{C}{\lambda} = 0.5$ (b) $\frac{C}{\lambda} = 1.25$ (c) $\frac{C}{\lambda} = 3$

FIGURE 3.21 Loop radiation intensity.

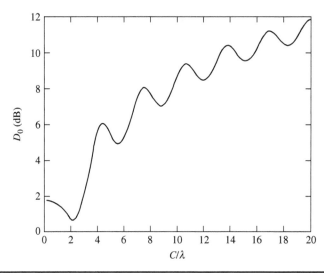

FIGURE 3.22 Maximum directivity of a circular loop.

the loop maximum directivity (D_0) versus the circumference C in wavelengths to yield Fig. 3.22.

3.9 References

1. Balanis, C., *Antenna Theory Analysis and Design*, 2d ed., Wiley, New York, 1997.
2. Kraus, J., and R. Marhefka, *Antennas for All Applications*, 3d ed., McGraw-Hill, New York, 2002.
3. Stutzman, W. L., and G. A. Thiele, *Antenna Theory and Design*, Wiley, New York, 1981.
4. Haykin, S., *Communication Systems*, Wiley, New York, 2001, p. 723.
5. King, R. W. P., "The Linear Antenna—Eighty Years of Progress," *Proceedings of the IEEE*, Vol. 55, pp. 2–16, Jan. 1967.
6. Gradshteyn, I. S., and I. M. Ryzhik, *Table of Integrals, Series, and Products*, Academic Press, New York, 1980.

3.10 Problems

3.1 For an antenna in free space creating an electric field intensity given by

$$\bar{E}_s = \frac{4\cos\theta}{r} e^{-jkr} \hat{\theta} \text{ V/m}$$

(a) What is the average power density?
(b) What is the radiation intensity?
(c) What is the total power radiated from the antenna?
(d) Use MATLAB to plot the normalized power pattern.

3.2 Use the help given in Example 3.4 and plot the following power patterns using MATLAB for $-90° < \theta < 90°$:
(a) $\cos^4(\theta)$
(b) $\sin^2(\theta)$
(c) $\sin^4(\theta)$

3.3 What is the 3-dB beamwidth for the patterns given in Prob. 3.2?

3.4 For the following radiation intensities, use MATLAB to plot the elevation plane polar pattern when $\phi = 0°$ and the azimuthal plane polar pattern when $\theta = 90°$. What are the boresight directions for these two patterns? Plot the first pattern for $0° < \theta < 180°$ and the second pattern for $-90° < \phi < 90°$.
 (a) $U(\theta,\phi) = \sin^2(\theta)\cos^2(\phi)$
 (b) $U(\theta,\phi) = \sin^6(\theta)\cos^6(\phi)$

3.5 Calculate the maximum directivity for the following radiation intensities:
 (a) $U(\theta,\phi) = 4\cos^2(\theta)$
 (b) $U(\theta,\phi) = 2\sin^4(\theta)$
 (c) $U(\theta,\phi) = 2\sin^2(\theta)\cos^2(\phi)$
 (d) $U(\theta,\phi) = 6\sin^2(2\theta)$

3.6 Using Eq. (3.19), find the beam solid angle for the following radiation intensities:
 (a) $U(\theta,\phi) = 2\cos^2(\theta)$
 (b) $U(\theta,\phi) = 4\sin^2(\theta)\cos^2(\phi)$
 (c) $U(\theta,\phi) = 4\cos^4(\theta)$

3.7 For the figure shown below calculate the receive power P_2 at antenna 2 if the transmit power $P_1 = 5$ kW. The transmitter gain is $G_1(\theta_1, \phi_1) = 4\sin^4(\theta_1)$ and the receiver gain is $G_2(\theta_2, \phi_2) = 2\sin^2(\theta_2)$. The operating frequency is 10 GHz. Use the Fig. 3P.1 to solve the problem. The y-z and y'-z' planes are coplanar.

3.8 Create three normalized plots in one polar plot figure using MATLAB for the finite length dipole, where $\frac{L}{\lambda} = 0.5, 1,$ and 1.5. Normalize each plot before plotting. One can use the "hold on" command to overlay the plots in the same figure.

3.9 What is the maximum directivity in dB for a finite length dipole whose length is $\frac{L}{\lambda} = 1.25$?

3.10 Create three normalized plots in one polar plot figure using MATLAB for the loop antenna where $\frac{C}{\lambda} = 0.5, 1.25,$ and 3. Normalize each plot before plotting. One can use the "hold on" command to overlay the plots in the same figure.

3.11 What is the maximum directivity in dB for the loop antenna whose circumference in wavelengths is $\frac{C}{\lambda} = 10$?

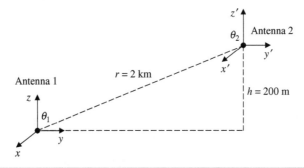

FIGURE 3P.1 Two antennas separated by a distance r.

CHAPTER 4
Array Fundamentals

Smart antennas are composed of a collection of two or more antennas working in concert to establish a unique radiation pattern for the electromagnetic environment at hand. The antenna elements are allowed to work in concert by means of array element phasing, which is accomplished with hardware or is performed digitally. In Chap. 3 we considered individual antenna elements such as the dipole or loop. In this chapter we look at generic collections of antennas. These collections could be composed of dipole or loop antennas, but it is not necessary to restrict ourselves to any particular antenna elements. As we will discover, the behavior of arrays transcends the specific elements used and the subject of arrays in and of itself has generated an extensive body of work. In fact the subject of arrays has merited entire textbooks devoted to the subject. Some very useful texts include Haykin [1], Johnson and Dudgeon [2], and Van Trees [3]. Extensive material can be found in the phased array text by Brookner [4] or in the article by Dudgeon [5].

Arrays of antennas can assume any geometric form. The various array geometries of common interest are linear arrays, circular arrays, planar arrays, and conformal arrays. We begin with a discussion of linear arrays. A thorough treatment of linear arrays is found in the texts by Balanis [6] and Kraus and Marhefka [7].

4.1 Linear Arrays

The simplest array geometry is the linear array. Thus, all elements are aligned along a straight line and generally have a uniform interelement spacing. Linear arrays are the simplest to analyze and many valuable insights can be gained by understanding their behavior. The minimum length linear array is the two-element array.

4.1.1 Two-Element Array

The most fundamental and simplest array to analyze is the two-element array. The two-element array demonstrates the same general behavior as much larger arrays and is a good starting point in order to understand the phase relationship between adjacent array elements. Figure 4.1 shows two vertically polarized infinitesimal

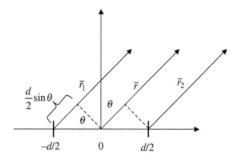

Figure 4.1 Two infinitesimal dipoles.

dipoles aligned along the y axis and separated by a distance d. The field point is located at a distance r from the origin such that $r \gg d$. We can therefore assume that the distance vectors $\bar{r}_1, \bar{r}$, and $\bar{r}_2$ are all approximately parallel to each other.

We can therefore make the following approximations:

$$r_1 \approx r + \frac{d}{2}\sin\theta \tag{4.1}$$

$$r_2 \approx r - \frac{d}{2}\sin\theta \tag{4.2}$$

Let us additionally assume that the electrical phase of element 1 is $-\delta/2$ such that the phasor current in element 1 is $I_0 e^{-j\frac{\delta}{2}}$. The electrical phase of element 2 is $+\delta/2$ such that the phasor current in element 2 is $I_0 e^{-j\frac{\delta}{2}}$. We can now find the distant electric field by using superposition as applied to these two dipole elements. Using Eq. (3.36) and Eqs. (4.1) and (4.2) and assuming that $r_1 \approx r_1 \approx r$ in the denominator, we can now find the total electric field.

$$E_\theta = \frac{jk\eta I_0 e^{-j\frac{\delta}{2}} L \sin\theta}{4\pi r_1} e^{-jkr_1} + \frac{jk\eta I_0 e^{j\frac{\delta}{2}} L \sin\theta}{4\pi r_2} e^{-jkr_2}$$

$$= \frac{jk\eta I_0 L \sin\theta}{4\pi r} e^{-jkr} \left[e^{-j\frac{(kd\sin\theta + \delta)}{2}} + e^{j\frac{(kd\sin\theta + \delta)}{2}} \right] \tag{4.3}$$

where δ = electrical phase difference between the two adjacent elements
L = dipole length
θ = angle as measured from the z axis in spherical coordinates
d = element spacing

We can further simplify Eq. (4.3) such that

$$E_\theta = \underbrace{\frac{jk\eta I_0 L e^{-jkr}}{4\pi r} \sin\theta}_{\text{Element factor}} \cdot \underbrace{\left(2\cos\left(\frac{(kd\sin\theta + \delta)}{2}\right) \right)}_{\text{Array factor}} \tag{4.4}$$

where the element factor is the far-field equation for one dipole and the array factor is the pattern function associated with the array geometry.

The distant field from an array of identical elements can always be broken down into the product of the element factor (EF) and the array factor (AF). The very fact that the antenna pattern can be multiplied by the array factor pattern demonstrates a property called *pattern multiplication*. Thus, the far-field pattern of any array of antennas is always given by (EF) × (AF). The AF is dependent on the geometric arrangement of the array elements, the spacing of the elements, and the electrical phase of each element.

The normalized radiation intensity can be found by substituting Eq. (4.4) into Eq. (3.13) to get

$$U_n(\theta) = [\sin\theta]^2 \cdot \left[\cos\left(\frac{(kd\sin\theta + \delta)}{2}\right)\right]^2$$
$$= [\sin\theta]^2 \cdot \left[\cos\left(\frac{\pi d}{\lambda}\sin\theta + \frac{\delta}{2}\right)\right]^2$$

(4.5)

We can demonstrate pattern multiplication by plotting Eq. (4.5) for the case where $d/\lambda = .5$ and $\delta = 0$. This is shown in Fig. 4.2. Figure 4.2*a* shows the power pattern for the dipole element alone. Figure 4.2*b* shows the array factor power pattern alone. And Fig. 4.2*c* shows the multiplication of the two patterns.

The overriding principle demonstrated by the two-element array is that we can separate the element factor from the array factor. The array factor can be calculated for any array regardless of the individual elements chosen as long as all elements are the same. Thus, it is easier to first analyze arrays of isotropic elements. When the general array design is complete, one can implement the design by inserting the specific antenna elements required. Those antenna elements can include, but are not restricted to, dipoles, loops, horns, waveguide apertures, and patch antennas. A more exact representation of array radiation must always include the effects of coupling between adjacent antenna elements. However, that topic is beyond the scope of this book and is left for treatment by more advanced texts. The reader may refer to the text by Balanis [6] for more information on element coupling.

4.1.2 Uniform N-Element Linear Array

The more general linear array is the *N*-element array. For simplification purposes, we assume that all elements are equally spaced and have equal amplitudes. Later we may allow the antenna elements to have any arbitrary amplitude. Figure 4.3 shows an *N*-element linear array composed of isotropic radiating antenna elements. It is assumed

66 Chapter Four

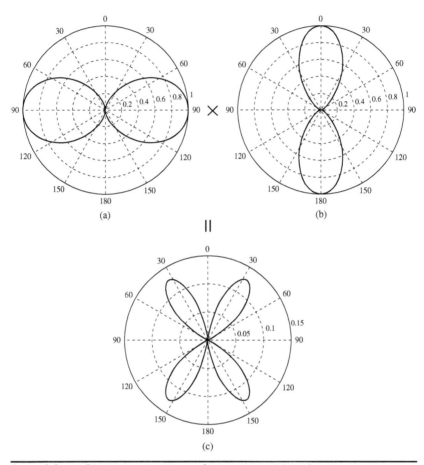

FIGURE 4.2 (a) Dipole pattern, (b) array factor pattern, (c) total pattern.

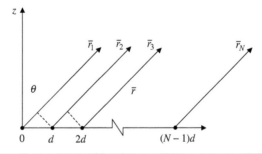

FIGURE 4.3 N-element linear array.

Array Fundamentals

that the nth element leads the $(n-1)$ element by an electrical phase shift of δ radians. This phase shift can easily be implemented by shifting the phase of the antenna current for each element.

Assuming far-field conditions such that $r \gg d$, we can derive the array factor as follows:

$$\text{AF} = 1 + e^{j(kd\sin\theta+\delta)} + e^{j2(kd\sin\theta+\delta)} + \cdots + e^{j(N-1)(kd\sin\theta+\delta)} \quad (4.6)$$

where δ is the phase shift from element to element.

This series can more concisely be expressed by

$$\text{AF} = \sum_{n=1}^{N} e^{j(n-1)(kd\sin\theta+\delta)} = \sum_{n=1}^{N} e^{j(n-1)\psi} \quad (4.7)$$

where $\psi = kd\sin\theta + \delta$.

It should be noted that if the array is aligned along the z axis, $\psi = kd\cos\theta + \delta$.

Because each isotropic element has unity amplitude, the entire behavior of this array is dictated by the phase relationship between the elements. The phase is directly proportional to the element spacing in wavelengths.

The array processing and array beamforming textbooks have taken an alternative approach to expressing Eq. (4.7). Let us begin by defining the array vector.

$$\bar{a}(\theta) = \begin{bmatrix} 1 \\ e^{j(kd\sin\theta+\delta)} \\ \vdots \\ e^{j(N-1)(kd\sin\theta+\delta)} \end{bmatrix} = [1 \quad e^{j(kd\sin\theta+\delta)} \quad \cdots \quad e^{j(N-1)(kd\sin\theta+\delta)}]^T \quad (4.8)$$

where $[\ ]^T$ signifies the transpose of the vector within the brackets.

The vector $\bar{a}(\theta)$ is a Vandermonde vector because it is in the form $[1\ z \cdots z^{(N-1)}]$. In the literature the array vector has been alternatively called: the *array steering vector* [2], the *array propagation vector* [8, 9], the *array response vector* [10], and the *array manifold vector* [3]. For simplicity's sake, we call $\bar{a}(\theta)$ the array vector. Therefore, the array factor, in Eq. (4.7), can alternatively be expressed as the sum of the elements of the array vector.

$$\text{AF} = \text{sum}(\bar{a}(\theta)) \quad (4.9)$$

The utility of the vector notation in Eq. (4.8) is more readily seen in Chaps. 7 and 8 when we study angle-of-arrival estimation and smart antennas. In our current development, it is sufficient to use the notation of Eq. (4.7).

We may simplify the expression in Eq. (4.6) by multiplying both sides by $e^{j\psi}$ such that

$$e^{j\psi}\text{AF} = e^{j\psi} + e^{j2\psi} + \cdots + e^{jN\psi} \quad (4.10)$$

Subtracting Eq. (4.6) from Eq. (4.10) yields

$$(e^{j\psi} - 1)AF = (e^{jN\psi} - 1) \tag{4.11}$$

The array factor can now be rewritten.

$$AF = \frac{(e^{jN\psi} - 1)}{(e^{j\psi} - 1)} = \frac{e^{j\frac{N}{2}\psi}\left(e^{j\frac{N}{2}\psi} - e^{-j\frac{N}{2}\psi}\right)}{e^{j\frac{\psi}{2}}\left(e^{j\frac{\psi}{2}} - e^{-j\frac{\psi}{2}}\right)}$$

$$= e^{j\frac{(N-1)}{2}\psi} \frac{\sin\left(\frac{N}{2}\psi\right)}{\sin\left(\frac{\psi}{2}\right)} \tag{4.12}$$

The $e^{j\frac{(N-1)}{2}\psi}$ term accounts for the fact that the physical center of the array is located at $(N-1)d/2$. This array center produces a phase shift of $(N-1)\psi/2$ in the array factor. If the array is centered about the origin, the physical center is at 0 and Eq. (4.12) can be simplified to become

$$AF = \frac{\sin\left(\frac{N}{2}\psi\right)}{\sin\left(\frac{\psi}{2}\right)} \tag{4.13}$$

The maximum value of AF is when the argument $\psi = 0$. In that case $AF = N$. This is intuitively obvious as an array of N elements should have a gain of N over a single element. We may normalize the AF to be reexpressed as

$$AF_n = \frac{1}{N} \frac{\sin\left(\frac{N}{2}\psi\right)}{\sin\left(\frac{\psi}{2}\right)} \tag{4.14}$$

In the cases where the argument $\psi/2$ is very small, we can invoke the small argument approximation for the $\sin(\psi/2)$ term to yield an approximation

$$AF_n \approx \frac{\sin\left(\frac{N}{2}\psi\right)}{\frac{N}{2}\psi} \tag{4.15}$$

It should be noted that the array factor of Eq. (4.15) takes the form of a $\sin(x)/x$ function. This is because the uniform array itself presents a finite sampled rectangular window through which to radiate or receive a signal. The spatial Fourier transform of a rectangular window yields a $\sin(x)/x$ function. The Fourier transform relationship between an antenna array and its radiation pattern is explained in [6].

Let us now determine the array factor nulls, maxima, and the mainlobe beamwidth.

Nulls

From Eq. (4.15), the array nulls occur when the numerator argument $N\psi/2 = \pm n\pi$. Thus, the array nulls are given when

$$\frac{N}{2}(kd\sin\theta_{null} + \delta) = \pm n\pi$$

or

$$\theta_{null} = \sin^{-1}\left(\frac{1}{kd}\left(\pm\frac{2n\pi}{N} - \delta\right)\right) \qquad n = 1, 2, 3\ldots \qquad (4.16)$$

Because the $\sin(\theta_{null}) \le 1$, for real angles, the argument in Eq. (4.16) must be ≤ 1. Thus, only a finite set of n values satisfies the equality.

Example 4.1 Find all of the nulls for an $N = 4$ element array with $d = .5\lambda$ and $\delta = 0$.

Solution Substituting N, d, and δ into Eq. (4.16)

$$\theta_{null} = \sin^{-1}\left(\pm\frac{n}{2}\right)$$

Thus, $\theta_{null} = \pm 30°, \pm 90°$.

Maxima

The mainlobe maximum in Eq. (4.15) occurs when the denominator term $\psi/2 = 0$. Thus

$$\theta_{max} = -\sin^{-1}\left(\frac{\delta\lambda}{2\pi d}\right) \qquad (4.17)$$

The sidelobe maxima occur approximately when the numerator is a maximum. This occurs when the numerator argument $N\psi/2 = \pm(2n+1)\pi/2$. Thus

$$\theta_s = \sin^{-1}\left(\frac{1}{kd}\left(\pm\frac{(2n+1)\pi}{N} - \delta\right)\right)$$

$$= \pm\frac{\pi}{2} + \cos^{-1}\left(\frac{1}{kd}\left(\pm\frac{(2n+1)\pi}{N} - \delta\right)\right) \qquad (4.18)$$

Example 4.2 Find the mainlobe maximum and the sidelobe maxima for the case where $N = 4$, $d = .5\lambda$, and $\delta = 0$.

Solution Using Eq. (4.17), the mainlobe maximum can be found to be $\theta_{max} = 0$ or π. π is a valid solution because the array factor is symmetric about the $\theta = \pi/2$ plane. The sidelobe maxima can be found from Eq. (4.18) to be

$$\theta_s = \pm 48.59°, \pm 131.4°$$

Beamwidth

The beamwidth of a linear array is determined by the angular distance between the half-power points of the mainlobe. The mainlobe

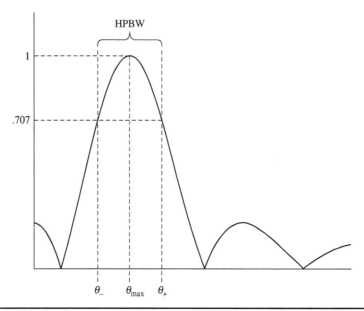

FIGURE 4.4 Half-power beamwidth of a linear array.

maximum is given by Eq. (4.17). Figure 4.4 depicts a typical normalized array radiation pattern with beamwidth as indicated.

The two half-power points (θ_+ and θ_-) are found when the normalized $AF_n = .707 (AF_n^2 = .5)$. If we use the array approximation given in Eq. (4.15), we can simplify the calculation of the beamwidth.

It is well known that $\sin(x)/x = .707$ when $x = \pm 1.391$. Thus, the normalized array factor is at the half-power points when

$$\frac{N}{2}(kd \sin \theta_\pm + \delta) = \pm 1.391 \tag{4.19}$$

Rearranging to solve for $\theta_\pm$, we get

$$\theta_\pm = \sin^{-1}\left(\frac{1}{kd}\left(\frac{\pm 2.782}{N} - \delta\right)\right) \tag{4.20}$$

The half-power beamwidth is now easily shown to be

$$\text{HPBW} = |\theta_+ - \theta_-| \tag{4.21}$$

For large arrays, the beamwidth is narrow enough such that the HPBW can be approximated as

$$\text{HPBW} = 2|\theta_+ - \theta_{max}| = 2|\theta_{max} - \theta_-| \tag{4.22}$$

θ_{max} is given by Eq. (4.17) and $\theta_\pm$ is given by Eq. (4.20).

Example 4.3 For the four-element linear array with $\delta = -2.22$ and $d = .5\lambda$, what is the half-power beamwidth?

Solution θ_{max} is first found using Eq. (4.17). Thus, $\theta_{max} = 45°$. θ_+ is found using Eq. (4.20). Thus, $\theta_+ = 68.13°$. The half-power beamwidth is then approximated by using Eq. (4.21) to be HPBW = 46.26°.

Broadside Linear Array

The most common mode of operation for a linear array is in the broadside mode. This is the case where $\delta = 0$ such that all element currents are in phase. Figure 4.5 shows three polar plots for a four-element array for element distances $d/\lambda = .25, .5,$ and $.75$.

This array is called a *broadside array* because the maximum radiation is broadside to the array geometry. Two major lobes are seen because the broadside array is symmetric about the $\theta = \pm\pi/2$ line. As the array element spacing increases, the array physically is longer,

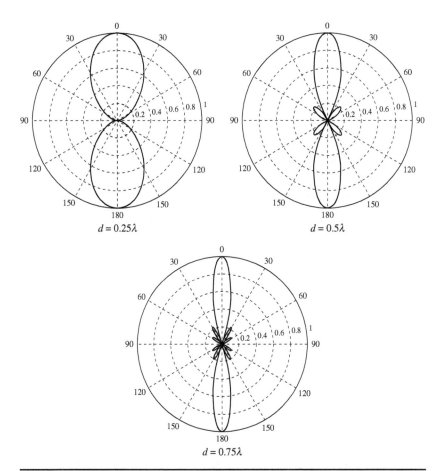

FIGURE 4.5 The four-element broadside array with $\delta = 0$ and $d = .25\lambda, .5\lambda,$ and $.75\lambda$.

Chapter Four

thereby decreasing the mainlobe width. The general rule for array radiation is that the mainlobe width is inversely proportional to the array length.

End-Fire Linear Array

The name *end-fire* indicates that this array's maximum radiation is along the axis containing the array elements. Thus, maximum radiation is "out the end" of the array. This case is achieved when $\delta = -kd$. Figure 4.6 shows three polar plots for the end-fire four-element array for the distances $d/\lambda = .25, .5,$ and $.75$.

It should be noted that the mainlobe width for the ordinary end-fire case is much greater than that for the broadside case. Thus, ordinary end-fire arrays do not afford the same beamwidth efficiency as the broadside array. The beamwidth efficiency in this context is the beamwidth available relative to the overall array length.

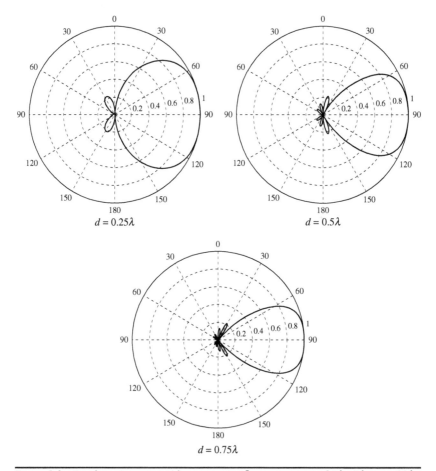

FIGURE 4.6 The four-element end-fire array with $\delta = -kd$ and $d = .25\lambda, .5\lambda,$ and $.75\lambda$.

An increased directivity end-fire array has been developed by Hansen and Woodyard [11] where the phase shift is modified such that $\delta = -(kd + \frac{\pi}{N})$. This provides a dramatic improvement in the beamwidth and directivity. Because the Hansen and Woodyard 1938 paper is not generally accessible, a detailed derivation can be found in [5].

Beamsteered Linear Array

A beamsteered linear array is an array where the phase shift δ is a variable, thus allowing the mainlobe to be directed toward any direction of interest. The broadside and end-fire conditions are special cases of the more generalized beamsteered array. The beamsteering conditions can be satisfied by defining the phase shift $\delta = -kd \sin\theta_0$. We may rewrite the array factor in terms of beamsteering such that

$$AF_n = \frac{1}{N} \frac{\sin\left(\frac{Nkd}{2}(\sin\theta - \sin\theta_0)\right)}{\sin\left(\frac{kd}{2}(\sin\theta - \sin\theta_0)\right)} \quad (4.23)$$

Figure 4.7 shows polar plots for the beamsteered eight-element array for the $d/\lambda = .5$, and $\theta_0 = 20°$, $40°$, and $60°$. Major lobes exist above and below the horizontal because of array is symmetry.

The beamwidth of the beamsteered array can be determined by using Eqs. (4.20) and (4.21) such that

$$\theta_\pm = \sin^{-1}\left(\pm\frac{2.782}{Nkd} + \sin\theta_0\right) \quad (4.24)$$

where $\delta = -kd \sin\theta_0$
  θ_0 = steering angle

The beamsteered array beamwidth is now given as

$$\text{HPBW} = |\theta_+ - \theta_-| \quad (4.25)$$

For the $N = 6$-element array, where $\theta_0 = 45°$, $\theta_+ = 58.73°$, and $\theta_- = 34.02°$. Thus the beamwidth can be calculated to be HPBW = 24.71°.

4.1.3 Uniform N-Element Linear Array Directivity

Antenna directivity was previously defined in Eq. (3.16). Directivity is a measure of the antennas ability to preferentially direct energy in certain directions. The directivity equation is repeated as follows:

$$D(\theta,\phi) = \frac{4\pi U(\theta,\phi)}{\int_0^{2\pi}\int_0^{\pi} U(\theta,\phi)\sin\theta\,d\theta\,d\phi} \quad (4.26)$$

Our previous derivation for the array factor assumed the array was aligned along the horizontal axis. This derivation helped us visualize the array performance relative to a broadside reference angle.

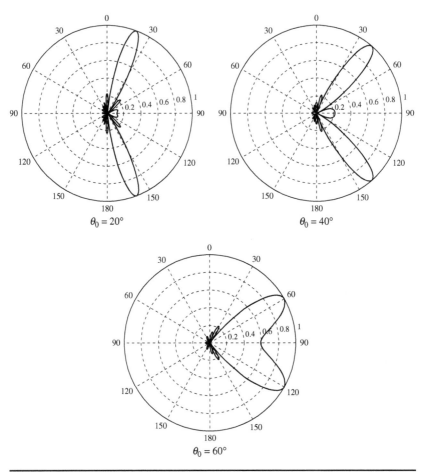

FIGURE 4.7 Beamsteered linear array with $\theta_0 = 45°$.

However, the horizontal array does not fit symmetrically into spherical coordinates. In order to simplify the calculation of the directivity, let us align the linear array along the z axis as shown in Fig. 4.8.

Because we have now rotated the array by 90° to make it vertical, we can modify the AF by allowing $\psi = kd\cos\theta + \delta$. Now the broadside angle is when $\theta = 90°$. Because the array factor is proportional to the signal level and not the power, we must square the array factor to yield the array radiation intensity $U(\theta)$. We now substitute the normalized approximate $(AF_n)^2$ into Eq. (4.26).

$$D(\theta) = \frac{4\pi \left(\frac{\sin\left(\frac{N}{2}(kd\cos\theta + \delta)\right)}{\frac{N}{2}(kd\cos\theta + \delta)} \right)^2}{\int_0^{2\pi} \int_0^{\pi} \left(\frac{\sin\left(\frac{N}{2}(kd\cos\theta + \delta)\right)}{\frac{N}{2}(kd\cos\theta + \delta)} \right)^2 \sin\theta \, d\theta \, d\phi} \qquad (4.27)$$

Array Fundamentals

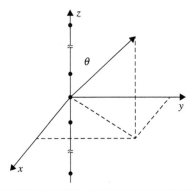

FIGURE 4.8 The N-element linear array along z axis.

The maximum value of the normalized array factor is unity. Thus the maximum directivity is given as

$$D_0 = \frac{4\pi}{\int_0^{2\pi}\int_0^{\pi}\left(\frac{\sin\left(\frac{N}{2}(kd\cos\theta+\delta)\right)}{\frac{N}{2}(kd\cos\theta+\delta)}\right)^2 \sin\theta\, d\theta\, d\phi} \tag{4.28}$$

Solving for the maximum directivity is now simply the matter of solving the denominator integral although the integral itself is not trivial.

Broadside Array Maximum Directivity

As was noted before, the case of broadside maximum directivity requires that $\delta = 0$. We can simplify the directivity equation by integrating over the ϕ variable. Thus, Eq. (4.28) can be simplified to be

$$D_0 = \frac{2}{\int_0^{\pi}\left(\frac{\sin\left(\frac{N}{2}kd\cos\theta\right)}{\frac{N}{2}kd\cos\theta}\right)^2 \sin\theta\, d\theta} \tag{4.29}$$

We can define the variable $x = \frac{N}{2}kd\cos\theta$. Then $dx = -\frac{N}{2}kd\sin\theta\, d\theta$. Substituting the new variable x into Eq. (4.29) yields

$$D_0 = \frac{Nkd}{\int_{-Nkd/2}^{Nkd/2}\left(\frac{\sin(x)}{x}\right)^2 dx} \tag{4.30}$$

As $Nkd/2 \gg \pi$, the limits can be extended to infinity without a significant loss in accuracy. The integral solution can be found in integral tables. Thus

$$D_0 \approx 2N\frac{d}{\lambda} \tag{4.31}$$

End-Fire Array Maximum Directivity

End-fire radiation conditions are achieved when the electrical phase between elements is $\delta = -kd$. This is equivalent to writing the overall phase term as $\psi = kd(\cos\theta - 1)$. Rewriting the maximum directivity, we get

$$D_0 = \frac{2}{\int_0^\pi \left(\frac{\sin\left(\frac{N}{2}kd(\cos\theta - 1)\right)}{\frac{N}{2}kd(\cos\theta - 1)}\right)^2 \sin\theta\, d\theta} \tag{4.32}$$

We may again make a change of variable such that $x = \frac{N}{2}kd(\cos\theta - 1)$. Then $dx = -\frac{N}{2}kd\sin\theta\, d\theta$. Upon substitution of x into Eq. (4.32), we have

$$D_0 = \frac{Nkd}{\int_0^{Nkd}\left(\frac{\sin(x)}{x}\right)^2 dx} \tag{4.33}$$

As $Nkd/2 \gg \pi$, the upper limit can be extended to infinity without a significant loss in accuracy. Thus

$$D_0 \approx 4N\frac{d}{\lambda} \tag{4.34}$$

The end-fire array has twice the directivity of the broadside array. This is true because the end-fire array only has one major lobe, whereas the broadside array has two symmetric major lobes.

Beamsteered Array Maximum Directivity

The most general case for the array directivity is found by defining the element-to-element phase shift δ in terms of the steering angle θ_0. Let us rewrite Eq. (4.27) by substituting $\delta = -kd\cos\theta_0$.

$$D(\theta, \theta_0) = \frac{4\pi\left(\frac{\sin\left(\frac{N}{2}(kd(\cos\theta - \cos\theta_0))\right)}{\frac{N}{2}(kd(\cos\theta - \cos\theta_0))}\right)^2}{\int_0^{2\pi}\int_0^\pi \left(\frac{\sin\left(\frac{N}{2}(kd(\cos\theta - \cos\theta_0))\right)}{\frac{N}{2}(kd(\cos\theta - \cos\theta_0))}\right)^2 \sin\theta\, d\theta\, d\phi} \tag{4.35}$$

It is instructive to plot the linear array directivity versus angle for a few different steering angles using MATLAB. We would expect the end-fire and broadside maximum directivities, calculated previously, to be two points on the more general plot of Eq. (4.35). Allowing $N = 4$ and $d = .5\lambda$, the directivity is shown plotted in Fig. 4.9. The maximum values are slightly higher than predicted by Eqs. (4.31) and (4.34) because no approximations were made to simplify the integration limits.

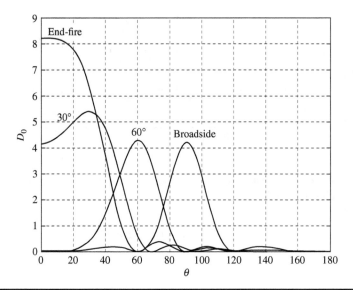

FIGURE 4.9 Family of steered array maximum directivity curves.

4.2 Array Weighting

The previous derivation for the array factor assumed that all of the isotropic elements had unity amplitude. Because of this assumption, the AF could be reduced to a simple series and a simple $\sin(x)/x$ approximation.

It was apparent from Figs. 4.5 and 4.6 that the array factor has sidelobes. For a uniformly weighted linear array, the largest sidelobes are down approximately 24 percent from the peak value. The presence of sidelobes means that the array is radiating energy in unintended directions. Additionally, because of reciprocity, the array is receiving energy from unintended directions. In a multipath environment, the sidelobes can receive the same signal from multiple angles. This is the basis for fading experienced in communications. If the direct transmission angle is known, it is best to steer the beam toward the desired direction and to shape the sidelobes to suppress unwanted signals. The sidelobes can be suppressed by *weighting*, *shading*, or *windowing* the array elements. These terms are taken from the EM, underwater acoustics, and array signal processing communities, respectively. Array element weighting has numerous applications in areas such as digital signal processing (DSP), radio astronomy, radar, sonar, and communications. Two excellent foundational articles on array weighting are written by Harris [12] and Nuttall [13].

Figure 4.10 shows a symmetric linear array with an even number of elements N. The array is symmetrically weighted with weights as indicated:

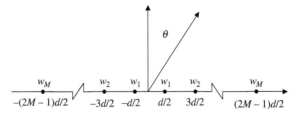

FIGURE 4.10 Even array with weights.

The array factor is found by summing the weighted outputs of each element such that

$$AF_{even} = w_M e^{-j\frac{(2M-1)}{2}kd\sin\theta} + \cdots + w_1 e^{-j\frac{1}{2}kd\sin\theta}$$
$$+ w_1 e^{j\frac{1}{2}kd\sin\theta} + \cdots + w_M e^{j\frac{(2M-1)}{2}kd\sin\theta} \quad (4.36)$$

where $2M = N =$ total number of array elements. Each apposing pair of exponential terms in Eq. (4.36) forms complex conjugates. We can invoke Euler's identity for the cosine to recast the even array factor given as follows:

$$AF_{even} = 2\sum_{n=1}^{M} w_n \cos\left(\frac{(2n-1)}{2}kd\sin\theta\right) \quad (4.37)$$

Without loss of generality, the 2 can be eliminated from the expression in Eq. (4.37) to produce a quasi-normalization.

$$AF_{even} = \sum_{n=1}^{M} w_n \cos((2n-1)u) \quad (4.38)$$

where $u = \frac{\pi d}{\lambda}\sin\theta$.

The array factor is maximum when the argument is zero, implying $\theta = 0$. The maximum is then the sum of all of the array weights. Thus, we may completely normalize AF_{even} to be

$$AF_{even} = \frac{\sum_{n=1}^{M} w_n \cos((2n-1)u)}{\sum_{n=1}^{M} w_n} \quad (4.39)$$

It is easiest to express the array factor in the form of Eq. (4.38). However, for plotting purposes, it is best to use the normalized array factor given in Eq. (4.39).

An odd array is depicted in Fig. 4.11 with the center element at the origin.

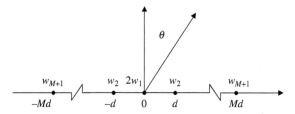

FIGURE 4.11 Odd array with weights.

We may again sum all of the exponential contributions from each array element to get the quasi-normalized odd array factor.

$$\text{AF}_{\text{odd}} = 2\sum_{n=1}^{M+1} w_n \cos(2(n-1)u) \tag{4.40}$$

where $2M + 1 = N$.

In order to normalize Eq. (4.40), we must again divide by the sum of the array weights to get

$$\text{AF}_{\text{odd}} = \frac{\sum_{n=1}^{M+1} w_n \cos(2(n-1)u)}{\sum_{n=1}^{M+1} w_n} \tag{4.41}$$

We may alternatively express Eqs. (4.38) and (4.40) using the array vector nomenclature previously addressed in Eq. (4.8). Then, the array factor can be expressed in vector terms as

$$\text{AF} = \bar{w}^T \cdot \bar{a}(\theta) \tag{4.42}$$

where $\bar{a}(\theta)$ = array vector

$$\bar{w}^T = [w_M \quad w_{M-1} \quad \cdots \quad w_1 \quad \cdots \quad w_{M-1} \quad w_M]$$

The weights w_n can be chosen to meet any specific criteria. Generally the criterion is to minimize the sidelobes or possibly to place nulls at certain angles. However, symmetric scalar weights can only be utilized to shape sidelobes.

There are a vast number of possible window functions available that can provide weights for use with linear arrays. Some of the more common window functions, along with corresponding plots, are explained next. Unless otherwise noted, we assume that the array plotted has $N = 8$ weighted isotropic elements.

Binomial

Binomial weights create an array factor with no sidelobes, provided that the element spacing $d \leq \frac{\lambda}{2}$. The binomial weights are chosen from the rows of Pascal's triangle. The first nine rows are shown in Table 4.1.

TABLE 4.1 Pascal's Triangle

N = 1						1					
N = 2					1		1				
N = 3					1	2	1				
N = 4				1	3	3	1				
N = 5			1	4	6	4	1				
N = 6			1	5	10	10	5	1			
N = 7		1	6	15	20	15	6	1			
N = 8		1	7	21	35	35	21	7	1		
N = 9	1	8	28	56	70	56	28	8	1		

If we choose an $N = 8$-element array, the array weights are taken from row 8 to be $w_1 = 35$, $w_2 = 21$, $w_3 = 7$, and $w_4 = 1$. The normalized binomial weights are $w_1 = 1$, $w_2 = 6$, $w_3 = .2$, and $w_4 = .0286$. The eight array weights can more conveniently be found using the MATLAB command diag(rot90(pascal(N))). The normalized array weights are shown plotted in Fig. 4.12 using the stem command. The weighted array factor is superimposed over the unweighted array factor as shown in Fig. 4.13. The price paid for suppressing the sidelobes is seen in the broadening of the main beamwidth.

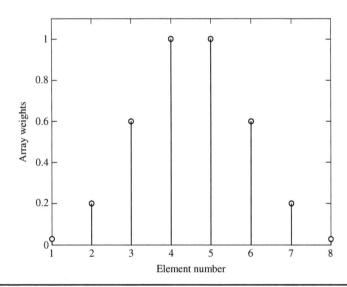

FIGURE 4.12 Binomial array weights.

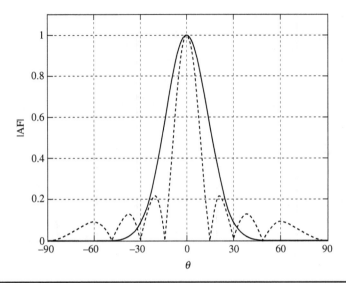

Figure 4.13 Array factor with binomial weights.

Blackman

The Blackman weights are defined by

$$w(k+1) = .42 - .5\cos(2\pi k/(N-1)) \\ + .08\cos(4\pi k/(N-1)) \quad k = 0, 1, \ldots, N-1 \quad (4.43)$$

For the $N = 8$ element array, the normalized Blackman weights are $w_1 = 1$, $w_2 = .4989$, $w_3 = .0983$, and $w_4 = 0$. The eight array weights can be found using the Blackman(N) command in MATLAB. The normalized array weights are shown plotted in Fig. 4.14 using the stem command. The weighted array factor is shown in Fig. 4.15.

Hamming

The Hamming weights are given by

$$w(k+1) = .54 - .46\cos[2\pi k/(N-1)] \quad k = 0, 1, \ldots, N-1 \quad (4.44)$$

The normalized Hamming weights are $w_1 = 1$, $w_2 = .673$, $w_3 = .2653$, and $w_4 = .0838$. The eight array weights can be found using the hamming(N) command in MATLAB. The normalized array weights are shown plotted in Fig. 4.16 using the stem command. The weighted array factor is shown in Fig. 4.17.

Gaussian

The Gaussian weights are determined by the Gaussian function to be

$$W(k+1) = e^{-\frac{1}{2}\left(\alpha\frac{k-N/2}{N/2}\right)^2} \quad k = 0, 1, \ldots, N \quad \alpha \geq 2 \quad (4.45)$$

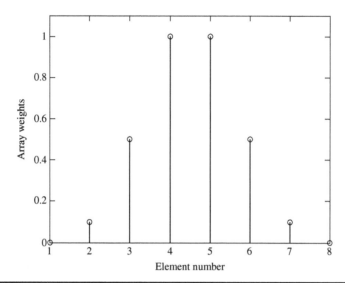

Figure 4.14 Blackman array weights.

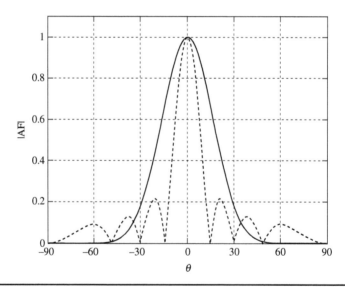

Figure 4.15 Array factor with Blackman weights.

The normalized Gaussian weights for $\alpha = 2.5$ are $w_1 = 1$, $w_2 = .6766$, $w_3 = .3098$, and $w_4 = .0960$. The eight array weights can be found using the gausswin(N) command in MATLAB. The normalized array weights are shown plotted in Fig. 4.18 using the stem command. The weighted array factor is shown in Fig. 4.19.

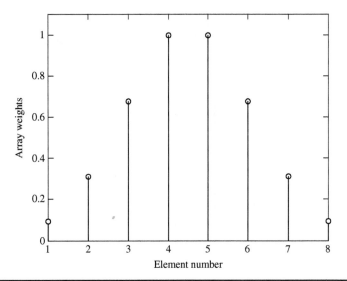

FIGURE 4.16 Hamming array weights.

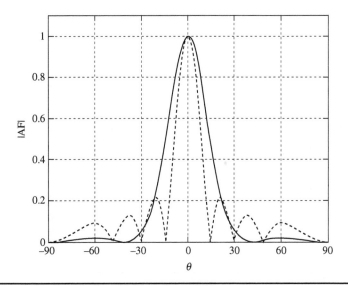

FIGURE 4.17 Array factor with Hamming weights.

Kaiser-Bessel

The Kaiser-Bessel weights are determined by

$$w(k) = \frac{I_0\left[\pi\alpha\sqrt{1-\left(\frac{k}{N/2}\right)^2}\right]}{I_0[\pi\alpha]} \quad k = 0, 1, \ldots, N/2 \quad \alpha > 1 \quad (4.46)$$

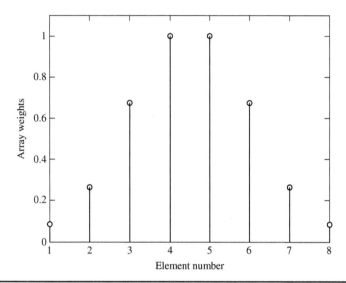

FIGURE 4.18 Gaussian array weights.

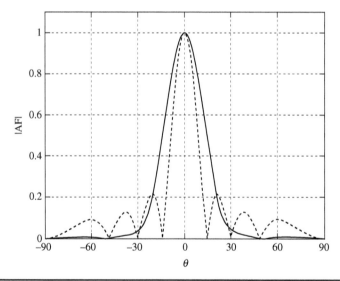

FIGURE 4.19 Array factor with Gaussian weights.

The normalized Kaiser-Bessel weights for $\alpha = 3$ are $w_1 = 1$, $w_2 = .8136$, $w_3 = 5137$, and $w_4 = .210$. The eight array weights can be found using the Kaiser (N, α) command in MATLAB. The normalized array weights are shown plotted in Fig. 4.20 using the stem command. The weighted array factor is shown in Fig. 4.21.

It should be noted that the Kaiser-Bessel weights provide one of the lowest array sidelobe levels while still maintaining nearly the

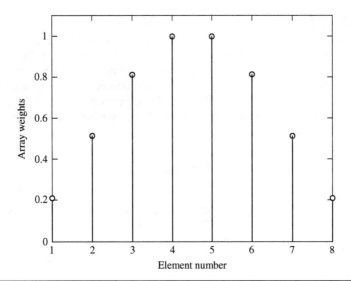

Figure 4.20 Kaiser-Bessel array weights.

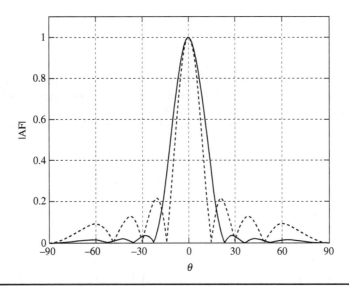

Figure 4.21 Array factor with Kaiser-Bessel weights.

same beamwidth as uniform weights. Additionally, the Kaiser-Bessel weights for $\alpha = 1$ yield the uniform set of weights.

Other potential weight functions are the Blackman-Harris, Bohman, Hanning, Bartlett, Dolph-Chebyshev, and Nuttall. A detailed description of these functions can be found in [12, 13]. In addition, many of these weight functions are available in MATLAB.

4.2.1 Beamsteered and Weighted Arrays

Previously, in Sec. 4.1.3 we discussed the beamsteered uniformly weighted array. We could steer the mainlobe to any desired direction, but we still experienced the problem of relatively large minor lobes. The nonuniformly weighted array can also be modified in order to steer the beam to any direction desired and with suppressed sidelobe levels.

We can repeat Eqs. (4.38) and (4.40), but we can modify them to include beamsteering.

$$AF_{even} = \sum_{n=1}^{M} w_n \cos((2n-1)u) \quad (4.47)$$

$$AF_{odd} = \sum_{n=1}^{M+1} w_n \cos(2(n-1)u) \quad (4.48)$$

where

$$u = \frac{\pi d}{\lambda}(\sin\theta - \sin\theta_0)$$

As an example, we can use Kaiser-Bessel weights and steer the mainlobe to three separate angles. Let $N = 8$, $d = \lambda/2$, $\alpha = 3$, $w_1 = 1$, $w_2 = .8136$, $w_3 = .5137$, and $w_4 = .210$. The beamsteered array factor for the weighted even element array is shown in Fig. 4.22.

In general, any array can be steered to any direction by either using phase shifters in the hardware or by digitally phase shifting the data at the back end of the receiver. If the received signal is digitized and processed, this signal processing is often called *digital beamforming* (DBF) [8]. Current technologies are making it more feasible

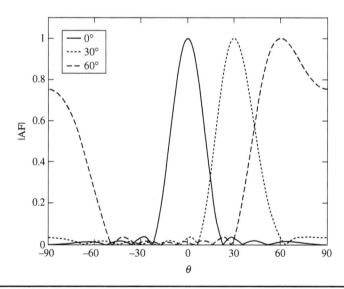

FIGURE 4.22 Beamsteered Kaiser-Bessel weighted array.

to perform DBF and therefore allow the array designer to bypass the need for hardware phase shifters. The DBF performed can be used to steer the antenna beam according to any criteria specified by the user.

4.3 Circular Arrays

Linear arrays are very useful and instructive, but there are occasions where a linear array is not appropriate for the building, structure, or vehicle upon which it is mounted. Other array geometries may be necessary to appropriately fit into a given scenario. Such additional arrays can include the circular array. Just as the linear array was used for increased gain and beamsteering, the circular array can also be used. Figure 4.23 shows a circular array of N elements in the x-y plane. The array has N elements and the array radius is a.

The nth array element is located at the radius a with the phase angle ϕ_n. Additionally, each element can have an associated weight w_n and phase δ_n. As before, with the linear array, we assume far-field conditions and will assume that the observation point is such that the position vectors $\bar{r}$ and $\bar{r}_n$ are parallel. We can now define the unit vector in the direction of each array element n.

$$\hat{\rho}_n = \cos\phi_n \hat{x} + \sin\phi_n \hat{y} \tag{4.49}$$

We can also define the unit vector in the direction of the field point.

$$\hat{r} = \sin\theta \cos\phi \hat{x} + \sin\theta \sin\phi \hat{y} + \cos\theta \hat{z} \tag{4.50}$$

It can be shown that the distance r_n is less than the distance r by the scalar projection of $\hat{\rho}_n$ onto $\hat{r}$. (This is indicated by the dotted line in Fig. 4.23). Thus,

$$r_n = r - a\hat{\rho}_n \cdot \hat{r} \tag{4.51}$$

with

$$\hat{\rho}_n \cdot \hat{r} = \sin\theta \cos\phi \cos\phi_n + \sin\theta \sin\phi \sin\phi_n = \sin\theta \cos(\phi - \phi_n)$$

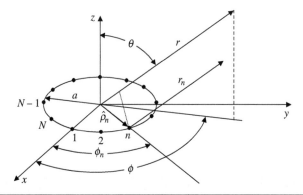

FIGURE 4.23 Circular array of N elements.

The array factor can now be found in a similar fashion as was done with the linear array. With some effort, it can be shown that

$$\text{AF} = \sum_{n=1}^{N} w_n e^{-j(ka\hat{\rho}\cdot\hat{r}+\delta_n)} = \sum_{n=1}^{N} w_n e^{-j[ka\sin\theta\cos(\phi-\phi_n)+\delta_n]} \quad (4.52)$$

where

$$\phi_n = \frac{2\pi}{N}(n-1) = \text{angular location of each element.}$$

4.3.1 Beamsteered Circular Arrays

The beamsteering of circular arrays is identical in form to the beamsteering of linear arrays. If we beamsteer the circular array to the angles (θ_0, ϕ_0), we can determine that the element to element phase angle is $\delta_n = -ka\sin\theta_0\cos(\phi_0 - \phi_n)$. We can thus rewrite the array factor as

$$\text{AF} = \sum_{n=1}^{N} w_n e^{-j\{ka[\sin\theta\cos(\phi-\phi_n)-\sin\theta_0\cos(\phi_0-\phi_n)]\}} \quad (4.53)$$

The circular array AF can be plotted in two or three dimensions. Let us assume that all weights are uniform and that the array is steered to the angles $\theta_0 = 30°$ and $\phi_0 = 0°$. With $N = 10$ and $a = \lambda$, we can plot the elevation pattern in the $\phi = 0°$ plane as shown in Fig. 4.24.

We can also plot the array factor in three dimensions as a mesh plot. Using the same parameters as above, we see the beamsteered circular array pattern in Fig. 4.25.

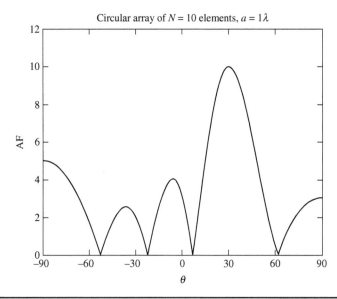

Figure 4.24 AF elevation pattern for beamsteered circular array ($\theta_0 = 30°$, $\phi_0 = 0°$).

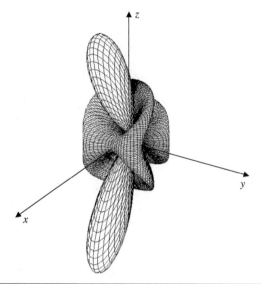

FIGURE 4.25 3D AF pattern for beamsteered circular array ($\theta_0 = 30°$, $\phi_0 = 0°$).

4.4 Rectangular Planar Arrays

Having explored the linear and circular arrays, we can move on to slightly more complex geometries by deriving the pattern for rectangular planar arrays. The following development is similar to that found in both Balanis [6] and in Johnson and Jasik [14].

Figure 4.26 shows a rectangular array in the x-y plane. There are M elements in the x-direction and N elements in the y-direction creating an $M \times N$ array of elements. The m-nth element has weight w_{mn}. The x-directed elements are spaced d_x apart and the y-directed elements are spaced d_y apart. The planar array can be viewed as M linear arrays of N elements or as N linear arrays of M elements. Because we already know the array factor for an M or N element array acting alone, we can use pattern multiplication to find the pattern of the entire $M \times N$ element array. Using pattern multiplication, we have

$$\text{AF} = \text{AF}_x \cdot \text{AF}_y = \sum_{m=1}^{M} a_m e^{j(m-1)(kd_x \sin\theta\cos\phi + \beta_x)} \sum_{n=1}^{N} b_n e^{j(n-1)(kd_y \sin\theta\sin\phi + \beta_y)}$$

$$= \sum_{m=1}^{M}\sum_{n=1}^{N} w_{mn} e^{j[(m-1)(kd_x \sin\theta\cos\phi + \beta_x)+(n-1)(kd_y \sin\theta\sin\phi + \beta_y)]} \quad (4.54)$$

where $w_{mn} = a_m \cdot b_n$.

The weights a_m and b_n can be uniform or can be in any form according to the designer's needs. This could include the various weights discussed in Sec. 4.2 such as the binomial, Kaiser-Bessel, Hamming, or Gaussian weights. The a_m weights do not have to be identical to the b_n weights.

90   Chapter Four

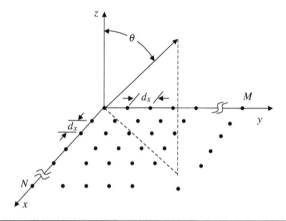

FIGURE 4.26 N × M rectangular planar array.

Thus, we might choose the a_m weights to be binomial weights while the b_n weights are Gaussian. Any combination of weighting can be used and w_{mn} is merely the consequence of the product $a_m \cdot b_n$.

If beamsteering is desired, the phase delays β_x and β_y are given by

$$\beta_x = -kd_x \sin\theta_0 \cos\phi_0 \qquad \beta_y = -kd_y \sin\theta_0 \sin\phi_0 \qquad (4.55)$$

Example 4.4 Design and plot a 8 × 8 element array with equal element spacing such that $d_x = d_y = .5\lambda$. Let the array be beamsteered to $\theta_0 = 45°$ and $\phi_0 = 45°$. The element weights are chosen to be the Kaiser-Bessel weights given in Sec. 4.2. Plot the pattern for the range $0 \le \theta \le \pi/2$ and $0 \le \phi \le 2\pi$.

Solution Taking the weights from Sec. 4.2, we have $a_1 = a_4 = b_1 = b_4 = .2352$, and $a_2 = a_3 = b_2 = b_3 = 1$. Thus, $w_{1n} = w_{m1} = .0055$ and all other w_{mn}'s = 1. We can substitute the weights into Eq. (4.51) to calculate the array factor. MATLAB can be used to plot Fig. 4.27.

FIGURE 4.27
Beamsteered planar array pattern.

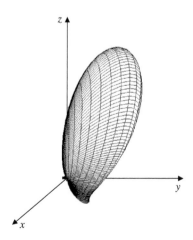

4.5 Fixed Beam Arrays

Fixed beam arrays are designed such that the array pattern consists of several simultaneous *spot* beams transmitting in fixed angular directions. Normally these directions are in equal angular increments so as to ensure a relatively uniform coverage of a region in space. However, this is not a necessary restriction. These fixed beams can be used in satellite communications to create spot beams toward fixed earth-based locations. As an example, the *Iridium*[1] (low earth orbit) satellite constellation system has 48 spot beams per satellite. Spot beams are sometimes also called *pincushion* beams because of the similarity to pins in a pin cushion. Figure 4.28 shows an example of a planar array creating three spot beams.

Fixed beams can also be used for mobile communication base stations in order to provide space division multiple access (SDMA) capabilities. Several have treated the subject of fixed beam systems such as Mailloux [15], Hansen [16], and Pattan [17].

4.5.1 Butler Matrices

One method for easily creating fixed beams is through the use of Butler matrices. Details of the derivation can be found in Butler and Lowe [18], and Shelton and Kelleher [19]. The Butler matrix is an analog means of producing several simultaneous fixed beams through the use of phase shifters. As an example, let us assume a

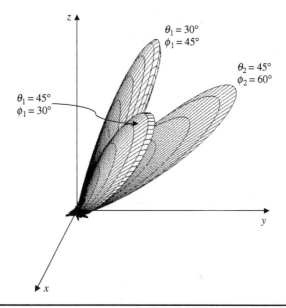

FIGURE 4.28 Three spot beams created by a 16 × 16 planar array.

---

[1] Iridium Satellite LLC.

linear array of N elements. If $N = 2^n$ elements, the array factor can be given as

$$\text{AF}(\theta) = \frac{\sin\left(N\pi\frac{d}{\lambda}\sin\theta - \beta_\ell\right)}{N\pi\frac{d}{\lambda}\sin\theta - \beta_\ell} = \frac{\sin\left[N\pi\frac{d}{\lambda}(\sin\theta - \sin\theta_\ell)\right]}{N\pi\frac{d}{\lambda}(\sin\theta - \sin\theta_\ell)} \quad (4.56)$$

with

$$\sin\theta_\ell = \frac{\ell\lambda}{Nd}$$

$$\beta_\ell = \ell\pi$$

$$\ell = \pm\frac{1}{2}, \pm\frac{3}{2}, \cdots, \pm\frac{(N-1)}{2}$$

The ℓ values create evenly spaced contiguous beams about $\theta = 0°$. If the element spacing is $d = \lambda/2$, the beams are evenly distributed over the span of 180°. If $d > \lambda/2$, the beams span an ever-decreasing range of angles. Grating lobes are not created provided that the only phase shifts used are those defined by β_ℓ.

As an example, let us choose an $N = 4$ element array with $d = \lambda/2$. We can use Eq. (4.56) to produce N fixed beams and plot the results using MATLAB. These beams are sometimes referred to as *scalloped beams* because of the resemblance to the scallop shell. Because $N = 4$, $\ell = -\frac{3}{2}, -\frac{1}{2}, \frac{1}{2}, \frac{3}{2}$. Substituting these values into the equation for $\sin\theta_\ell$, we get the polar plot as shown in Fig. 4.29.

These beams can be created using fixed phase shifters by noting that $\beta_\ell = \ell\pi = \pm\frac{\pi}{2}, \pm\frac{3\pi}{2}$. Figure 4.30 depicts a Butler matrix labyrinth of phase shifters for $N = 4$ array elements.

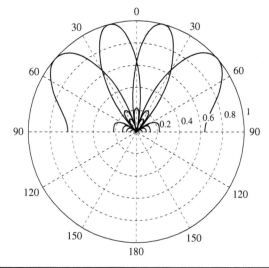

FIGURE 4.29 Scalloped beams using Butler approach.

Array Fundamentals

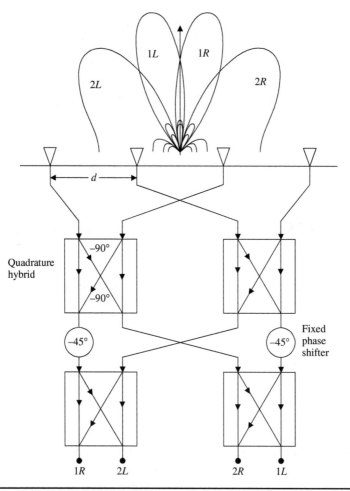

FIGURE 4.30 Butler matrix labyrinth for $N = 4$.

As can be noted from the figure, the 1R port will yield the first beam to the right of broadside to the array. The 2R port yields the second beam to the right, and so on. Thus, through the use of the Butler matrix labyrinth, one can simultaneously look in N directions with an N-element array.

4.6 Fixed Sidelobe Canceling

The basic goal of a fixed *sidelobe canceller* (SLC) is to choose array weights such that a null is placed in the direction of interference while the mainlobe maximum is in the direction of interest. The concept of an SLC was first presented by Howells in 1965 [20].

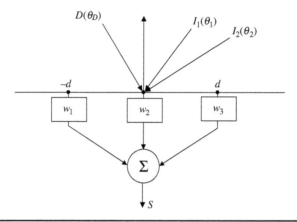

Figure 4.31 The three-element array with desired and interfering signals.

Because that time adaptive SLCs have been extensively studied. A detailed description for an adaptive SLC will be reserved for Chap. 8.

In our current development, we discuss fixed sidelobe canceling for one fixed known desired source and two fixed undesired interferers. All signals are assumed to operate at the same carrier frequency. Let us assume a three-element array with the desired signal and interferers as shown in Fig. 4.31.

The array vector is given by

$$\bar{a}(\theta) = [e^{-jkd\sin\theta} \quad 1 \quad e^{jkd\sin\theta}]^T \tag{4.57}$$

The, as yet to be determined, array weights are given by

$$\bar{w}^T = [w_1 \quad w_2 \quad w_3] \tag{4.58}$$

Therefore, the total array output from the summer is given as

$$S = \bar{w}^T \cdot \bar{a} = w_1 e^{-jkd\sin\theta} + w_2 + w_3 e^{jkd\sin\theta} \tag{4.59}$$

The array output for the desired signal will be designated by S_D whereas the array output for the interfering signals will be designated by S_1 and S_2. Because there are three unknown weights, there must be three conditions satisfied.

Condition 1: $S_D = w_1 e^{-jkd\sin\theta_D} + w_2 + w_3 e^{jkd\sin\theta_D} = 1$

Condition 2: $S_1 = w_1 e^{-jkd\sin\theta_1} + w_2 + w_3 e^{jkd\sin\theta_1} = 0$

Condition 2: $S_2 = w_1 e^{-jkd\sin\theta_2} + w_2 + w_3 e^{jkd\sin\theta_2} = 0$

Condition 1 demands that $S_D = 1$ for the desired signal, allowing the desired signal to be received without modification. Conditions 2 and 3 reject the undesired signals. These conditions can be recast in matrix form as

$$\begin{bmatrix} e^{-jkd \sin \theta_D} & 1 & e^{jkd \sin \theta_D} \\ e^{-jkd \sin \theta_1} & 1 & e^{jkd \sin \theta_1} \\ e^{-jkd \sin \theta_2} & 1 & e^{jkd \sin \theta_2} \end{bmatrix} \cdot \begin{bmatrix} w_1 \\ w_2 \\ w_3 \end{bmatrix} = \begin{bmatrix} 1 \\ 0 \\ 0 \end{bmatrix} \qquad (4.60)$$

One can invert the matrix to find the required complex weights w_1, w_2, and w_3. As an example, if the desired signal is arriving from $\theta_D = 0°$ as while $\theta_1 = -45°$ and $\theta_2 = 60°$, the necessary weights can be calculated to be

$$\begin{bmatrix} w_1 \\ w_2 \\ w_2 \end{bmatrix} = \begin{bmatrix} 0.748 + 0.094i \\ -0.496 \\ 0.748 - 0.094i \end{bmatrix} \qquad (4.61)$$

The array factor is shown plotted in Fig. 4.32.

There are some limitations to this scheme. The number of nulls cannot exceed the number of array elements. In addition, the array maximum cannot be closer to a null than the array resolution allowed. The array resolution is inversely proportional to the array length.

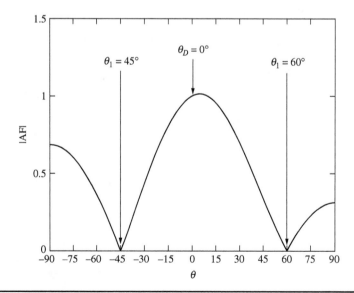

Figure 4.32 Sidelobe cancellation.

4.7 Retrodirective Arrays

Retrodirective arrays are the array equivalent of a corner reflector. Corner reflectors can be studied in any standard antenna book. It was Van Atta [21, 22] who invented a scheme to convert a linear array into a reflector. In this case, the array redirects the incident field in the direction of arrival. Thus, the use of the term "retrodirective" is appropriate. Retrodirective arrays are sometimes also called *self-phasing* arrays, *self-focusing* arrays, *conjugate matched* arrays, or *time-reversal mirrors* [23–25]. A conjugate matched array is retrodirective because it retransmits the signal with the phases conjugated. If the phases are conjugated, it is the same as reversing time in the time domain. This is why retrodirective arrays are sometimes called "time-reversal" arrays. The acoustic community has aggressively pursued time-reversal methods in underwater acoustics as a means for addressing the multipath challenge [26].

It is not necessary that the array is linear in order for the array to be retrodirective. In fact, the Van Atta array is a special case of the more general subject of self-phasing arrays. However, in this development, we restrict our discussion to linear arrays.

One of the obvious advantages of a retrodirective array is the fact that if the array can redirect energy in the direction of arrival then the array will work extremely well in a multipath environment. Some suggested ways of using a retrodirective array in mobile communications are given by Fusco and Karode [24]. If the same signal arrives from multiple directions, a retrodirective array will retransmit at the same angles and the signal will return to the source as if multipath did not exist. Figure 4.33 shows a linear retrodirective array in a multipath environment.

If indeed the retrodirective array can retransmit along the angles of arrival, the retransmitted signal will retrace the multiple paths back to the transmitter.

4.7.1 Passive Retrodirective Array

One possible way of implementing a retrodirective array is shown in Fig. 4.34 for an $N = 6$-element array. A plane wave is incident on the array at the angle θ_0.

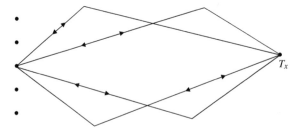

Figure 4.33 Retrodirective array with multipath.

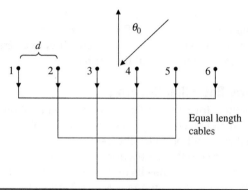

FIGURE 4.34 Retrodirective array.

The array vector for this N-element array is given as

$$\bar{a} = \begin{bmatrix} e^{-j\frac{5}{2}kd\sin\theta} & e^{-j\frac{3}{2}kd\sin\theta} & \cdots & e^{j\frac{3}{2}kd\sin\theta} & e^{j\frac{5}{2}kd\sin\theta} \end{bmatrix}^T \quad (4.62)$$

The received array vector, at angle θ_0, is given as

$$\bar{a}_{rec} = \begin{bmatrix} e^{-j\frac{5}{2}kd\sin\theta_0} & e^{-j\frac{3}{2}kd\sin\theta_0} & \cdots & e^{j\frac{3}{2}kd\sin\theta_0} & e^{j\frac{5}{2}kd\sin\theta_0} \end{bmatrix}^T \quad (4.63)$$

The input to element 6, $e^{j\frac{5}{2}kd\sin\theta_0}$, propagates down the transmission line to element 1 and is retransmitted. The same process is repeated for all elements. Thus the transmitted signal for element i was the received signal for element $N - i$. This can be shown to be the same as multiplying the array vector $\bar{a}$ in Eq. (4.62) by the reverse of the vector $\bar{a}_{rec}$ of Eq. (4.63). Reversing the vector of Eq. (4.63) is the same as reversing or flipping the individual elements. One method of reversing vector elements is through the use of the permutation matrix. In MATLAB, this function can be accomplished through the use of the fliplr( ) command. The array transmission can be now calculated to be

$$\text{AF} = \begin{bmatrix} e^{-j\frac{5}{2}kd\sin\theta} & e^{-j\frac{3}{2}kd\sin\theta} & \cdots & e^{j\frac{3}{2}kd\sin\theta} & e^{j\frac{5}{2}kd\sin\theta} \end{bmatrix}^T \begin{bmatrix} e^{j\frac{5}{2}kd\sin\theta_0} \\ e^{j\frac{3}{2}kd\sin\theta_0} \\ \vdots \\ e^{-j\frac{3}{2}kd\sin\theta_0} \\ e^{-j\frac{5}{2}kd\sin\theta_0} \end{bmatrix} \quad (4.64)$$

Based on our derivation in Eq. (4.13), this is equivalent to a beam-steered array factor.

$$\text{AF} = \frac{\sin\left(\frac{Nkd}{2}(\sin\theta - \sin\theta_0)\right)}{\sin\left(\frac{kd}{2}(\sin\theta - \sin\theta_0)\right)} \quad (4.65)$$

Thus, this retrodirective array has successfully retransmitted the signal back toward the θ_0 direction. This process works regardless of the angle of arrival (AOA). Therefore, the retrodirective array serves to focus the reflected signal back at the source.

4.7.2 Active Retrodirective Array

A second method for self-phasing or phase conjugation is achieved through mixing the received signal with a local oscillator. In this case, the local oscillator will be twice the carrier frequency. Although it is not necessary to make this restriction, the analysis is easiest for this case. Each antenna output has its own mixer as shown in Fig. 4.35.

The output of the nth antenna is $R_n(t, \theta_n)$, given by

$$R_n(t, \theta_n) = \cos(w_0 t + \theta_n) \tag{4.66}$$

The mixer output is given by

$$S_{mix} = \cos(w_0 t + \theta_n) \cdot \cos(2w_0 t)$$

$$= \frac{1}{2}[\cos(-w_0 t + \theta_n) + \cos(3w_0 t + \theta_n)] \tag{4.67}$$

After passing through the low pass filter and selecting the lower sideband, the transmitted signal, for element n, is given by

$$T_n(t, \theta_n) = \cos(w_0 t - \theta_n) \tag{4.68}$$

Thus, the phase has been conjugated from the phase imposed at arrival. This array will then redirect the signal back toward the AOA θ_0. Since, the term in Eq. (4.68) could have been written as $\cos(-w_0 t + \theta_n)$.

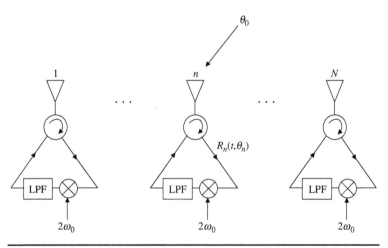

Figure 4.35 Phase conjugation through heterodyne mixing.

It can thus be seen how this procedure is alternatively called time-reversal and how the array is called a time-reversal mirror. If we choose a different local oscillator frequency, it is sufficient to choose the lower sideband for retransmission.

4.8 References

1. Haykin, S., ed., *Array Signal Processing*, Prentice Hall, New York, 1985.
2. Johnson, D., and D. Dudgeon, *Array Signal Processing—Concepts and Techniques*, Prentice Hall, New Jersey, 1993.
3. Trees, H. V., *Optimum Array Processing—Part IV of Detection, Estimation, and Modulation Theory*, Wiley Interscience, New York, 2002.
4. Brookner, E., *Practical Phased-Array Antenna Systems*, Artech House, Boston, MA, 1991.
5. Dudgeon, D. E., "Fundamentals of Digital Array Processing," *Proceedings of IEEE*, Vol. 65, pp. 898–904, June 1977.
6. Balanis, C., *Antenna Theory: Analysis and Design*, 2d ed., Wiley, New York, 1997.
7. Kraus, J. D., and R. J. Marhefka, *Antennas for All Applications*, 3d ed., McGraw-Hill, New York, 2002.
8. Litva, J., and T. K-Y. Lo, *Digital Beamforming in Wireless Communications*, Artech House, 1996.
9. Monzingo, R. A., and T. W. Miller, *Introduction to Adaptive Arrays*, Wiley, New York, 1980.
10. Ertel, R. B., P. Cardieri, K. W. Sowerby, et al., "Overview of Spatial Channel Models for Antenna Array Communication Systems," *IEEE Personal Commun. Mag.*, Vol. 5, No. 1, pp. 10–22, Feb. 1998.
11. Hansen, W. W., and J. R.Woodyard, "A New Principle in Directional Antenna Design," *Proceedings IRE*, Vol. 26, No. 3, pp. 333–345, March 1938.
12. Harris, F. J. "On the Use of Windows for Harmonic Analysis with the DFT," *IEEE Proceedings*, pp. 51–83, Jan. 1978.
13. Nuttall, A. H., in "Some Windows with Very Good Sidelobe Behavior," *IEEE Transactions on Acoustics, Speech, and Signal Processing*, Vol. ASSP-29, No. 1, Feb. 1981.
14. Johnson, R. C., and H. Jasik, *Antenna Engineering Handbook*, 2d ed., McGraw-Hill, New York, pp. 20–16, 1984.
15. Mailloux, R. J., *Phased Array Antenna Handbook*, Artech House, Norwood, MA, 1994.
16. Hansen, R. C., *Phased Array Antennas*, Wiley, New York, 1998.
17. Pattan, B., *Robust Modulation Methods and Smart Antennas in Wireless Communications*, Prentice Hall, New York, 2000.
18. Butler, J., and R. Lowe, "Beam-Forming Matrix Simplifies Design of Electrically Scanned Antennas," *Electronic Design*, April 12, 1961.
19. Shelton, J. P., and K. S. Kelleher, "Multiple Beams from Linear Arrays," *IRE Transactions on Antennas and Propagation*, March 1961.
20. Howells, P. W., "Intermediate Frequency Sidelobe Canceller," U.S. Patent 3202990, Aug. 24, 1965.
21. Van Atta, L. C., "Electromagnetic Reflector," U.S. Patent 2908002, Oct. 6, 1959.
22. Sharp, E. D., and M. A. Diab, "Van Atta Reflector Array," *IRE Transactions on Antennas and Propagation Communications*, Vol. AP-8, pp. 436–438, July 1960.
23. Skolnik, M. I., and D. D. King, "Self-Phasing Array Antennas," *IEEE Transactions on Antennas and Propagation*, Vol. AP-12, No. 2, pp. 142–149, March 1964.
24. Fusco, V. F., and S. L. Karode, "Self-Phasing Antenna Array Techniques for Mobile Communications Applications," *Electronics and Communication Engineering Journal*, Dec. 1999.
25. Blomgren, P., G. Papanicolaou, and H. Zhao, "Super-Resolution in Time-Reversal Acoustics," *Journal Acoustical Society of America*, Vol. 111, No. 1, Pt. 1, Jan. 2002.
26. Fink, M., "Time-Reversed Acoustics," *Scientific American*, pp. 91–97, Nov. 1999.

4.9 Problems

4.1 Use Eq. (4.14) and MATLAB and plot the array factor in rectangular coordinate form for the following broadside array cases:
  (a) $N = 4, d = \lambda/2$
  (b) $N = 8, d = \lambda/2$
  (c) $N = 8, d = \lambda$

4.2 For the three arrays given in Prob. 4.1, calculate θ_{null} and θ_s for each of the arrays given above.

4.3 Repeat Prob. 4.1 above but for end-fire arrays.

4.4 Use MATLAB and the command trapz( ) to calculate the maximum directivity for the following two arrays:
  (a) Broadside with $N = 8, d = \lambda/2$
  (b) End-fire with $N = 8, d = \lambda/2$

4.5 Use MATLAB and the command trapz( ) to calculate the maximum directivity for the beamsteered array where $d = \lambda/2, N = 8$.
  (a) $\theta_0 = 30°$
  (b) $\theta_0 = 45°$

4.6 What is the beamwidth for the following array parameters?
  (a) $\theta_0 = 0°, N = 8, d = \lambda/2$
  (b) $\theta_0 = 45°, N = 8, d = \lambda/2$
  (c) $\theta_0 = 90°, N = 8, d = \lambda/2$

4.7 For an $N = 6, d = \lambda/2$ uniformly weighted broadside array, plot the array factor. Superimpose plots of the same array with the following weights. (Create one new plot for each new set of weights.)
  (a) Kaiser-Bessel for $\alpha = 2$ using Kaiser (N, α)
  (b) Blackman-Harris using blackmanharris(N)
  (c) Nuttall using nuttallwin(N)
  (d) Chebyshev window for $R = 30$ dB using chebwin(N, R)

4.8 Repeat Prob. 4.7 for $N = 9, d = \lambda/2$.

4.9 Using MATLAB, create and superimpose three normalized array factor plots using the chebwin( ) function for $R = 20, 40$, and 60 dB. $N = 9, d = \lambda/2$.

4.10 Using MATLAB, create and superimpose three normalized array factor plots using the chebwin( ) function for $R = 40$ and beamsteer the array to three angles such that $\theta_0 = 0°, 30°, 60°, N = 9, d = \lambda/2$.

4.11 For $d = \lambda/2$, use MATLAB and plot the broadside array beamwidth versus element number N for $2 < N < 20$.

4.12 For $d = \lambda/2$, use MATLAB and plot the $N = 8$ element array beamwidth for a range of steering angles such that $0 < \theta_0 < 90°$.

4.13 For the $N = 40$ element circular array with radius $a = 2\lambda$. Use MATLAB to plot the elevation plane pattern in the x-z plane when the array is beamsteered to $\theta_0 = 20°$, $40°$, and $60°$, $\phi_0 = 0$. Plot for $(0 < \theta < 180°)$.

4.14 For the $N = 40$ element circular array with radius $\alpha = 2\lambda$. Use MATLAB to plot the azimuth plane pattern in the x-y plane when the array is beamsteered to $\theta_0 = 90°$, $\phi_0 = 0°$, $40°$, and $60°$. Plot for $(-90° < \phi < 90°)$.

4.15 Design and plot a 5×5 element array with equal element spacing such that $d_x = d_y = .5\lambda$. Let the array be beamsteered to $\theta_0 = 45°$ and $\phi_0 = 90°$.

The element weights are chosen to be the Blackman-Harris weights using the blackmanharris( ) command in MATLAB. Plot the pattern for the range $0 \le \theta \le \pi/2$ and $0 \le \phi \le 2\pi$.

4.16 Use Eq. (4.56) to create scalloped beams for the $N = 6$-element array with $d = \lambda/2$.
   (a) What are the l values?
   (b) What are the angles of the scalloped beams?
   (c) Plot and superimpose all beams on a polar plot similar to Fig. 4.29.

4.17 For the fixed beam sidelobe canceller in Sec. 4.6, with $N = 3$-antenna elements, calculate the array weights to receive the desired signal at $\theta_D = 30°$, and to suppress the interfering signals arriving at $\theta_1 = -30°$ and $\theta_2 = -60°$.

CHAPTER 5
Principles of Random Variables and Processes

Every wireless communication system or radar system must take into account the noise-like nature of the arriving signals as well as the internal system noise. The arriving intelligent signals are usually altered by propagation, spherical spreading, absorption, diffraction, scattering, and/or reflection from various objects. This being the case, it is important to know the statistical properties of the propagation channel as well as the statistical properties of the noise internal to the system. Chapter 6 addresses the subject of multipath propagation, which will be seen to be a random process. Chapters 7 and 8 deal with system noise as well as the statistical properties of arriving signals. In addition, methodologies used in the remaining chapters require computations based on the assumption that the signals and noise are random.

It is thus assumed that students or researchers using this book have a working familiarity with random processes. Typically this material is covered in any undergraduate course in statistical topics or communications. However, for the purposes of consistency in this text, we perform a brief review of some fundamentals of random processes. Subsequent chapters apply these principles to specific problems.

Several books are devoted to the subject of random processes, including texts by Papoulis [1], Peebles [2], Thomas [3], Schwartz [4], and Haykin [5]. The treatment of the subject in this chapter is based on the principles discussed in depth in these references.

5.1 Definition of Random Variables

In the context of communication systems, the received voltages, currents, phases, time delays, and angles of arrival tend to be random variables. As an example, if one were to conduct measurements of the receiver phase every time the receiver is turned on, the numbers measured would tend to be randomly distributed between 0 and 2π. One could not say with certainty what value the next measurement will produce but one could state the probability of getting a certain measured value. A random variable is a function that describes all possible outcomes of an experiment. In general, some values of a random variable are more likely to be measured than other values. The probability of getting a specific number when rolling a die is equal to the probability of any other number. However, most random variables in communications problems do not have equally likely probabilities. Random variables can either be discrete or continuous variables. A random variable is discrete if the variable can only take on a finite number of values during an observation interval. An example of a discrete random variable might be the arrival angle for indoor multipath propagation. A random variable is continuous if the variable can take on a continuum of values during an observation interval. An example of a continuous random variable might be the voltage associated with receiver noise or the phase of an arriving signal. Since random variables are the result of random phenomena, it is often best to describe the behavior of random variables using probability density functions.

5.2 Probability Density Functions

Every random variable x is characterized by a probability density function $p(x)$. The *probability density function* (pdf) is established after a large number of measurements have been performed, which determine the likelihood of all possible values of x. A discrete random variable possesses a discrete pdf. A continuous random variable possesses a continuous pdf. Figure 5.1 shows a typical pdf for a discrete random variable. Figure 5.2 shows a typical pdf for a continuous random variable.

The probability that x will take on a range of values between two limits x_1 and x_2 is defined by

$$P(x_1 \leq x \leq x_2) = \int_{x_1}^{x_2} p(x)\,dx \tag{5.1}$$

There are two important properties for pdfs. First, no event can have a negative probability. Thus

$$p(x) \geq 0 \tag{5.2}$$

Second, the probability that an x value exists somewhere over its range of values is certain. Thus

$$\int_{-\infty}^{\infty} p(x)\,dx = 1 \tag{5.3}$$

Both properties must be satisfied by any pdf. Since the total area under the pdf is equal to 1, the probability of x existing over a finite range of possible values is always less than 1.

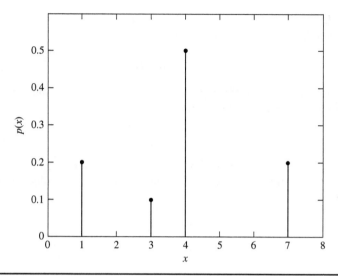

FIGURE 5.1 Probability density function for discrete x values.

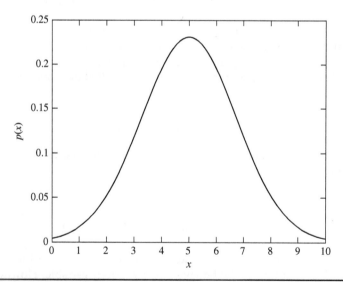

FIGURE 5.2 Probability density function for continuous x values.

5.3 Expectation and Moments

It is valuable to understand various properties of the random variable x or various properties of functions of the random variable x. The most obvious property is the statistical average. The *statistical average* is defined as the expected value denoted by E. Thus, the expected value of x is defined as

$$E[x] = \int_{-\infty}^{\infty} xp(x)dx \tag{5.4}$$

Not only can we find the expected value of x, but we can also find the expected value of any function of x. Thus

$$E[f(x)] = \int_{-\infty}^{\infty} f(x)p(x)dx \tag{5.5}$$

The function of x could be x^2, x^3, $\cos(x)$, or any other operation on the random variable. The expected value of x is typically called the *first moment* denoted as m_1.

$$m_1 = \int_{-\infty}^{\infty} xp(x)dx \tag{5.6}$$

The *n*th *moment* is defined as the expected value of x^n, thus

$$m_n = \int_{-\infty}^{\infty} x^n p(x)dx \tag{5.7}$$

The concept of moments is borrowed from the terminology of moments in mechanics.

If the random variable was expressed in volts, the first moment would correspond to the average, mean, or dc voltage. The second moment would correspond to the average power.

The spreading about the first moment is called the *variance* and is defined as

$$E[(x-m_1)^2] = \mu_2 = \int_{-\infty}^{\infty} (x-m_1)^2 p(x)dx \tag{5.8}$$

The *standard deviation* is denoted by σ and is defined as the spread about the mean, thus

$$\sigma = \sqrt{\mu_2} \tag{5.9}$$

By expanding the squared term in Eq. (5.8), it can be shown that $\sigma = m_2 - m_1^2$.

The first moment and standard deviation tend to be the most useful descriptors of the behavior of a random variable. However, other moments may need to be calculated to understand more fully the behavior of a given random variable x. Since the calculation of each new moment requires a reevaluation Eq. (5.7), sometimes it is useful to utilize the moment generating function that will simplify the calculation of multiple moments. The *moment generating function* is defined as

$$E[e^{sx}] = \int_{-\infty}^{\infty} e^{sx} p(x)dx = F(s) \tag{5.10}$$

Principles of Random Variables and Processes

The moment generating function resembles the Laplace transform of the pdf. This now brings us to the moment theorem. If we differentiate Eq. (5.10) n times with respect to s, it can be shown that

$$F^n(s) = E[x^n e^{sx}] \tag{5.11}$$

Thus, when $s = 0$, we can derive the nth moment as shown below

$$F^n(0) = E[x^n] = m_n \tag{5.12}$$

Example 5.1 If the discrete pdf of the random variable is given by $p(x) = .5[\delta(x+1) + \delta(x-1)]$, what are the first three moments using the moment generating function $F(s)$?

Solution Finding the moment generating function, we have

$$F(s) = \int_{-\infty}^{\infty} e^{sx}(.5)[\delta(x+1) + \delta(x-1)]dx = .5[e^{-s} + e^{s}]$$

$$= \cosh(s)$$

The first moment is given as

$$m_1 = F^1(s)\big|_{s=0} = \frac{d\cosh(s)}{ds}\bigg|_{s=0} = \sinh(0) = 0$$

The second moment is given as

$$m_2 = F^2(s)\big|_{s=0} = \cosh(0) = 1$$

The third moment is given as

$$m_3 = F^3(s)\big|_{s=0} = \sinh(0) = 0$$

5.4 Common Probability Density Functions

There are numerous pdfs commonly used in both radar, sonar, and communications. These pdfs describe the characteristics of the receiver noise, the arriving signal from multipaths, the distribution of the phase, envelope, and power of arriving signals. A quick summary of these pdfs and their behavior will be useful in supporting concepts addressed in Chaps. 6 to 8.

5.4.1 Gaussian Density

The Gaussian or normal probability density is perhaps the most common pdf. The Gaussian distribution generally defines the behavior of noise in receivers and also the nature of the random amplitudes of arriving multipath signals. According to the Central Limit Theorem, the sum of numerous continuous random variables as the number increases, tends toward a Gaussian distribution. The *Gaussian density* is defined as

$$p(x) = \frac{1}{\sqrt{2\pi\sigma^2}} e^{-\frac{(x-x_o)^2}{2\sigma^2}} \quad -\infty \leq x \leq \infty \tag{5.13}$$

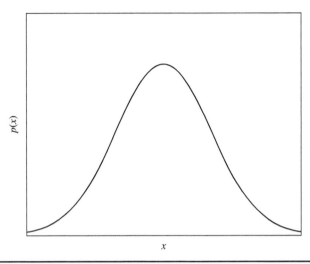

Figure 5.3 Gaussian density function.

This is a bell-shaped curve, which is symmetric about the mean value x_0, and has a standard deviation of σ. A plot of a typical Gaussian distribution is shown in Fig. 5.3.

Example 5.2 For the Gaussian probability density function with $x_0 = 0$ and $\sigma = 2$, calculate the probability that x exists over the range $0 \leq x \leq 4$.

Solution Invoking Eqs. (5.1) and (5.13), we can find the probability

$$P(0 \leq x \leq 4) = \int_0^4 \frac{1}{\sqrt{8\pi}} e^{-x^2/8} = .477$$

5.4.2 Rayleigh Density

The Rayleigh probability density generally results when one finds the envelope of two independent Gaussian processes. This envelope can be found at the output of a linear filter where the inputs are Gaussian random variables. Rayleigh distributions are normally attributed to the envelope of multipath signals when there is no direct path. The *Rayleigh distribution* is defined as

$$p(x) = \frac{x}{\sigma^2} e^{-x^2/2\sigma^2} \qquad x \geq 0 \qquad (5.14)$$

The standard deviation can be shown to be σ. A plot of a typical Rayleigh distribution is shown in Fig. 5.4.

Example 5.3 For the Rayleigh probability density function with $\sigma = 2$, calculate the probability that x exists over the range $0 \leq x \leq 4$.

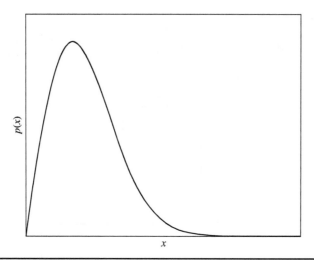

FIGURE 5.4 Rayleigh density function.

Solution Invoking Eqs. (5.1) and (5.14), we can find the probability

$$P(0 \leq x \leq 4) = \int_0^4 \frac{x}{4} e^{-x^2/8} = .865$$

5.4.3 Uniform Density

The uniform distribution is normally attributed to the distribution of the random phase for propagating signals. Not only does the phase delay tend to be uniformly distributed, but often the angles of arrival for diverse propagating waves can also take on a uniform distribution. The *uniform distribution* is defined as

$$P(x) = \frac{1}{b-a}[u(x-a) - u(x-b)] \quad a \leq x \leq b \quad (5.15)$$

This mean value can be shown to be $(a + b)/2$. A plot of a typical uniform distribution is shown in Fig. 5.5.

Example 5.4 For the uniform distribution where $a = -2$, and $b = 2$, use the moment generating function to find the first three moments.

Solution By substituting Eq. (5.15) into Eq. (5.10), we have

$$F(s) = \frac{1}{4}\int_{-\infty}^{\infty} [u(x+2) - u(x-2)] e^{sx}\, dx = \frac{1}{4}\int_{-2}^{2} e^{sx}\, dx$$

$$= \frac{1}{2s}\sinh(2s)$$

The first moment is found as

$$m_1 = F^1(s)\Big|_{s=0} = \frac{\cosh(2s)}{s} - \frac{\sinh(2s)}{2s^2} = 0$$

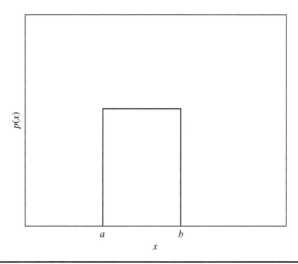

FIGURE 5.5 Uniform density function.

The second moment is found as

$$m_2 = F^2(s)\big|_{s=0} = \sinh(2s)\left[\frac{2}{s} + \frac{1}{s^3}\right] - \frac{2\cosh(2s)}{s^2} = \frac{4}{3}$$

The third moment is found as

$$m_3 = F^3(s)\big|_{s=0} = \cosh(2s)\left[\frac{4}{s} + \frac{6}{s^3}\right] - \sinh(2s)\left[\frac{6}{s^2} + \frac{3}{s^4}\right] = 0$$

5.4.4 Exponential Density

The exponential density function is sometimes used to describe the angles of arrival for incoming signals. It can also be used to describe the power distribution for a Rayleigh process. The *Exponential density* is the Eralang density when $n = 1([1])$ and is defined by

$$P(x) = \frac{1}{\sigma}e^{-x/\sigma} \qquad x \geq 0 \qquad (5.16)$$

The mean value can be shown to be σ. The standard deviation can also be shown to be σ. The literature sometimes replaces σ with $2\sigma^2$. in Eq. (5.16). A plot of a typical exponential distribution is shown in Fig. 5.6.

Example 5.5 Using the exponential density with $\sigma = 2$, calculate the probability that $2 \leq x \leq 4$ and find the first two moments using the moment generating function.

Solution The probability is given as

$$P(2 \leq x \leq 4) = \int_2^4 \frac{1}{\sigma}e^{-x(1/\sigma)}dx = .233$$

The moment generating function is derived as

$$F(s) = \int_0^\infty \frac{1}{\sigma}e^{-x\left(\frac{1}{\sigma}-s\right)}dx = \frac{1}{1-\sigma s}$$

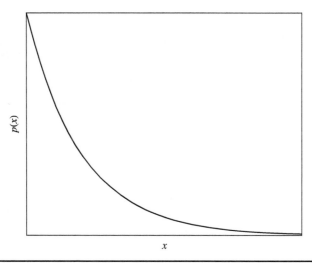

FIGURE 5.6 Exponential density function.

Finding the first moment

$$F^1(s) = \frac{\sigma}{(1-\sigma s)^2}\bigg|_{s=0} = \sigma = 2$$

Finding the second moment

$$F^{21}(s) = \frac{2\sigma^2}{(1-\sigma s)^3}\bigg|_{s=0} = 2\sigma^2 = 8$$

5.4.5 Rician Density

The Rician distribution is common for propagation channels where there is a direct path signal added to the multipath signals. The direct path inserts a nonrandom carrier, thereby modifying the Rayleigh distribution. Details of the derivation of the Rician distribution can be found in [1, 4]. The *Rician distribution* is defined as

$$p(x) = \frac{x}{\sigma^2} e^{-\frac{(x^2+A^2)}{2\sigma^2}} I_0\left(\frac{xA}{\sigma^2}\right) \quad x \geq 0, A \geq 0 \quad (5.17)$$

where $I_0(\ )$ is the Modified Bessel function of first kind and zero-order. A plot of a typical Rician distribution is shown in Fig. 5.7.

Example 5.6 For the Rician distribution with $\sigma = 2$ and $A = 2$, what is the probability that $x \geq 5$?

Solution Using Eqs. (5.17) and (5.1), we have the probability as

$$P(x \geq 5) = \int_5^\infty \frac{x}{4} e^{-\frac{(x^2+4)}{8}} I_0\left(\frac{x}{2}\right) dx = .121$$

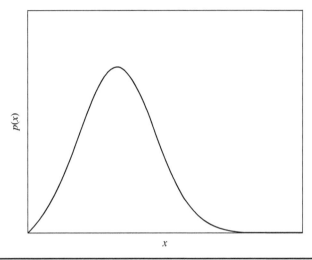

FIGURE 5.7 Rician density function.

5.4.6 Laplace Density

The Laplace density function is generally attributed to the distribution of indoor or congested urban angles of arrival. The Laplace distribution is given as

$$P(x) = \frac{1}{\sqrt{2}\sigma} e^{-\left|\frac{\sqrt{2}x}{\sigma}\right|} \quad -\infty \leq x \leq \infty \quad (5.18)$$

Because the Laplace distribution is symmetric about the origin, the first moment is zero. The second moment can be shown to be σ^2. A plot of the Laplace distribution is shown in Fig. 5.8.

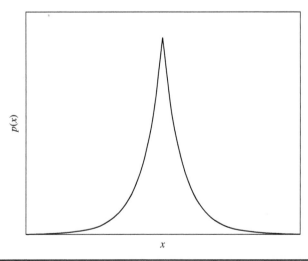

FIGURE 5.8 Laplace density function.

There are many other probability density functions described in the literature, but the six functions mentioned in this chapter are the most common distribution functions applied to wireless communication problems.

5.5 Stationarity and Ergodicity

In realistic applications, we may know the statistical properties of the signals and noise, but we are often confronted with the challenge of performing operations on a limited block of sampled data. If the statistical mean m_1 is the average value of the random variable x, one might intuitively assume that the time average would be equal to the statistical average. We can estimate the statistical average by using the time average over a block length T. The time average for the random variable x can be written as

$$\hat{x} = \frac{1}{T}\int_0^T x(t)\,dt \qquad (5.19)$$

where $\hat{x}$ is the estimate of the statistical average of x.

If the data is sampled data, Eq. (5.19) can be rewritten as a series to be expressed as

$$\hat{x} = \frac{1}{K}\sum_{k=1}^{K} x(k) \qquad (5.20)$$

Because the random variable $x(t)$ is changing with time, one might expect that the estimate $\hat{x}$ might also vary with time depending on the block length T. Because we are performing a linear operation on the random variable x, we produce a new random variable $\hat{x}$. One might expect that the time average and the statistical average would be similar if not identical. We can take the expected value of both sides of Eq. (5.19)

$$E(\hat{x}) = \frac{1}{T}\int_0^T E[x(t)]\,dt \qquad (5.21)$$

If all statistics of the random variable x do not change with time, the random process is said to be *strict-sense stationary* [1]. A strict sense stationary process is one in which the statistical properties are invariant to a shift in the time origin. If the mean value of a random variable does not change with time, the process is said to be *wide-sense stationary*. If x is wide-sense stationary, Eq. (5.21) simplifies to

$$E[\hat{x}] = E[x(t)] = m_1 \qquad (5.22)$$

In reality, the statistics might change for short blocks of time T but stabilize over longer blocks of time. If by increasing T (or K) we can force the time average estimate to converge to the statistical average, the process is said to be *ergodic* in the mean or *mean-ergodic*. This can be written as

$$\lim_{T\to\infty} \hat{x} = \lim_{T\to\infty} \frac{1}{T}\int_0^T x(t)\,dt = m_1 \qquad (5.23)$$

or

$$\lim_{T\to\infty} \hat{x} = \lim_{T\to\infty} \frac{1}{K}\sum_{k=1}^{K} x(k) = m_1 \qquad (5.24)$$

In a similar way, we can also use a time average to estimate the *variance* of x defined as

$$\hat{\sigma}_x^2 = \frac{1}{T}\int_0^T (x(t)-\hat{x})^2\,dt \qquad (5.25)$$

If the data is sampled data, Eq. (5.25) can be rewritten as a series to be expressed as

$$\hat{\sigma}_x^2 = \frac{1}{K}\sum_{k=1}^{K}(x(k)-\hat{x})^2 \qquad (5.26)$$

If by increasing T (or K) we can force the variance estimate to converge to the statistical variance, the process is said to be *ergodic* in the variance or *variance-ergodic*. This can be written as

$$\lim_{T\to\infty} \hat{\sigma}_x^2 = \lim_{T\to\infty} \frac{1}{T}\int_0^T (x(t)-\hat{x})\,dt = \sigma_x^2 \qquad (5.27)$$

or

$$\lim_{T\to\infty} \hat{\sigma}_x^2 = \lim_{T\to\infty} \frac{1}{K}\sum_{k=1}^{K}(x(k)-\hat{x})^2 = \sigma_x^2 \qquad (5.28)$$

In summary, stationary processes are ones in which the statistics of the random variables do not change at different times. Ergodic processes are ones where it is possible to estimate the statistics, such as mean, variance, and autocorrelation, from the measured values in time. Stationarity and ergodicity will prove valuable in practical communication systems because, under certain conditions, one can reliably estimate the mean, variance, and other parameters based on computing the time averages.

5.6 Autocorrelation and Power Spectral Density

It is valuable to know how well a random variable correlates with itself at different points in time. That is, how does x at the time t_1 correlate with x at the time t_2? This correlation is defined as an *autocorrelation* as we are correlating x with itself. The autocorrelation is normally written as

$$R_x(t_1, t_2) = E[x(t_1)x(t_2)] \qquad (5.29)$$

If the random variable x is wide-sense stationary, the specific values of t_1 and t_2 are not as important as the time interval between these two values defined by τ. Thus, for a wide-sense stationary process, the autocorrelation can be rewritten as

$$R_x(t) = E[x(t)x(t+\tau)] \tag{5.30}$$

It should be noted that the autocorrelation value at $\tau = 0$ is the second moment. Thus

$$R_x(0) = E[x^2] = m_2$$

Again, in practical systems where we are constrained to process limited blocks of data, one is forced to estimate the autocorrelation based upon using a time average. Therefore, the *estimate* of the autocorrelation can be defined as

$$\hat{R}_x(\tau) = \frac{1}{T}\int_0^T x(t)x(t+\tau)dt \tag{5.31}$$

If the data is sampled data, Eq. (5.31) can be rewritten as a series to be expressed as

$$\hat{R}_x(n) = \frac{1}{K}\sum_{k=1}^{K} x(k)x(k+n) \tag{5.32}$$

If by increasing T (or K) we can force the autocorrelation estimate to converge to the statistical autocorrelation, the process is said to be *ergodic* in the autocorrelation or *autocorrelation-ergodic*. This can be written as

$$\lim_{T\to\infty}\hat{R}_x(\tau) = \lim_{T\to\infty}\frac{1}{T}\int_0^T x(t)x(t+\tau)dt = R_x(\tau) \tag{5.33}$$

It should be noted that the units of the autocorrelation function for electrical systems are normally expressed in watts. Thus $R_x(0)$ yields the average power of the random variable x.

As with normal signals and linear systems, it is instructive to understand the behavior of the spectrum of the random variable x. Such parameters as bandwidth and center frequency help the system designer to understand how to best process the desired signal. The autocorrelation itself is a function of the time delay between two time-separated random variables. Thus, the autocorrelation is subject to Fourier analysis. Let us define the power spectral density as the Fourier transform of the autocorrelation function.

$$S_x(f) = \int_{-\infty}^{\infty} R_x(\tau)e^{-j2\pi f\tau}d\tau \tag{5.34}$$

$$R_x(f) = \int_{-\infty}^{\infty} S_x(f)e^{j2\pi f\tau}df \tag{5.35}$$

The Fourier transform pair in Eqs. (5.34) and (5.35) is frequently referred to as the *Wiener-Khinchin pair* [6].

5.7 Covariance Matrix

In the previous treatment of random variables, we assumed that only one random variable x existed and we performed expectation operations on these scalar values. Several circumstances arise when a collection of random variables exist. One such example is the output of each element of an antenna array. If an incoming planewave induces a random voltage on all M-array elements, the received signal x is a vector. Using the notation of Chap. 4, we can describe the array element output voltages for one incident planewave.

$$\bar{x}(t) = \bar{a}(\theta) \cdot s(t) \tag{5.36}$$

where $s(t)$ = incident monochromatic signal at time t
$\bar{a}(\theta)$ = M-element array steering vector for the θ direction of arrival

Let us now define the $M \times M$-array covariance matrix $\bar{R}_{xx}$ as

$$\begin{aligned}\bar{R}_{xx} &= E[\bar{x} \cdot \bar{x}^H] = E[(\bar{a}s)(s^*\bar{a}^H)] \\ &= \bar{a}E[|s|^2]\bar{a}^H \\ &= S\bar{a} \cdot \bar{a}^H\end{aligned} \tag{5.37}$$

where $(\ )^H$ indicates the Hermitian transpose and $S = E[|s|^2]$.

The covariance matrix in Eq. (5.37) assumes that we are calculating the ensemble average using the expectation operator $E[\ ]$. It should be noted that this is not a vector autocorrelation because we have imposed no time delay in the vector $\bar{x}$. Sometimes the correlation matrix is called the covariance matrix.

For realistic systems where we have a finite data block, we must resort to estimating the covariance matrix using a time average. Therefore, we can reexpress the operation in Eq. (5.37) as

$$\hat{R}_{xx} = \frac{1}{T}\int_0^T \bar{x}(t) \cdot \bar{x}(t)^H\, dt = \frac{\bar{a} \cdot \bar{a}^H}{T}\int_0^T |s(t)|^2\, dt \tag{5.38}$$

If the data is sampled data, Eq. (5.38) can be rewritten as a series to be expressed as

$$\hat{R}_{xx} = \frac{\bar{a} \cdot \bar{a}^H}{K}\sum_{k=1}^{K}|s(k)|^2 \tag{5.39}$$

If by increasing T (or K) we can force the covariance matrix estimate to converge to the statistical correlation matrix, the process is said to be *ergodic* in the covariance matrix. This can be written as

$$\lim_{T \to \infty}\hat{R}_{xx}(\tau) = \lim_{T \to \infty}\frac{1}{T}\int_0^T x(t)x(t)^H\, dt = \bar{R}_{xx} \tag{5.40}$$

The covariance matrix will be heavily used in Chaps. 7 and 8.

5.8 References

1. Papoulis, A., *Probability, Random Variables, and Stochastic Processes*, 2d ed., McGraw-Hill, New York, 1984.
2. Peebles, P., *Probability, Random Variables, and Random Signal Principles*, McGraw-Hill, New York, 1980.
3. Thomas, J., *An Introduction to Statistical Communication Theory*, Wiley, New York, 1969.
4. Schwartz, M., *Information, Transmission, Modulation, and Noise*, McGraw-Hill, New York, 1970.
5. Haykin, S., *Communication Systems*, 2d ed., Wiley, New York, 1983.
6. Papoulis, A., and S. Pillai, *Probability, Random Variables, and Stochastic Processes*, 4th ed., McGraw-Hill, New York, 2002.

5.9 Problems

5.1 For the discrete probability density function given as

$$p(x) = \frac{1}{3}[\delta(x) + \delta(x-1) + \delta(x-2)]$$

(a) What is the moment generating function $F(s)$?
(b) Calculate the first two moments using Eq. (5.7).
(c) Calculate the first two moments using the moment generating function and Eq. (5.12).

5.2 For the Gaussian density with $\sigma = 1$ and $x_0 = 3$
(a) Use MATLAB and plot the function for $-10 \le x \le 10$.
(b) What is the probability $P(x \ge 2)$?

5.3 For the Rayleigh density with $\sigma = 2$
(a) Use MATLAB and plot the function for $0 \le x \le 10$.
(b) What is the probability $P(x \ge 2)$?

5.4 For the uniform density with $a = 0, b = 5$
(a) Find the first two moments using Eq. (5.7).
(b) Find the first two moments using Eq. (5.12).

5.5 For the exponential density with $\sigma = 2$
(a) Use MATLAB and plot the function for $0 \le x \le 5$.
(b) What is the probability $P(x \ge 2)$?

5.6 For the Rician density with $\sigma = 3$ and $A = 5$
(a) Use MATLAB and plot the function for $0 \le x \le 10$.
(b) What is the probability $P(x \ge 5)$?

5.7 For the Laplace density with $\sigma = 2$
(a) Use MATLAB and plot the function for $-6 \le x \le 6$.
(b) What is the probability $P(x \le -2)$?

Chapter Five

5.8 Create the 30 sample zero mean Gaussian random variable x using MATLAB such that the $\sigma = 2$. This is done by the command $x = \sigma *$ randn (1, 30). This is a discrete time series of block length $K = 30$.
  (a) Use Eq. (5.20) to estimate the mean value.
  (b) Use Eq. (5.26) to estimate the standard deviation σ_x.
  (c) What is the percent error between these estimates and the mean and standard deviation for the true Gaussian process?

5.9 Use the same sequence from Prob. 5.8.
  (a) Use the MATLAB xcorr( ) command to calculate and plot the autocorrelation $R_x(n)$ of x.
  (b) Use the FFT command and fftshift to calculate the power spectral density of $R_x(n)$. Plot the absolute value.

5.10 For the $N = 2$ element array with elements spaced $\lambda/2$ apart
  (a) What is the array steering vector for the $\theta = 30°$?
  (b) Define the time signal impinging upon the array as $s(t) = 2\exp(j\pi t/T)$. Use Eq. (5.38) and calculate the array covariance matrix.

CHAPTER 6
Propagation Channel Characteristics

Free-space transmission occurs when the received signal is exclusively the result of direct path propagation. In this case there is no interference at the receiver caused by multipath signals. The received signal strength calculations are straightforward and deterministic. The free-space transmission model is a useful construct that can be used to understand fundamental propagation behavior. However, the free-space model is unrealistic because it fails to account for the numerous terrestrial effects of multipath propagation. It is normally assumed in this chapter that the propagation channel includes at least two propagation paths.

A *channel* is defined as the communication path between transmit and receive antennas. The channel accounts for all possible propagation paths as well as the effects of absorption, spherical spreading, attenuation, reflection losses, Faraday rotation, scintillation, polarization dependence, delay spread, angular spread, Doppler spread, dispersion, interference, motion, and fading. It may not be necessary that any one channel has all of the above effects, but often channels have multiple influences on communication waveforms. Obviously, the complexity of the channel increases as the number of available propagation paths increases. It also becomes more complex if one or more variables vary with time such as the receiver or transmitter position. Several excellent texts exist, which describe channel characteristics [1–9].

Indoor propagation modeling can be a formidable challenge. This is partly due to the regular and periodic location of structures such as windows, doors, wall studs, ceiling tiles, electrical conduits, ducts, and plumbing. This is also partly due to the very close proximity of scattering objects relative to the transmitter and/or receiver. An excellent treatment of indoor propagation channels can be found in Sarkar et al. [5], Shankar [3], and Rappaport [9]. This chapter previews channel characterization basics but will assume outdoor propagation conditions.

6.1 Flat Earth Model

In Chap. 2 we discussed propagation over flat earth. This propagation model was simplistic, but it demonstrated the elements of more complex propagation scenarios. We will repeat Fig. 2.17 as Fig. 6.1. The transmitter is at height h_1 while the receiver is at height h_2. The direct path term has path length r_1. There is also a reflected term due to the presence of a ground plane, this is called the *indirect path*. Its overall path length is r_2. The ground causes a reflection at the point y with a reflection coefficient R. The reflection coefficient R is normally complex and can alternatively be expressed as $R = |R|e^{j\psi}$. The reflection coefficient can be found from the Fresnel reflection coefficients as described in Eq. (2.63) or (2.69). The reflection coefficient is polarization dependent. One expression is used for parallel polarization where the E field is parallel to the plane of incidence (E is perpendicular to the ground). The other expression is for perpendicular polarization where the E field is perpendicular to the plane of incidence (E is parallel to the ground).

Through simple algebra, it can be shown that

$$r_1 = \sqrt{d^2 + (h_2 - h_1)^2} \tag{6.1}$$

$$r_2 = \sqrt{d^2 + (h_2 + h_1)^2} \tag{6.2}$$

The total received phasor field can be found to be

$$E_{rs} = \frac{E_0 e^{-jkr_1}}{r_1} + \frac{E_0 R e^{-jkr_2}}{r_2} \tag{6.3}$$

where E_{rs} = phasor representation of the received electric field E_r.

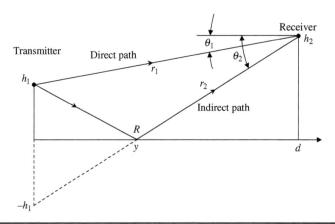

FIGURE 6.1 Propagation over flat earth.

Propagation Channel Characteristics

The reflection point is given by $y = dh_1/(h_1 + h_2)$. Therefore, the angles of arrival may be calculated to be

$$\theta_1 = \tan^{-1}\left(\frac{h_2 - h_1}{d}\right) \tag{6.4}$$

$$\theta_2 = \tan^{-1}\left(\frac{h_2 + h_1}{d}\right) \tag{6.5}$$

Last, the time delays of arrival are given by $\tau_1 = \frac{r_1}{c}$ and $\tau_2 = \frac{r_2}{c}$, where c is the speed of propagation.

If we make the assumption that the antenna heights $h_1, h_2 \ll r_1, r_2$, we can use a binomial expansion on Eqs. (6.1) and (6.2) to simplify the distance expressions.

$$r_1 = \sqrt{d^2 + (h_2 - h_1)^2} \approx d + \frac{(h_2 - h_1)^2}{2d} \tag{6.6}$$

$$r_2 = \sqrt{d^2 + (h_2 + h_1)^2} \approx d + \frac{(h_2 + h_1)^2}{2d} \tag{6.7}$$

If we additionally assume that $r_1 \approx r_2$, we can substitute Eqs. (6.6) and (6.7) into Eq. (6.3) to get

$$\begin{aligned}
E_{rs} &= \frac{E_0 e^{-jkr_1}}{r_1}\left[1 + \mathrm{Re}^{-jk(r_2 - r_1)}\right] \\
&= \frac{E_0 e^{-jkr_1}}{r_1}\left[1 + |R|e^{-j\left(k\frac{2h_1 h_2}{d} - \psi\right)}\right] \\
&= \frac{E_0 e^{-jkr_1}}{r_1}\left[1 + |R|\left(\cos\left(k\frac{2h_1 h_2}{d} - \psi\right) - j\sin\left(k\frac{2h_1 h_2}{d} - \psi\right)\right)\right]
\end{aligned} \tag{6.8}$$

Equation (6.8) is in phasor form. We can transform this phasor back into the time domain by finding $\mathrm{Re}\{E_{rs} e^{j\omega t}\}$. Through the use of Euler identities and some manipulation, we can write the time domain expression as

$$E_r(t) = \frac{E_0}{r_1}\left[\begin{array}{l}\left(1 + |R|\cos\left(k\frac{2h_1 h_2}{d} - \psi\right)\right)\cos(\omega t - kr_1) \\ + |R|\sin\left(k\frac{2h_1 h_2}{d} - \psi\right)\sin(\omega t - kr_1)\end{array}\right] \tag{6.9}$$

This solution is composed of two quadrature sinusoidal signals harmonically interacting with each other. If $R = 0$, Eq. (6.9) reverts to the direct path solution. Equation (6.9) is of the general form

$$X\cos(\omega t - kr_1) + Y\sin(\omega t - kr_1) = A\cos(\omega t - kr_1 + \phi) \tag{6.10}$$

where $X = \dfrac{E_0}{r_1}\left(1 + |R|\cos\left(k\dfrac{2h_1 h_2}{d} - \psi\right)\right)$

$Y = \dfrac{E_0}{r_1}|R|\sin\left(k\dfrac{2h_1 h_2}{d} - \psi\right)$

$A = \sqrt{X^2 + Y^2} =$ signal envelope

$\phi = \tan^{-1}\left(\dfrac{Y}{X}\right) =$ signal phase

Thus, the envelope and phase of Eq. (6.9) is given by

$$A = \dfrac{E_0}{r_1}\sqrt{\left(1 + |R|\cos\left(\dfrac{2kh_1 h_2}{d} - \psi\right)\right)^2 + \left(|R|\sin\left(\dfrac{2kh_1 h_2}{d} - \psi\right)\right)^2} \quad (6.11)$$

$$\phi = \tan^{-1}\left(\dfrac{\dfrac{E_0}{r_1}|R|\sin\left(k\dfrac{2h_1 h_2}{d} - \psi\right)}{\dfrac{E_0}{r_1}\left(1 + |R|\cos\left(k\dfrac{2h_1 h_2}{d} - \psi\right)\right)}\right) \quad (6.12)$$

where $\dfrac{E_0}{r_1} =$ direct path amplitude.

Example 6.1 Assuming the following values: $|R| = .3, .6, .9$, $\psi = 0$, $E_0/r_1 = 1$, $h_1 = 5$ m, $h_2 = 20$ m, $d = 100$ m. What are the time delays τ_1 and τ_2? Also, plot a family of curves for the envelope and the phase versus $(2kh_1 h_2)/d$.

Solution Using Eqs. (6.6) and (6.7), we can calculate the distances $r_1 = 101.12$ m and $r_2 = 103.08$ m. The corresponding time delays are $\tau_1 = .337$ μs and $\tau_2 = .344$ μs. The plots of the envelope and phase are shown in Figs. (6.11) and (6.12).

Figures 6.2 and 6.3 demonstrate that if any one of the abscissa variables change, whether antenna heights (h_1, h_2), antenna distance (d), or the wavenumber k, the received signal will undergo significant interference effects. In general, the total received signal can fluctuate anywhere between zero and up to twice the amplitude of the direct path component. The phase angle can change anywhere between $-90°$ and $90°$. If antenna 1 represents a mobile phone in a moving vehicle, one can see how multipath can cause interference resulting in signal fading. Fading is defined as the *waxing* and *waning* of the received signal strength.

6.2 Multipath Propagation Mechanisms

The flat earth model discussed earlier is a deterministic model that is instructive in demonstrating the principles of basic interference. In the case of the flat earth model, we know the received signal voltage

Propagation Channel Characteristics   123

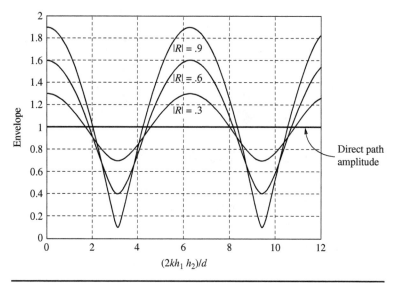

FIGURE 6.2 Envelope of direct and reflected signals over flat earth.

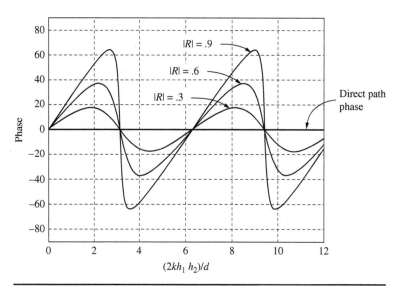

FIGURE 6.3 Phase of direct and reflected signals over flat earth.

at all times. In typical multipath/channel scenarios, the distribution of large numbers of reflecting, diffracting, refracting, and scattering objects becomes random. In this case, numerous multiple paths can be created and it becomes extremely difficult to attempt to model the channel deterministically. Thus, one must revert to a statistical model for estimating signal and channel behavior.

124   Chapter Six

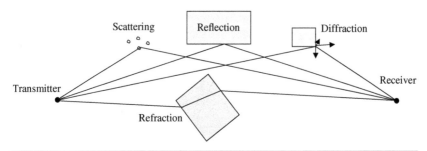

FIGURE 6.4 Different mechanisms creating multipath.

Figure 6.4 shows a propagation channel with several candidate mechanisms for creating multiple propagation paths. Let us define these mechanisms.

- *Scattering:* Scattering occurs when the electromagnetic signal strikes objects that are much smaller than a wavelength. These objects could be water droplets, clouds, or insects, for example. In the electromagnetic community this mechanism is often termed *Rayleigh scattering*. (This is not to be confused with Rayleigh fading although the two phenomena are interrelated.)
- *Refraction:* Refraction occurs when an electromagnetic signal propagates through a structure. The propagation path is diverted because of the difference in the electrical properties of the medium. Boundary conditions help determine the extent of the refraction.
- *Reflection:* Reflection occurs when an electromagnetic signal strikes a smooth surface at an angle and is reflected toward the receiver. The angle of reflection is equal to the angle of incidence under normal conditions.
- *Diffraction:* Diffraction occurs when the electromagnetic signal strikes an edge or corner of a structure that is large in terms of wavelength. The incident ray is diffracted in a cone of rays following Keller's laws of diffraction [10].

The various mechanisms such as scattering, refraction, reflection, and diffraction give rise to alternate propagation paths such that the received signal is a composite of numerous replicas all differing in phase, amplitude, and in time delay. Thus, the multipath signal amplitudes, phases, and time delays become random variables.

6.3   Propagation Channel Basics

In order to better understand the performance of a wireless signal propagating in a typical outdoor environment, it is necessary to define some terms. These terms are commonly used to describe properties or characteristics of the channel.

6.3.1 Fading

Fading is a term used to describe the fluctuations in a received signal as a result of multipath components. Several replicas of the signal arrive at the receiver, having traversed different propagation paths, adding constructively and destructively. The fading can be defined as fast or slow fading. Additionally, fading can be defined as *flat* or *frequency selective fading*.

Fast fading is propagation characterized by rapid fluctuations over very short distances. This fading is due to scattering from nearby objects and thus is termed *small-scale fading*. Typically fast fading can be observed up to half-wavelength distances. When there is no direct path (line-of-sight), a Rayleigh distribution tends to best fit this fading scenario. Thus fast fading is sometimes referred to as *Rayleigh fading*. When there is a direct path or a dominant path, fast fading can be modeled with a Rician distribution.

Slow fading is propagation characterized by slow variations in the mean value of the signal. This fading is due to scattering from the more distant and larger objects, and thus is termed *large-scale fading*. Typically slow fading is the trend in signal amplitude as the mobile user travels over large distances relative to a wavelength. The slow fading mean value is generally found by averaging the signal over 10 to 30 wavelengths [11]. A log-normal distribution tends to best fit this fading scenario; thus slow fading is sometimes referred to as *lognormal fading*. Figure 6.5 shows a superimposed plot of fast and slow fading.

Flat fading is when the frequency response of the channel is flat relative to the frequency of the transmit signal; that is, the channel bandwidth B_c is greater than the signal bandwidth B_s ($B_c > B_s$).

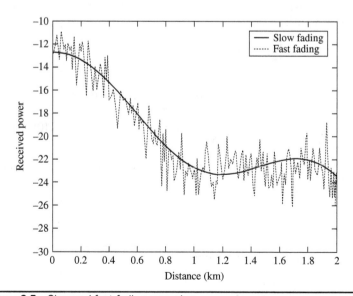

FIGURE 6.5 Slow and fast fading examples.

Thus, the multipath characteristics of the channel preserve the signal quality at the receiver.

Frequency selective fading is when the channel bandwidth B_C is less than the signal bandwidth B_S ($B_C < B_S$). In this case, the multipath delays start to become a significant portion of the transmit signal time duration and dispersion occurs.

Fast fading is of particular interest to the electrical engineer because the resulting rapid fluctuations can cause severe problems in reliably maintaining communication.

6.3.2 Fast Fading Modeling

Multipath with No Direct Path

Based on the scenario as given in Fig. 6.4, we assume that no direct path exists but that the entire received electric field is based on multipath propagation. We can express the received voltage phasor as the sum of all the possible multipath component voltages within the receiver.

$$v_{rs} = \sum_{n=1}^{N} a_n e^{-j(kr_n - \alpha_n)} = \sum_{n=1}^{N} a_n e^{j\phi_n} \qquad (6.13)$$

where a_n = random amplitude of the nth path
α_n = random phase associated with the nth path
r_n = length of nth path
$\phi_n = -kr_n + \alpha_n$

If we assume a large number of scattering structures N, which are randomly distributed, we can assume that the phases ϕ_n are uniformly distributed. We can express the time-domain version of the received voltage as

$$v_r = \sum_{n=1}^{N} a_n \cos(\omega_0 t + \phi_n)$$

$$= \sum_{n=1}^{N} a_n \cos(\phi_n)\cos(\omega_0 t) - \sum_{n=1}^{N} a_n \sin(\phi_n)\sin(\omega_0 t) \qquad (6.14)$$

We may further simplify Eq. (6.14) as we did in Eq. (6.10) using a simple trigonometric identity.

$$v_r = X\cos(\omega_0 t) - Y\sin(\omega_0 t) = r\cos(\omega_0 t + \phi) \qquad (6.15)$$

where $X = \sum_{n=1}^{N} a_n \cos(\phi_n)$

$Y = \sum_{n=1}^{N} a_n \sin(\phi_n)$

$r = \sqrt{X^2 + Y^2}$ = envelope

$\phi = \tan^{-1}\left(\dfrac{Y}{X}\right)$

In the limit, as $N \to \infty$, the Central Limit Theorem dictates that the random variables X and Y will follow a Gaussian distribution with zero mean and standard deviation σ. The phase ϕ can also be modeled as a uniform distribution such that $p(\phi) = \frac{1}{2\pi}$ for $0 \le \phi \le 2\pi$. The envelope r is the result of a transformation of the random variables X and Y and can be shown to follow a Rayleigh distribution as given by Schwarz [12] or Papoulis [13]. The *Rayleigh probability density function* is defined as

$$p(r) = \frac{r}{\sigma^2} e^{-\frac{r^2}{2\sigma^2}} \qquad r \ge 0 \qquad (6.16)$$

where σ^2 is the variance of the Gaussian random variables X or Y.

Figure 6.6 shows a Rayleigh distribution for two different standard deviations. Although the mean value of X or Y is zero, the mean value of the Rayleigh distribution is $\sigma\sqrt{\pi/2}$.

Example 6.2 For the Rayleigh-fading channel where $\sigma = .003$ V, what is the probability that the received voltage envelope will exceed a threshold of 5 mV?

Solution The probability of the envelope as exceeding 5 mV is given by

$$P(r \ge .005) = \int_{.005}^{\infty} \frac{r}{\sigma^2} e^{-\frac{r^2}{2\sigma^2}} dr = .249$$

This can be shown as the shaded area under the curve in the Fig. 6.7.

It can be shown that if the envelope is Rayleigh distributed, the power p (watts) will have an exponential distribution [3, 13] (also called an *Erlang distribution* with $n = 1$.)

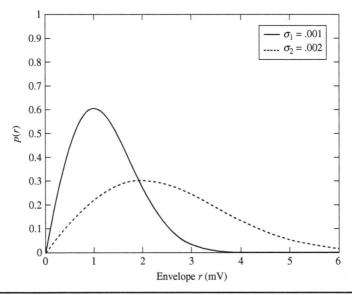

FIGURE 6.6 Rayleigh probability density.

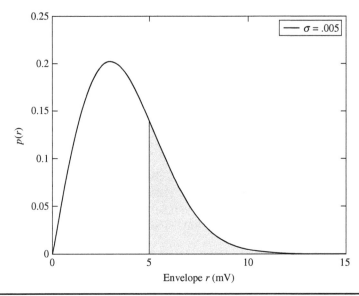

Figure 6.7 Rayleigh probability density with indicated threshold.

$$p(p) = \frac{1}{2\sigma^2} e^{-\frac{p}{2\sigma^2}} \quad p \geq 0 \tag{6.17}$$

The average value of the power is given by

$$E[p] = p_0 = \int_0^\infty p \cdot p(p) dp$$

$$= \int_0^\infty \frac{p}{2\sigma^2} e^{-\frac{p}{2\sigma^2}} dp = 2\sigma^2 \tag{6.18}$$

Thus $p_0 = 2\sigma^2$ and can be substituted in Eq. (6.17). The power distribution is plotted in Fig. 6.8 for $p_0 = 2\mu W$, 4 µW.

The minimum detectable threshold power in a receiver is P_{th}. This power level is dictated by the receiver noise floor, noise figure, and detector thresholds. If the received power falls below the threshold, the receiver goes into "outage" because the backend signal-to-noise ratio is insufficient. The *outage probability* is the probability that the received power is too small for detection. This is given by

$$P(p \leq P_{th}) = \int_0^{P_{th}} \frac{1}{p_0} e^{-\frac{p}{p_0}} dp \tag{6.19}$$

Example 6.3 What is the outage probability of a Rayleigh channel if the average power is 2 µW and the threshold power is 1 µW?

Solution The outage probability is given by

$$P(p \leq 1\mu W) = \int_0^{1\mu W} \frac{1}{2\mu W} e^{-\frac{p}{2\mu W}} dp = 0.393$$

This can be shown as the shaded area under the curve in the Fig. 6.9.

Propagation Channel Characteristics

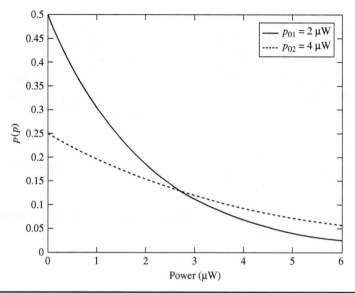

FIGURE 6.8 Exponential probability density.

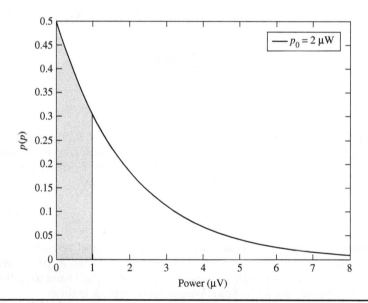

FIGURE 6.9 Outage probability shown as area under curve.

Multipath with Direct Path

Let us now consider that a direct path is present as indicated in Fig. 6.10. If a direct path is allowed in the received voltage, we must modify

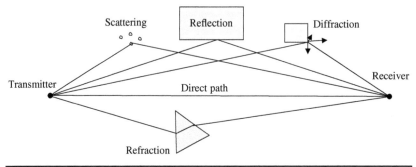

Figure 6.10 A multipath channel with a direct path.

Eqs. (6.14) and (6.15) by adding the direct path term with the direct path amplitude A (volts).

$$v_r = A\cos(\omega_0 t) + \sum_{n=1}^{N} a_n \cos(\omega_0 t + \phi_n)$$

$$= \left[A + \sum_{n=1}^{N} a_n \cos(\phi_n)\right]\cos(\omega_0 t) - \sum_{n=1}^{N} a_n \sin(\phi_n)\sin(\omega_0 t) \quad (6.20)$$

Again, the envelope $r = \sqrt{X^2 + Y^2}$. We now must accordingly revise the random variables X and Y.

$$X = A + \sum_{n=1}^{N} a_n \cos(\phi_n)$$

$$Y = \sum_{n=1}^{N} a_n \sin(\phi_n) \quad (6.21)$$

The random variable X is Gaussian with mean of A and standard deviation of σ. Random variable Y is Gaussian with zero mean and standard deviation of σ. The probability density function for the envelope is now a Rician distribution and is given by

$$p(r) = \frac{r}{\sigma^2} e^{-\frac{(r^2 + A^2)}{2\sigma^2}} I_0\left(\frac{rA}{\sigma^2}\right) \quad r \geq 0 \; A \geq 0 \quad (6.22)$$

where $I_0() =$ Modified Bessel function of first kind and zero-order.

We can characterize the Rician distribution by a parameter $K = A^2/(2\sigma^2)$. K is the direct signal power to multipath variance ratio. K is also called the *Rician factor*. We can also express K in dB as

$$K(\text{dB}) = 10 \log_{10}\left(\frac{A^2}{2\sigma^2}\right) \quad (6.23)$$

In the case where $A = 0$, the Rician distribution reverts to a Rayleigh distribution. A plot of the Rician distribution for three K values is shown in Fig. 6.11.

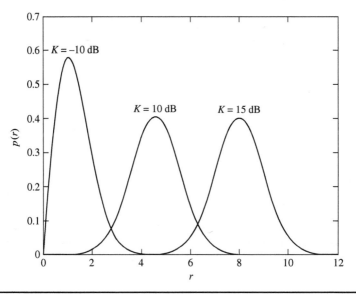

FIGURE 6.11 Rician distribution.

Example 6.4 For the Rician-fading channel where $\sigma = 3$ mV, the direct path amplitude $A = 5$ mV, what is the probability that the received voltage envelope will exceed a threshold of 5 mV?

Solution The probability of the envelope as exceeding 5 mV is given by

$$P(r \geq .005) = \int_{.005}^{\infty} \frac{r}{\sigma^2} e^{-\frac{(r^2+A^2)}{2\sigma^2}} I_0\left(\frac{rA}{\sigma^2}\right) dr = .627$$

This can be shown as the shaded area under the curve in Fig. 6.12.

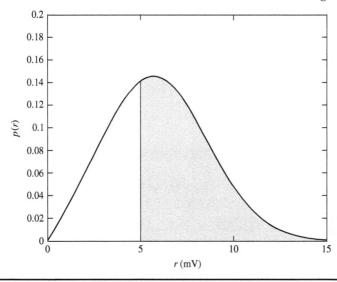

FIGURE 6.12 Rician probability density with indicated threshold.

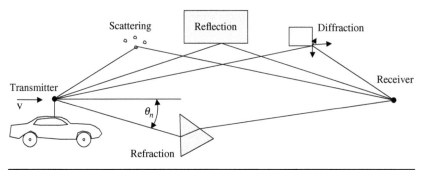

Figure 6.13 A multipath channel with no direct path.

As compared to Example 6.2 with no direct path, Example 6.4 demonstrates that the likelihood of having a detectable signal dramatically increases when the direct path is present.

Motion in a Fast Fading Channel

In the above examples of *line-of-sight* (LOS) and *non–line-of-sight* (NLOS) propagation, it was assumed that there was no motion in the channel. Motion will change the channel behavior by changing the location of the transmitter or receiver. In addition, motion introduces many discrete Doppler shifts in the received signal. Figure 6.13 shows a moving transmitter in a multipath environment with no direct path.

As the vehicle moves at a constant velocity in Fig. 6.13, many factors change with time. The angles (θ_n) of each multipath signal are time dependent. Each multipath experiences a different Doppler shift because the angle of scattering with respect to the moving vehicle is different for each scattering object. Also, the overall phase shift (α_n) changes with time because the propagation delays are changing.

The maximum possible Doppler shift is given by

$$f_d = f_0 \frac{v}{c} \qquad (6.24)$$

where f_d = Doppler frequency
f_0 = carrier frequency
v = vehicle velocity
c = speed of light

Because the direction of vehicle travel is at an angle θ_n with the nth multipath, the Doppler shift is modified accordingly. The Doppler shift for the nth path is given by

$$f_n = f_d \cos\theta_n = f_0 \frac{v}{c} \cos\theta_n \qquad (6.25)$$

We can now rewrite Eq. (6.14) accounting for the Doppler frequency shifts f_n.

Propagation Channel Characteristics

$$v_r = \sum_{n=1}^{N} a_n \cos(2\pi f_n t + \phi_n)\cos(\omega_0 t) - \sum_{n=1}^{N} a_n \sin(2\pi f_n t + \phi_n)\sin(\omega_0 t)$$

$$= \sum_{n=1}^{N} a_n \cos(2\pi f_d \cos(\theta_n)t + \phi_n)\cos(\omega_0 t) \qquad (6.26)$$

$$-\sum_{n=1}^{N} a_n \sin(2\pi f_d \cos(\theta_n)t + \phi_n)\sin(\omega_0 t)$$

We now have three random variables a_n, ϕ_n, and θ_n. The amplitude coefficients are Gaussian distributed whereas the phase coefficients are presumed to have a uniform distribution such that $0 \le \phi_n$ and $\theta_n \le 2\pi$. The envelope of v_r again has a Rayleigh distribution. The envelope r is given by

$$r = \sqrt{X^2 + Y^2} \qquad (6.27)$$

where $X = \sum_{n=1}^{N} a_n \cos(2\pi f_d \cos(\theta_n)t + \phi_n)$

$Y = \sum_{n=1}^{N} a_n \sin(2\pi f_d \cos(\theta_n)t + \phi_n)$

This model is called the *Clarke flat fading model* [7, 14].

Example 6.5 Use MATLAB to plot the envelope in Eq. (6.27) where the carrier frequency is 2 GHz, the vehicle velocity is 50 mph, the phase angles ϕ_n and θ_n are uniformly distributed, the coefficient a_n has a Gaussian distribution with zero mean and standard deviation of $\sigma = .001$. Let $N = 10$.

Solution We must first convert the velocity to meters/second. Therefore, 50 mph = 22.35 m/s. Thus, the maximum Doppler shift is

$$f_d = 2 \cdot 10^9 \cdot \frac{22.35}{3 \cdot 10^8} = 149 \text{ Hz}$$

Using the following short MATLAB program, we can plot the results as demonstrated in Fig. 6.14.

```
% Fast fading with velocity Example 6.5
N = 10;                  % number of scatterers
a=.001*randn(N,1);       % create Gaussian amplitude
                           coefficients
th=rand(N,1)*2*pi;       % create uniform phase angles
ph=rand(N,1)*2*pi;
fd=149;                  % Doppler
tmax = 10/fd;            % Maximum time
omega=2*pi*fd;
t=[0:1000]*tmax/1000;    % generate timeline
X=[zeros(1,length(t))];
Y=[zeros(1,length(t))];
for n=1:N                % generate the sums for X and
                           Y
X=X+a(n)*cos(omega*cos(th(n))*t+ph(n));
Y=Y+a(n)*sin(omega*cos(th(n))*t+ph(n));
end
```

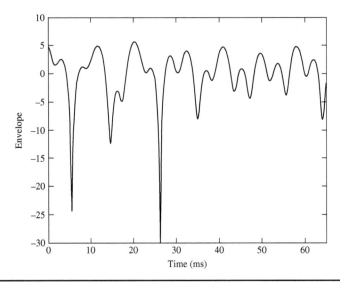

Figure 6.14 Doppler fading channel with N = 10 Paths.

```
r=sqrt(X.^2+Y.^2);    % Calculate the Rayleigh enve-
                        lope
rdb=20*log10(r);      % Calculate the envelope in dB
figure;
plot(t*1000,rdb,'k') % plot
xlabel('time (ms)')
ylabel('envelope')
axis([0 65 -30 10])
```

Example 6.5 is representative of Doppler fading but makes assumptions that are not realistic. In the example, the scattering from objects is angularly dependent. Thus, the coefficients a_n will be a function of time. Additionally, the phase angles ϕ_n and θ_n change with time. The Clarke model can be modified to reflect the time dependence of a_n, ϕ_n, and θ_n.

If we assume a large number of paths, the uniform distribution of angles θ_n results in a sinusoidal variation in the Doppler frequencies f_n. This transformation of the random variable results in a Doppler power spectrum derived by Gans [15] and Jakes [16].

$$S_d(f) = \frac{\sigma^2}{\pi f_d \sqrt{1 - \left(\frac{f}{f_d}\right)}} \qquad |f| \leq f_d \qquad (6.28)$$

where $\sigma^2 = \sum_{n=1}^{N} E[a_n^2]$ = average power of signal.

A plot of the Doppler power spectrum is shown in Fig. 6.15.

As a test of the validity of Eq. (6.28), we can rerun the program of Example 6.5 by increasing the scatterers to 100 and padding the amplitude

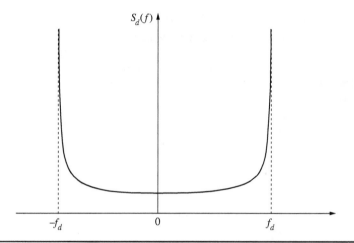

FIGURE 6.15 Doppler power density spectrum.

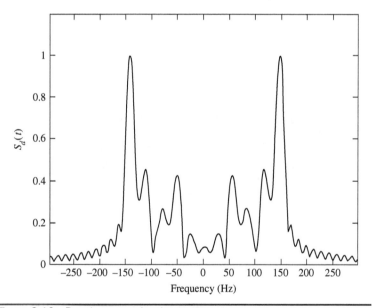

FIGURE 6.16 Doppler power density spectrum.

X with zeros. Again, the maximum Doppler shift is 149 Hz. After taking the fast Fourier transform, we can plot the results as seen in Fig. 6.16. The spectrum bears a similarity to the theoretical spectrum. However, the difference lies in the fact that the theoretical spectrum assumes that the number of scatterers is large enough to apply the Central Limit Theorem. In that case, there is a true Gaussian distribution on the scattering amplitudes and a true uniform distribution on the angles.

6.3.3 Channel Impulse Response

If we assume that the radio channel can be modeled as a linear filter, the characteristics of the channel can be modeled by finding the impulse response of the channel. Thus, all signal responses to the channel are dictated by the impulse response. The impulse response also gives an indication of the nature and number of the multiple paths. If it is assumed that the channel characteristics can change with time (i.e., the mobile user is moving), the channel impulse response will also be function of time. The generalized channel impulse response is given as follows:

$$h_c(t,\tau) = \sum_{n=1}^{N} a_n(t) e^{j\psi_n(t)} \delta(\tau - \tau_n(t)) \qquad (6.29)$$

where $a_n(t)$ = time varying amplitude of path n
$\psi_n(t)$ = time varying phase of path n which can include the effects of Doppler
$\tau_n(t)$ = time varying delay of path n

An example of an impulse response magnitude, $|h_c(t,\tau)|$, is shown plotted in Fig. 6.17. The amplitudes generally diminish with increasing delay because the spherical spreading increases in proportion to the delay. Differences in reflection coefficients can cause an exception to this rule for similar path lengths because slightly more distant sources of scattering might have larger reflection coefficients. The discrete time-delayed impulses are sometimes called *fingers*, *taps*, or *returns*.

If we additionally assume that the channel is *wide-sense stationary* (WSS) over small-scale times and distances, the impulse response will remain stable for short time/fast fading. Thus, we can further simplify the impulse response to be approximated as

$$h_c(\tau) = \sum_{n=1}^{N} a_n e^{j\psi_n} \delta(\tau - \tau_n) \qquad (6.30)$$

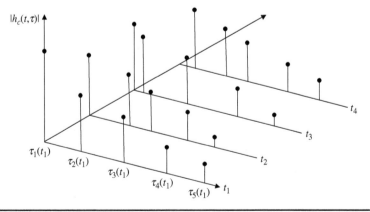

FIGURE 6.17 Channel impulse response at four instants in time.

6.3.4 Power Delay Profile

In the case of small-scale fading, a channel metric can be defined to aid in understanding channel behavior. This metric is called the *power delay profile* (PDP). (The PDP is essentially identical to the *multipath intensity profile* or MIP [17, 18].) The PDP can be defined as ([8, 9, 19])

$$P(\tau) = E[\,|h_c(t,\tau)|^2\,] = \sum_{n=1}^{N} P_n \delta(\tau - \tau_n) \qquad (6.31)$$

where $P_n = \langle |a_n(t)|^2 \rangle = a_n^2$

$\tau_n = \langle \tau_n(t) \rangle$

$\langle x \rangle$ = estimate of the random value x

A typical power delay profile for an urban area is shown in Fig. 6.18, where the fingers are the result of scattering from buildings.

The PDP is partly defined by the trip time delays (τ_n). The nature and characteristics of the delays helps to define the expected channel performance. Thus, it is important to define some terms regarding trip delays. Several valuable texts define these statistics [3, 5, 6, 8].

- *First Arrival Delay* (τ_A): This is measured as the delay of the earliest arriving signal. The earliest arriving signal is either the shortest multipath or the direct path if a direct path is present. All other delays can be measured relative to τ_A. Channel analysis can be simplified by defining the first arrival delay as zero (i.e., $\tau_A = 0$).

- *Excess Delay:* This is the additional delay of any received signal relative to the first arrival delay τ_A. Typically, all delays are defined as excess delays.

- *Maximum Excess Delay* (τ_M): This is the maximum excess delay where the PDP is above a specified threshold. Thus, $P(\tau_M) = P_{th}$ (dB).

- *Mean Excess Delay* (τ_0): The mean value or first moment of all excess delays.

$$\tau_0 = \frac{\sum_{n=1}^{N} P_n \tau_n}{\sum_{n=1}^{N} P_n} = \frac{\sum_{n=1}^{N} P_n \tau_n}{P_T} \qquad (6.32)$$

where $P_T = \sum_{n=1}^{N} P_n$ = multipath power gain.

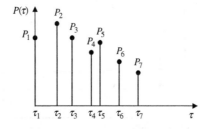

FIGURE 6.18 Typical urban power delay profile.

138   Chapter Six

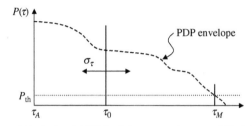

FIGURE 6.19   Depiction of PDP delay statistics.

- **RMS Delay Spread (σ_τ):** This is the standard deviation for all excess delays.

$$\sigma_\tau = \sqrt{\langle \tau^2 \rangle - \tau_0^2} = \sqrt{\frac{\sum_{n=1}^{N} P_n \tau_n^2}{P_T} - \tau_0^2} \qquad (6.33)$$

Figure 6.19 demonstrates the various terms defined earlier.

Example 6.6   Calculate the multipath power gain (P_T), the mean excess delay (τ_0), and the RMS delay spread (σ_τ) for the PDP given in the Fig. 6.20.

Solution   First we must convert all powers to a linear scale. Thus, $P_1 = .1$, $P_2 = .32$, $P_3 = .1$, $P_4 = .032$. The multipath power gain is

$$P_T = \sum_{n=1}^{N} P_n = .552 \quad \text{or} - 2.58 \text{ dB.}$$

The mean excess delay is

$$\tau_0 = \frac{\sum_{n=1}^{4} P_n \tau_n}{P_T} = \frac{.1 \times 0 + .32 \times 1 + .1 \times 3 + .032 \times 5}{.552} = 1.41 \text{ μs}$$

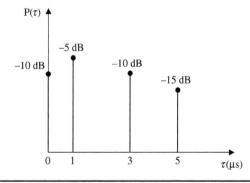

FIGURE 6.20   Power delay profile.

The RMS delay spread is

$$\sigma_\tau = \sqrt{\frac{\sum_{n=1}^{N} P_n \tau_n^2}{P_T} - \tau_0^2}$$

$$= \sqrt{\frac{.1 \times 0^2 + .32 \times 1^2 + .1 \times 3^2 + .032 \times 5^2}{.552} - (1.41)^2}$$

$$= 1.29 \text{ μs}$$

6.3.5 Prediction of Power Delay Profiles

It is sometimes informative to model the power delay profiles in order to: (1) understand channel behavior, (2) evaluate the performance of equalizers, and (3) estimate *bit error rate* (BER) performance. Numerous measurements have been performed on indoor and outdoor channels and three relatively useful models have been proposed Chuang [20], Feher [6]). The total received power is given by P_T. The three models are:

One-Sided Exponential Profile. This profile seems to most accurately describe both indoor and urban channels.

$$P(\tau) = \frac{P_T}{\sigma_\tau} e^{-\frac{\tau}{\sigma_\tau}} \qquad \tau \geq 0 \qquad (6.34)$$

Gaussian Profile

$$P(\tau) = \frac{P_T}{\sqrt{2\pi}\sigma_\tau} e^{-\frac{1}{2}\left(\frac{\tau}{\sigma_\tau}\right)^2} \qquad (6.35)$$

Equal Amplitude Two-Ray Profile

$$P(\tau) = \frac{P_T}{2}[\delta(\tau) + \delta(\tau - 2\sigma_\tau)] \qquad (6.36)$$

6.3.6 Power Angular Profile

The PDP and the path delays are instructive in helping one understand the dispersive characteristics of the channel and calculate the channel bandwidth. The PDP is especially relevant for *single-input single-output* (SISO) channels as one can view the impulse response as being for a SISO system. However, when an array is used at the receiver, the angles of arrival are of interest as well. Because the array has varying gain versus angle of arrival $(G(\theta))$, it is instructive to also understand the statistics of the angles of arrival such as angular spread and mean angle of arrival (AOA). Every channel has angular statistics as well as delay statistics. Basic concepts in modeling the AOA have been addressed by Rappaport [9], Gans [15], Fulghum, Molnar, Duel-Hallen [21, 22], Boujemaa and Marcos [23], and Klein and Mohr [24].

This author will define the angular equivalent of the PDP as the *power angular profile* (PAP). Earlier references discuss the concept of a *power angle density* (PAD) [23] or a *power azimuth spectrum* (PAS) [22]. However, the PAD or the PAS is more analogous to the traditional *power spectral density* (PSD) rather than to the PDP used previously. The concept of a PAP immediately conveys angular impulse response information which assists in channel characterization. Thus the PAP is given as

$$P(\theta) = \sum_{n=1}^{N} P_n \delta(\theta - \theta_n) \tag{6.37}$$

Along with the PAP, we can define some metrics as indicators of the angular characteristics of the propagation paths.

- *Maximum Arrival Angle* (θ_M): This is the maximum angle relative to the boresight (θ_B) of the receive antenna array. The boresight angle is often the broadside direction for a linear array. The maximum angle is restricted such that $|\theta_M| - \theta_B \leq 180°$.

- *Mean Arrival Angle* (θ_0): The mean value or first moment of all arrival angles

$$\theta_0 = \frac{\sum_{n=1}^{N} P_n \theta_n}{\sum_{n=1}^{N} P_n} = \frac{\sum_{n=1}^{N} P_n \theta_n}{P_T} \tag{6.38}$$

where $P_T = \sum_{n=1}^{N} P_n$ = multipath power gain.

- *RMS Angular Spread* (σ_θ): This is the standard deviation for all arrival angles

$$\sigma_\theta = \sqrt{\frac{\sum_{n=1}^{N} P_n \theta_n^2}{P_T} - \theta_0^2} \tag{6.39}$$

Figure 6.21 depicts a representative PAP.

Example 6.7 Calculate the multipath power gain (P_T), the mean arrival angle (θ_0), and the RMS angular spread (σ_θ) for the PAP given in the Fig. 6.22.

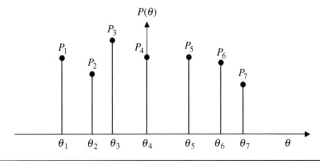

FIGURE 6.21 Power angular profile.

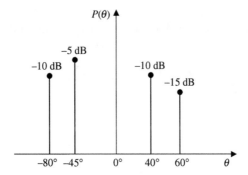

FIGURE 6.22 Power angular profile.

Solution First we must convert all powers to a linear scale. Thus, $P_1 = .1$, $P_2 = .32$, $P_3 = .1$, $P_4 = .032$. The multipath power gain is

$$P_T = \sum_{n=1}^{N} P_n = .552 \text{ W}$$

The mean arrival angle is

$$\theta_0 = \frac{\sum_{n=1}^{4} P_n \theta_n}{P_T} = \frac{.1 \times (-80) + .32 \times (-45) + .1 \times (40) + .032 \times (60)}{.552}$$
$$= -29.86°$$

The RMS angular spread is

$$\sigma_\theta = \sqrt{\frac{\sum_{n=1}^{N} P_n \theta_n^2}{P_T} - \theta_0^2}$$

$$= \sqrt{\frac{.1 \times (-80)^2 + .32 \times (-45)^2 + .1 \times (40)^2 + .032 \times (60)^2}{.552} - (-29.86)^2}$$

$$= 44°$$

An alternative approach for defining angular spread is given in Rappaport [9] where the angular spread is not found by defining first and second moments but rather the angular spread is determined through the use of a Fourier transform. This is akin to the use of the moment generating function in stochastic processes. The following description is slightly modified from the original Rappaport definition. We must first find the complex Fourier transform of the PAP. Thus

$$F_k = \int_0^{2\pi} P(\theta) e^{-jk\theta} d\theta \qquad (6.40)$$

where $F_k = k$th complex Fourier coefficient.

The angular spread is now defined as

$$\sigma_\theta = \theta_{\text{width}} \sqrt{1 - \frac{|F_1|^2}{F_0^2}} \qquad (6.41)$$

where θ_{width} is the angular width of the PAP.

If indeed we use the PAP defined in Eq. (6.34), we can use the sifting property of the delta function and can find the Fourier coefficients.

$$F_k = \int_0^{2\pi} \sum_{n=1}^{N} P_n \delta(\theta - \theta_n) e^{-jk\theta} d\theta = \sum_{n=1}^{N} P_n e^{-jk\theta_n} \quad (6.42)$$

where

$$F_0 = \sum_{n=1}^{N} P_n = P_T$$

$$F_1 = \sum_{n=1}^{N} P_n e^{-j\theta_n}$$

Example 6.8 Repeat Example 6.7 solving for the angular spread using the Rappaport method.

Solution The total angular width in Example 6.7 is 140°. The value of F_0 is the same as the total power, thus $F_0 = P_T = .552$. The value of F_1, found from Eq. (6.40), is

$$F_1 = \sum_1^4 P_n e^{-j\theta_n} = .1e^{j80°} + .32e^{j45°} + .1e^{-j40°} + .032e^{-j60°}$$
$$= .34 + j.23$$

Substituting into Eq. (6.41) we can find the angular spread thus

$$\sigma_\theta = 94.3°$$

This angular spread is larger than the angular spread in Example 6.7 by almost a factor of 2.

6.3.7 Prediction of Angular Spread

There are numerous potential models to describe angular spread under various conditions. An excellent overview is given by Ertel et al. [25]. Additionally, derivations are given for a ring of scatterers and a disk of scatterers in Pedersen et al. [26]. Several models are given next showing the angular distribution for a ring of scatterers, disk of scatterers, and an indoor distribution of scatterers.

Ring of Scatterers: Figure 6.23 shows the transmitter and receiver where the transmitter is circled by a ring of scatterers uniformly distributed about the ring.

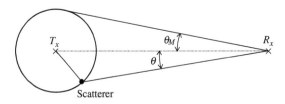

FIGURE 6.23 Ring of scatterers.

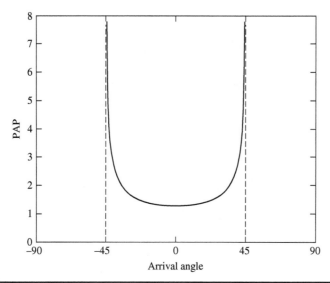

FIGURE 6.24 PAP for ring of scatterers.

If we assume that the scattering occurs from a ring of scatterers, at constant radius, surrounding the transmitting antenna, then it can be shown that the PAP can be modeled as

$$P(\theta) = \frac{P_T}{\pi \sqrt{\theta_M^2 - \theta^2}} \qquad (6.43)$$

where θ_M is the maximum arrival angle and P_T the total power in all angular paths.

A plot of this distribution is shown in Fig. 6.24 for the example where $P_T = \pi$ and $\theta_M = 45°$.

Disk of Scatterers: Figure 6.25 shows the transmitter and receiver where the transmitter is enveloped by a uniform disk of scatterers.

The disk diameter is chosen to encompass the *circle of influence*. That is, the region whose radius encompasses the significant

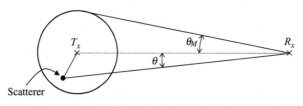

FIGURE 6.25 Disk of scatterers.

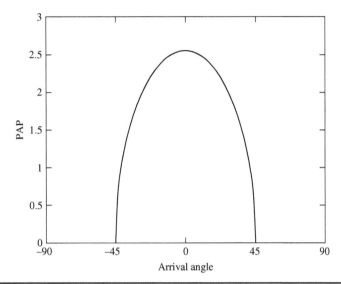

FIGURE 6.26 PAP for disk of scatterers.

scatterers that create the largest multipath signals. Because the scatterers are uniformly distributed within the disk, the PAP is given by

$$P(\theta) = \frac{2P_T}{\pi \theta_M^2} \sqrt{\theta_M^2 - \theta^2} \qquad (6.44)$$

A plot of this distribution is shown in Fig. 6.26 for the example where $P_T = \pi$ and $\theta_M = 45°$.

Indoor Distribution: Figure 6.27 shows a typical transmitter and receiver indoors.

Extensive field tests have demonstrated that both indoor scattering and congested urban scattering can most closely be modeled by a Laplace distribution [22, 26, 27]. Thus, the PAP can be modeled as

$$P(\theta) = \frac{P_T}{\sqrt{2}\sigma_\theta} e^{-\left|\frac{\sqrt{2}\theta}{\sigma_\theta}\right|} \qquad (6.45)$$

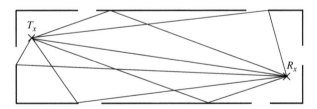

FIGURE 6.27 Indoor propagation.

Propagation Channel Characteristics   145

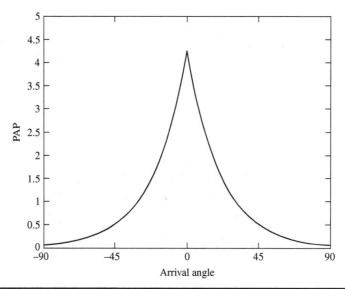

FIGURE 6.28   PAP for indoor scattering.

A plot of this distribution is shown in Fig. 6.28 for the example where $P_T = \pi$ and $\sigma_\theta = 30°$.

The exponential distribution is intuitive because the largest received signal is the direct path or line-of-sight signal at 0°. Because the largest reflection coefficients are for near grazing angles, the smallest angles of arrival represent signals reflecting at grazing angles from the closest structures to the line of sight. As the angles further increase, the additional paths are more the result of diffraction and less the result of reflection. Diffraction coefficients are generally much smaller than reflection coefficients because diffraction produces an additional signal spreading. Wider arrival angles tend to represent higher order scattering mechanisms.

6.3.8   Power Delay–Angular Profile

The next logical extension to the PDP and the PAP is the *power delay–angular profile* (PDAP). This can be viewed as the power profile resulting from the extended tap-delay-line method developed by Klein and Mohr [24], one that was further explored by Liberti and Rappaport [28]. This concept also can be derived from the work of Spencer et al. [27]. Thus, the PDAP is

$$P(\tau, \theta) = \sum_{n=1}^{N} P_n \delta(\tau - \tau_n) \delta(\theta - \theta_n) \qquad (6.46)$$

We can combine the excess delays of Example 6.7 along with the arrival angles of Example 6.8 to generate a three-dimensional plot of the PDAP as shown in Fig. 6.29.

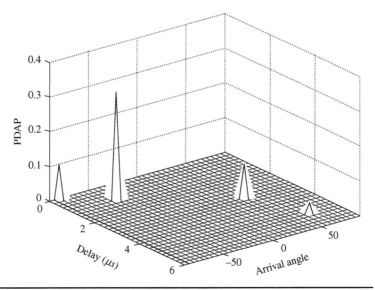

Figure 6.29 Power delay-angular profile.

The extent of correlation between the time delays and angles of arrival will depend on the scattering obstacles in the channel. If most scatterers lie between the transmitter and the receiver, the correlation will be high. If the transmitter and receiver are surrounded by scatterers, there may be little or no correlation.

6.3.9 Channel Dispersion

From an electromagnetics perspective, dispersion normally occurs when, in a medium, the propagation velocities are frequency dependent. Thus, the higher frequencies of a transmitted signal will propagate at different velocities than the lower frequencies. The high- and low-frequency components will produce different propagation delays even with no multipath present. This results in signal degradation at the receiver. However, a received signal can also be degraded by virtue of the fact that the time delays of the multipath components can become a reasonable fraction of the symbol period. Thus, the received signal is effectively degraded by time-delayed versions of itself. As the delay spread increases, the excess delays increase causing time dispersion. Figure 6.30 displays a direct path Gaussian pulse and the received signal for three increasing delay spreads (σ_1, σ_2, and σ_3). All signals are normalized. It is clear that the original pulse "broadens" as the delay spread increases. If the delay spread increases, this corresponds to a narrower channel bandwidth.

Because a channel can cause time-dispersion, it is necessary to define a channel bandwidth B_C, also referred to as a *coherence bandwidth*. This will help give an indication as to whether the channel

Propagation Channel Characteristics

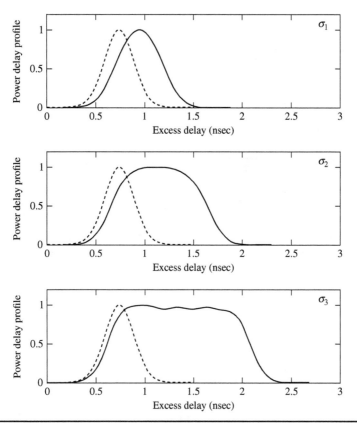

FIGURE 6.30 Dispersion caused by delay spread.

bandwidth is sufficient to allow a signal to be transmitted with minimal dispersion. The channel bandwidth is described in Shankar [3] and Stein [29] and is approximately defined by

$$B_C = \frac{1}{5\sigma_\tau} \qquad (6.47)$$

where σ_τ is the delay spread.

Therefore, if the signal chip rate or bandwidth (B_S) is less than the channel bandwidth (B_C), the channel undergoes flat fading. If the signal bandwidth is greater than the channel bandwidth, the channel will undergo frequency-selective fading. In this case dispersion occurs.

6.3.10 Slow-Fading Modeling

Figure 6.5 demonstrates the trend of slow fading (also known as *shadow fading*). This represents the average, about which fast fading occurs. Instead of a single scattering mechanism, the transmitted

148 Chapter Six

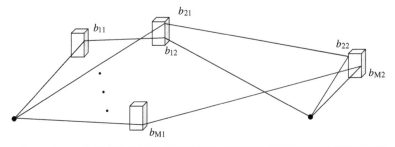

FIGURE 6.31 Slow fading with multiple scattering mechanisms.

signal can reflect, refract, diffract, and scatter multiple times before arriving at the receiver. Figure 6.31 demonstrates some of the candidate multipath mechanisms in the slow fading case along with coefficients representing each scattering location. It is assumed that no line-of-sight path exists.

The received signal can be represented as a sum of all multipath terms.

$$v_r(t) = \sum_{n=1}^{N} a_n e^{j\phi_n t} \qquad (6.48)$$

The coefficients a_n represent the cascade product of each reflection or diffraction coefficient along path n. Thus, we can write a separate expression for the amplitude coefficients as

$$a_n = \prod_{m=1}^{M} b_{mn} \qquad (6.49)$$

where b_{mn} = Rayleigh distributed random variables
M = number of scatterers along path n

These multiple scattering events will affect the mean value of the received power. As was discussed earlier, the total power received (multipath power gain) is the sum of the square of the coefficients a_n. However, the coefficients a_n are the consequence of the products of the b_{mn}. The logarithm of the power is the sum of these random variables. Using the Central Limit Theorem, this becomes a Gaussian (normal) distribution. Hence the name *log-normal*. Several references on the log-normal distribution are Shankar [3], Agrawal and Zheng [4], Saunders [8], Rappaport [9], Lee [30], and Suzuki [31]. The pdf of the power in dBm is given as

$$p(P) = \frac{1}{\sqrt{2\pi}\sigma} e^{-\frac{(P-P_0)^2}{2\sigma^2}} \qquad (6.50)$$

where P = power p in dBm = $10 \cdot \log 10(p)$
P_0 = average signal level in dBm
σ = standard deviation in dBm

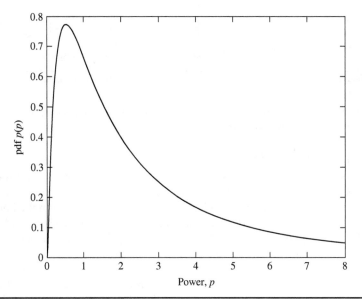

FIGURE 6.32 Log-normal density function.

Using a transformation of the random variable P, we can also express the pdf in terms of linear power p.

$$p(p) = \frac{1}{p\sqrt{2\pi}\sigma_0} e^{-\frac{\log_{10}^2(\frac{p}{p_0})}{2\sigma_0^2}} \qquad (6.51)$$

where p = power, mW
p_0 = average received signal level, mW
$\sigma_0 = \frac{\log_{10}(\sigma)}{10}$

A typical plot of the log-normal distribution given in Eq. (6.51) is shown in Fig. 6.32.

6.4 Improving Signal Quality

One of the drawbacks of the typical multipath channel is the fact that the signal quality is degraded by Doppler spread and dispersion. The negative effects of dispersion were briefly discussed in Sec. 6.3.9. If the signal bandwidth is greater than the channel bandwidth, we have frequency-selective fading. The consequence is an increasing *intersymbol interference* (ISI) leading to an unacceptable BER performance. Thus data transmission rates are limited by the delay spread of the channel. Because we understand the nature of dispersion and Doppler spreading, we can devise methods to compensate in order to improve signal quality at the receiver. An excellent discussion of compensation techniques is found in Rappaport [9]. The three basic

Chapter Six

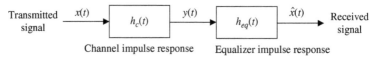

Figure 6.33 Communications system using an adaptive equalizer.

approaches to compensate for dispersion and Doppler spread are equalization, diversity, and channel coding.

6.4.1 Equalization

Channel equalization is the act of reducing amplitude, frequency, time, and phase distortion created by a channel. The goal of the equalizer is to correct for the frequency selective fading effects in a channel where the signal bandwidth (B_s) exceeds the channel coherence bandwidth (B_c). The equalizer can be a signal processing algorithm that seeks to minimize the ISI. Unless the channel characteristics are fixed with time, the equalizer must be an adaptive equalizer that compensates as the channel characteristics change with time. The ultimate goal of the equalizer is to completely neutralize the negative effects of the channel. Figure 6.33 demonstrates the ideal equalizer. The goal of the equalizer impulse response is to negate the channel impulse response such that the receive signal is nearly identical to the transmitted signal.

The ideal equalizer frequency response that negates the channel influence would be given as

$$H_{eq}(f) = \frac{1}{H_c^*(-f)} \tag{6.52}$$

If the channel frequency response can be characterized by $H_c(f) = |H_c(f)| e^{j\phi_c(f)}$, then the equalizer frequency response would be given by $H_{eq}(f) = \frac{e^{j\phi_c(-f)}}{|H_c(-f)|}$ such that the product of the two filters is unity. The channel frequency response and equalizer response are shown in Fig. 6.34.

Example 6.9 If the channel impulse response is given by $h_c(t) = .2\delta(t) + .3\delta(t-\tau)$, find the channel frequency response and the equalizer frequency response. Superimpose plots of the magnitude of $H_c(f)$ and $H_{eq}(f)$ for $0 \le f \le 1.5$.

Solution The channel frequency response is the Fourier transform of the channel impulse response. Thus

$$H_c(f) = \int_{-\infty}^{\infty} (.2\delta(t) + .3\delta(t-\tau)) e^{-j2\pi ft} dt$$
$$= .2 + .3 e^{-j2\pi f\tau}$$

The equalizer frequency response is given as

$$H_{eq}(f) = \frac{1}{H_c^*(-f)} = \frac{1}{.2 + .3 e^{-j2\pi f\tau}} = \frac{.2 + .3 e^{j2\pi f\tau}}{.13 + .12\cos(2\pi f\tau)}$$

Propagation Channel Characteristics

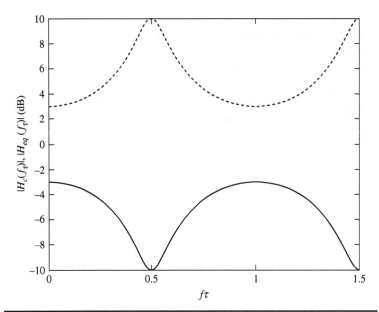

FIGURE 6.34 Channel and equalizer frequency response.

Since the channel characteristics are changing with time, it is necessary to design an adaptive equalizer that is able to respond to changing channel conditions. Figure 6.35 demonstrates a block diagram of an adaptive equalizer. The error signal is used as feedback to adjust the filter weights until the error is minimized.

The drawback of adaptive algorithms is that the algorithm must go through a *training* phase before it can *track* the received signal. The details of adaptive equalization are beyond the scope of this text but can be further explored in the Rappaport reference given previously.

6.4.2 Diversity

A second approach for mitigating the effects of fading is through diversity. Diversity means that a plurality of information is transmitted or

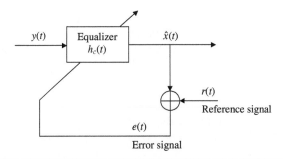

FIGURE 6.35 Adaptive equalizer.

received allowing one to reduce the depth and/or the duration of fades. This plurality can be achieved by having multiple receive antenna elements (antenna space diversity), different polarizations (polarization diversity), different transmit frequencies (frequency diversity), or different time characteristics such as Code division multiple access CDMA (time diversity). Thus, one can choose, amongst a collection of received signals, which signal provides the least amount of fading. This can be achieved by choosing perhaps the optimum antenna element, the optimum polarization, the optimum carrier frequency, or the optimum time diversity signal. Diversity has an advantage in that no adaptation or training is required in order to optimize the receiver.

Space diversity can be implemented through one of four basic approaches: selection diversity (select the largest signal from the antenna outputs), feedback diversity (scan all antenna outputs to find the first with a sufficient SNR for detection), maximal ratio combining (weight, co-phase, and sum all antenna outputs), and equal-gain combining (co-phase all received signals and combine with unity weights).

Polarization diversity is normally implemented through orthogonal polarizations. Orthogonal polarizations are uncorrelated and one can receive both polarizations with a dual polarized receive antenna. In addition, left- and right-hand circular polarizations can be used. One can select the polarization that maximizes the received signal-to-noise ratio.

Frequency diversity can be implemented by using multiple carrier frequencies. Presumably if a deep fade occurs at one frequency, the fade may be less pronounced at another frequency. It is suggested that the frequency separation be at least the channel coherence bandwidth (B_c) in order to ensure that the received signals are uncorrelated. Thus, the collection of possible transmit frequencies can be $f_0 \pm nB_c (n = 0, 1, 2 \ldots)$.

Time diversity can be implemented by transmitting the same information multiple times where the time delay exceeds the coherence time ($T_c = 1/B_c$). Thus, the times of transmission can be $T_0 + nT_c (n = 0, 1, 2 \ldots)$. Time diversity can also be implemented by modulating the transmission with CDMA pseudo-random codes. Typically these *pn* codes are uncorrelated if the different signal paths cause time delays exceeding a chip width. A RAKE receiver is a possible implementation of a time diversity scheme.

RAKE Receiver

A channelized correlation receiver can be used where each channel attempts a correlation with the received signal based on the anticipated path delays. One channel is assigned for each of the anticipated M strongest received components. The time delay τ_m is associated with the anticipated delay of the *m*th path. The correlating CDMA

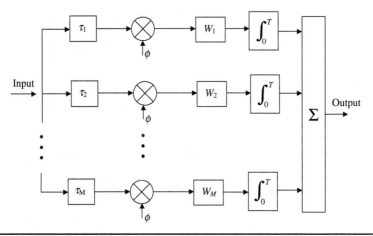

FIGURE 6.36 Structure of a RAKE receiver.

waveform is given by ϕ. Each channel can also have a weight W_n to allow the receiver to produce maximal ratio combining or equal-gain combining. Figure 6.36 demonstrates such a channelized receiver called a *RAKE receiver*. The RAKE receiver received its name based on its similarity to a garden rake (Price and Green [32]).

As an example of a two-channel RAKE receiver, we can transmit a 32-chip CDMA waveform of length $T = 2$ μs. Let there only be two paths to the receiver. The excess delay for path 1 is 0 μs and the excess delay for path 2 is 71.6 ns or 1.3 times the chip width ($\tau_{chip} = T/32 = 62.5$ ns. The received waveforms are shown in Fig. 6.37. The correlation of the first waveform with itself is greater than four times the correlation with the delayed copy of itself. Thus, the first channel has a high correlation with the first arriving waveform and the second channel has a high correlation with the second arriving waveform. The first channel excess delay $\tau_1 = 0$ μs. The second channel excess delay $\tau_2 = 71.6$ ns.

6.4.3 Channel Coding

Because of the adverse fading effects of the channel, digital data can be corrupted at the receiver. Channel coding deliberately introduces redundancies into the data to allow for correction of the errors caused by dispersion. These redundant symbols can be used to detect errors and/or correct the errors in the corrupted data. Thus, the channel codes can fall into two types: *error detection codes* and *error correction codes*. The error detection codes are called *automatic repeat request* (ARQ) *codes*. The error correction codes are *forward error correcting* (FEC) codes. The combination of both is called a *hybrid-ARQ code*. Two basic types of codes can be used to accomplish error detection and/or error correction. They are *block codes* and *convolutional codes*.

154 Chapter Six

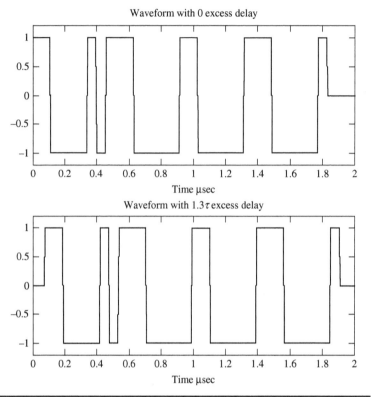

FIGURE 6.37 Channel 1 and 2 waveforms.

Turbo codes are a type of convolutional code. An excellent description of these codes and channel coding schemes is given by Rappaport [9], Sklar [10], Proakis [18], and Parsons [33].

6.4.4 MIMO

MIMO stands for multiple-input multiple-output communication. This is the condition under which the transmit and receive antennas have multiple antenna elements. This is also referred to as *volume-to-volume* or as a *multiple-transmit multiple-receive* (MTMR) communications link. MIMO has applications in broadband wireless, WLAN, 3G, and other related systems. MIMO stands in contrast to *single-input singleoutput* (SISO) systems. Because MIMO involves multiple antennas, it can be viewed as a space diversity approach to channel fading mitigation. Excellent sources for MIMO systems information can be found in the IEEE Journal on Selected Areas in Communications: MIMO Systems and Applications Parts 1 and 2 [34, 35], Vucetic and Yuan [36], Diggavi et al. [37], and Haykin and Moher [38]. A basic MIMO system is illustrated in Fig. 6.38, where $\bar{H}$ is the complex

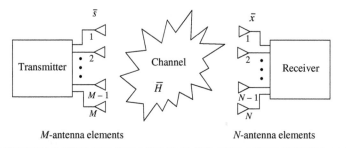

FIGURE 6.38 MIMO system.

channel matrix relating M inputs to N outputs, $\bar{s}$ is the complex transmit vector, and $\bar{x}$ is the complex receive vector.

The goal of MIMO is to combine signals on both the transmit and receive ends such that the data rate is increased and/or the ISI and BER are decreased. As was mentioned previously, one diversity option for a SISO system is time diversity. MIMO allows the user to combine time diversity and space diversity. It can be said that MIMO is more than a space diversity option but is a space-time signal processing solution to channel fading. Multipath propagation has been viewed as a nuisance in SISO systems, but it is actually an advantage for MIMO systems because the multipath information can be combined to produce a better received signal. Thus the goal of MIMO is not to mitigate channel fading but rather to take advantage of the fading process.

We may follow the development in Haykin and Moher [38] in order to understand the mathematical description of the MIMO system. Under flat-fading conditions, let us define the M-element complex transmit vector, created by the M-element transmit array, as

$$\bar{s} = [s_1 \quad s_2 \quad s_3 \quad \ldots \quad s_M]^T \tag{6.53}$$

The elements s_m are phased according to the array configuration and are assumed to be coded waveforms in the form of a data stream. If we assume that the transmit vector elements have zero mean and variance σ_s^2, the total transmit power is given by

$$P_t = M \cdot \sigma_s^2 \tag{6.54}$$

Each transmit antenna element, m, connects to a path (or paths) to the receive elements, n, creating a channel transfer function, h_{nm}. There are consequently $N \cdot M$ channel transfer functions connecting the transmit and receive array elements. Thus, we can define the $N \times M$ complex channel matrix as

$$\bar{H} = \begin{bmatrix} h_{11} & h_{12} & \cdots & h_{1M} \\ h_{21} & h_{22} & \cdots & h_{2M} \\ \vdots & \vdots & \ddots & \vdots \\ h_{N1} & h_{N2} & \cdots & h_{NM} \end{bmatrix} \tag{6.55}$$

Let us now define the N-element complex receive vector, created by the N-element receive array, as

$$\bar{x} = [x_1 \ x_2 \ x_3 \ \ldots \ x_N]^T \quad (6.56)$$

Let us also define the N-element complex channel noise vector as

$$\bar{n} = [n_1 \ n_2 \ n_3 \ \ldots \ n_N]^T \quad (6.57)$$

We can now describe the receive vector in matrix form as

$$\bar{x} = \bar{H} \cdot \bar{s} + \bar{n} \quad (6.58)$$

Assuming a Gaussian distribution for the transmitted signal, the channel, the received signal, and the noise, one can easily estimate the MIMO channel capacity. The correlation matrix for the transmitted signal is given by

$$\bar{R}_s = E[\bar{s} \cdot \bar{s}^H]$$
$$= \sigma_s^2 \bar{I}_M \quad (6.59)$$

where σ_s^2 = signal variance
$\bar{I}_M = M \times M$ identity matrix

$\bar{s}^H$ denotes the Hermitian transpose of $\bar{s}$. The correlation matrix for the noise is given by

$$\bar{R}_n = E[\bar{n} \cdot \bar{n}^H]$$
$$= \sigma_n^2 \bar{I}_N \quad (6.60)$$

where σ_n^2 = noise variance
$\bar{I}_N = N \times N$ identity matrix

The MIMO channel capacity is a random variable. Because the channel itself is random, we can define the ergodic (mean) capacity. If we assume that the sources are uncorrelated and of equal power, the ergodic capacity is given as [39–41]

$$C_{EP} = E\left[\log_2\left[\det\left(\bar{I}_N + \frac{\rho}{M}\bar{H} \cdot \bar{H}^H\right)\right]\right] \text{ bits/s/Hz} \quad (6.61)$$

where, the expectation is over the random channel matrix $\bar{H}$ and det = determinant, C_{EP} = equal power capacity, and ρ = SNR at each receive antenna = $\frac{P_t}{\sigma_n^2}$.

One possible algorithm for use in a MIMO system is the V-BLAST algorithm developed at Bell Labs [42]. The V-BLAST algorithm is an improvement over its predecessor D-BLAST. (V-BLAST stands for Vertically layered blocking structure, Bell Laboratories Layered

Space-Time). The V-BLAST algorithm demultiplexes a single data stream into M substreams that undergo a bit-to-symbol mapping. The mapped substreams are subsequently transmitted from the M transmit antennas. Thus, the total channel bandwidth used is a fraction of the original data stream bandwidth allowing for flat fading. The detection at the receiver can be performed by conventional adaptive beamforming. Each substream is considered to be the desired signal and all other substreams are deemed as interference and are therefore nulled. Thus, M simultaneous beams must be formed by the N-element receive array while nulling the unwanted substreams. The received substreams may then be multiplexed to recover the intended transmission.

A further exploration of MIMO concepts can be pursued in references [34–42].

6.5 References

1. Liberti, J. C., and T. S. Rappaport, *Smart Antennas for Wireless Communications: IS-95 and Third Generation CDMA Applications*, Prentice Hall, New York, 1999.
2. Bertoni, H. L., *Radio Propagation for Modern Wireless Systems*, Prentice Hall, New York, 2000.
3. Shankar, P. M., *Introduction to Wireless Systems*, Wiley, New York, 2002.
4. Agrawal, D. P., and Q. A. Zeng, *Introduction to Wireless and Mobile Systems*, Thomson Brooks/Cole, Toronto, Canada, 2003.
5. Sarkar, T. K., M. C. Wicks, M. Salazar-Palma, et al., *Smart Antennas*, IEEE Press & Wiley Interscience, New York, 2003.
6. Feher, K., *Wireless Digital Communications: Modulation and Spread Spectrum Applications*, Prentice Hall, New York, 1995.
7. Haykin, S. and M. Moher, *Modern Wireless Communications*, Prentice Hall, New York, 2005.
8. Saunders, S. R., *Antennas and Propagation for Wireless Communication Systems*, Wiley, New York, 1999.
9. Rappaport, T. S., *Wireless Communications: Principles and Practice*, 2d ed., Prentice Hall, New York, 2002.
10. Keller, J. B., "Geometrical Theory of Diffraction," *J. Opt. Soc. Amer.*, Vol. 52, pp. 116–130, 1962.
11. Sklar, B., *Digital Communications: Fundamentals and Applications*, 2d ed., Prentice Hall, New York, 2001.
12. Schwartz, M., *Information Transmission, Modulation, and Noise*, 4th ed., McGraw-Hill, New York, 1990.
13. Papoulis, A., *Probability, Random Variables, and Stochastic Processes*, 2d ed., McGraw-Hill, New York, 1984.
14. Clarke, R. H., "A Statistical Theory of Mobile-Radio Reception," *Bell Syst. Tech. J.*, Vol. 47, pp. 957–1000, 1968.
15. Gans, M. J., "A Power Spectral Theory of Propagation in the Mobile Radio Environment," *IEEE Trans. Veh. Technol.*, Vol. VT-21, No. 1, pp. 27–38, Feb. 1972.
16. Jakes, W. C., (ed.), *Microwave Mobile Communications*, Wiley, New York, 1974.
17. Ghassemzadeh, S. S., L. J. Greenstein, T. Sveinsson, et al., "A Multipath Intensity Profile Model for Residential Environments," *IEEE Wireless Communications and Networking*, Vol. 1, pp. 150–155, March 2003.
18. Proakis, J. G., *Digital Communications*, 2d ed., McGraw-Hill, New York, 1989.
19. Wesolowski, K., *Mobile Communication Systems*, Wiley, New York, 2004.
20. Chuang, J., "The Effects of Time Delay Spread on Portable Radio Communications Channels with Digital Modulation," *IEEE Journal on Selected Areas in Communications*, Vol. SAC-5, No. 5, June 1987.

21. Fulghum, T., and K. Molnar, "The Jakes Fading Model Incorporating Angular Sspread for a Disk of Scatterers," *IEEE 48th Vehicular Technology Conference*, Vol. 1, pp. 489–493, 18–21 May 1998.
22. Fulghum, T., K. Molnar, and A. Duel-Hallen, "The Jakes Fading Model for Antenna Arrays Incorporating Azimuth Spread," *IEEE Transactions on Vehicular Technology*, Vol. 51, No. 5, pp. 968–977, Sept. 2002.
23. Boujemaa, H., and S. Marcos, "Joint Estimation of Direction of Arrival and Angular Spread Using the Knowledge of the Power Angle Density," *Personal, Indoor and Mobile Radio Communications, 13th IEEE International Symposium*, Vol. 4, pp. 1517–1521, 15–18 Sept. 2002.
24. Klein, A., and W. Mohr, "A StatisticalWideband Mobile Radio Channel Model Including the Directions-of-Arrival," *Spread Spectrum Techniques and Applications Proceedings, IEEE 4th International Symposium on*, Vol. 1, Sept. 1996.
25. Ertel, R., P. Cardieri, K. Sowerby, et al., "Overview of Spatial Channel Models for Antenna Array Communication Systems," *IEEE Personal Communications*, pp. 10–22, Feb. 1998.
26. Pedersen, K., P. Mogensen, and B. Fleury, "A Stochastic Model of the Temporal and Azimuthal Dispersion Seen at the Base Station in Outdoor Propagation Environments," *IEEE Transactions on Vehicular Technology*, Vol. 49, No. 2, March 2000.
27. Spencer, Q., M. Rice, B. Jeffs, et al., "A Statistical Model for Angle of Arrival in Indoor Multipath Propagation," Vehicular Technology Conference, 1997 IEEE 47th , Vol. 3, pp. 1415–1419, 4–7 May 1997.
28. Liberti, J. C., and T. S. Rappaport, "A Geometrically Based Model for Line-of-Sight Multipath Radio Channels," *IEEE 46th Vehicular Technology Conference*, 1996. 'Mobile Technology for the Human Race', Vol. 2, pp. 844–848, 28 April-1 May 1996.
29. Stein, S., "Fading Channel Issues in System Engineering," *IEEE Journal on Selected Areas in Communications*, Vol. SAC-5, No. 2, pp. 68–89, Feb. 1987.
30. Lee, W. C. Y., "Estimate of Local Average Power of a Mobile Radio Signal," *IEEE Transactions Vehicular Technology*, Vol. 29, pp. 93–104, May 1980.
31. Suzuki, H., "A Statistical Model for Urban Radio Propagation," *IEEE Transactions on Communications*, Vol. COM-25, No. 7, 1977.
32. Price, R., and P. Green, "A Communication Technique for Multipath Channels," Proceedings of IRE, Vol. 46, pp. 555–570, March 1958.
33. Parsons, J., *The Mobile Radio Propagation Channel*, 2d ed., Wiley, New York, 2000.
34. "MIMO Systems and Applications Part 1," *IEEE Journal on Selected Areas in Communications*, Vol. 21, No. 3, April 2003.
35. "MIMO Systems and Applications Part 2," *IEEE Journal on Selected Areas in Communications*, Vol. 21, No. 5, June 2003.
36. Vucetic, B., and J. Yuan, *Space-Time Coding*, Wiley, New York, 2003.
37. Diggavi, S., N. Al-Dhahir, A. Stamoulis, et al., "Great Expectations: The Value of Spatial Diversity in Wireless Networks," *Proceedings of the IEEE*, Vol. 92, No. 2, Feb. 2004.
38. Haykin, S., and M. Moher, *Modern Wireless Communications*, Prentice Hall, New York, 2005.
39. Foschini, G., and M. Gans, "On Limits of Wireless Communications in a Fading Environment When Using Multiple Antennas," *Wireless Personal Communications* 6, pp. 311–335, 1998.
40. Gesbert, D., M. Shafi, D. Shiu, et al., "From Theory to Practice: An Overview of MIMO Space-Time Coded Wireless Systems," *Journal on Selected Areas in Communication*, Vol. 21, No. 3, April 2003.
41. Telatar, E., "Capacity of Multiantenna Gaussian Channels," *AT&T Bell Labs Technology Journal*, pp. 41–59, 1996.
42. Golden, G., C. Foschini, R. Valenzuela, et al., "Detection Algorithm and Initial Laboratory Results Using V-BLAST Space-Time Communication Architecture," *Electronic Lett.*, Vol. 35, No. 1, Jan. 1999.

6.6 Problems

6.1 Use the flat earth model. Assume the following values: $|R| = .5, .7, \psi = 0$, $E_0/r_1 = 1, h_1 = 5$ m, $h_2 = 20$ m, $d = 100$ m.
  (a) What are the arrival time delays τ_1 and τ_2?
  (b) Plot a family of curves for the envelope and the phase vs. $(2kh_1h_2)/d$.

6.2 Use the flat earth model. Assume the following values: $|R| = .5, \psi = 0$, $h_1 = 5$ m, $h_2 = 20$ m, $d = 100$ m. A baseband rectangular pulse is transmitted whose width is 10 ns. The transmit pulse is shown in Fig. 6P.1. The received voltage magnitude for the direct path $V_1 = 1$V and is proportional to E_0/r_1.
  (a) What are the arrival time delays τ_1 and τ_2?
  (b) Estimate the received voltage magnitude V_2 for the reflected path using spherical spreading and $|R|$.
  (c) Plot the received pulses for the direct and indirect paths showing pulse magnitudes and time delays.

6.3 Let us model an urban environment by using MATLAB to generate 100 values of the amplitude a_n, the phase α_n, and the distance r_n. Use Eqs. (6.13) through (6.15). Generate values for the amplitudes, a_n, using randn and assume that $\sigma = 1$ mV. Use rand to generate phase values for α_n to vary between 0 and 2π. Use rand to define $r_n = 500 + 50*$ rand(1,N). Find the phase ϕ_n. Plot the envelope r (mV) for 100 kHz $< f <$ 500 kHz. Let the x axis be linear and the y axis be log 10.

6.4 Calculate the outage probability for the Rayleigh process where $\sigma = 2$ mV and the threshold $p_{th} = 3$ μW.

6.5 Plot the probability density function for the Rician distribution where the Rician factor $K = -5$ dB, 5 dB, and 15 dB. Superimpose all curves on the same plot. Assume that $\sigma = 2$ mV.

6.6 Repeat Prob. 6.3 for the Rician channel where $A = 4\sigma$.

6.7 For the Rician-fading channel where $\sigma = 2$ mV, the direct path amplitude $A = 4$ mV, what is the probability that the received voltage envelope will exceed a threshold of 6 mV?

6.8 Use MATLAB to plot the envelope in Eq. (6.27) where the carrier frequency is 2 GHz, the vehicle velocity is 30 mph, the phase angles ϕ_n, θ_n are

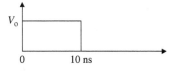

FIGURE 6P.1 Transmit pulse.

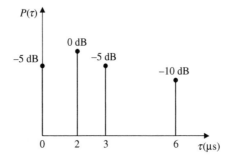

Figure 6P.2 Power delay profile.

uniformly distributed, the coefficient a_n has a Gaussian distribution with zero mean, and standard deviation of $\sigma = 0.002$. Let $N = 5$.

6.9 Calculate the multipath power gain P_T, the mean excess delay τ_0, and the RMS delay spread σ_τ for the PDP given in Fig. 6P.2.

6.10 Calculate the multipath power gain P_T, the mean excess delay τ_0, and the RMS delay spread σ_τ for the PDP given in Fig. 6P.3.

6.11 Calculate the multipath power gain P_T, the mean arrival angle θ_0, and the RMS angular spread σ_θ for the PAP given in Fig. 6P.4.

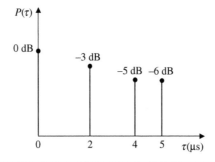

Figure 6P.3 Power delay profile.

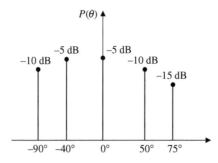

Figure 6P.4 Power angular profile.

Propagation Channel Characteristics   161

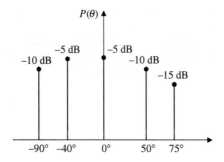

FIGURE 6P.5   Power angular profile.

6.12   Calculate the multipath power gain P_T, the mean arrival angle θ_o, and the RMS angular spread σ_θ for the PAP given in Fig. 6P.5.

6.13   For $P(\theta)$ in Prob. 6.11, use the Rapport method to calculate the angular spread.

6.14   For $P(\theta)$ in Prob. 6.12, use the Rapport method to calculate the angular spread.

6.15   For a transmitter transmitting 1 MW of power located at the rectangular coordinates (0, 250) and the receiver is located at (1 km, 250). Four scatterers are located at: (200, 400), (300, 150), (500, 100), (600, 400), and (800, 100). Each isotropic scatterer has a reflection coefficient of .7. Account for spherical spreading such that the power is inversely proportional to r^2.
   (a) Derive the time delays and powers associated with each path and plot the PDP.
   (b) Derive the angles of arrival and powers associated with each path and plot the PAP.
   (c) Plot the two-dimensional PDAP given in Eq. (6.46).

6.16   For a ring of scatterers where the circle radius $a = 20$ m and the distance between the transmit and receive antennas is 100 m. Allow $P_T = 1$ W
   (a) Solve for the PAP.
   (b) Plot the PAP.

6.17   For a disk of scatterers where the circle radius $a = 20$ m and the distance between the transmit and receive antennas is 100 m. Allow $P_T = 1$ W
   (a) Solve for the PAP.
   (b) Plot the PAP.

6.18   If the channel impulse response is given by $h_c(t) = .2\delta(t) + .4\delta(t - \tau) + .6\delta(t - 2\tau)$,
   (a) Derive and plot the channel frequency response for $.5 < f\tau < 1.5$.
   (b) Derive and plot the equalizer frequency response for $.5 < f\tau < 1.5$.
   (c) Are these plots conjugate inverses?

6.19 If the channel impulse response is given by $h_c(t) = .6\delta(t) + .4\delta(t-2\tau) + .2\delta(t-4\tau)$,

(a) Derive and plot the channel frequency response for $.5 < f\tau < 1.5$.
(b) Derive and plot the equalizer frequency response for $.5 < f\tau < 1.5$.
(c) Are these plots conjugate inverses?

CHAPTER 7
Angle-of-Arrival Estimation

In the propagation channel topics discussed in Chap. 6, it was apparent that even for one source there are many possible propagation paths and angles of arrival. If several transmitters are operating simultaneously, each source potentially creates many multipath components at the receiver. Therefore, it is important for a receive array to be able to estimate the angles of arrival in order to decipher which emitters are present and what are their possible angular locations. This information can be used to eliminate or combine signals for greater fidelity, suppress interferers, or both.

Angle-of-arrival (AOA) estimation has also been known as *spectral* estimation, *direction-of-arrival* (DOA) estimation, or *bearing* estimation. Some of the earliest references refer to spectral estimation as the ability to select various frequency components out of a collection of signals. This concept was expanded to include frequency-wavenumber problems and subsequently AOA estimation. *Bearing estimation* is a term more commonly used in the sonar community and is AOA estimation for acoustic problems. Much of the state of the art in AOA estimation has its roots in time *series analysis, spectrum analysis, periodograms, eigenstructure methods, parametric methods, linear prediction methods, beamforming, array processing,* and *adaptive array methods.* Some of the more useful materials include a survey paper by Godara [1], spectrum analysis by Capon [2], a review of spectral estimation by Johnson [3], an exhaustive text by Van Trees [4], and a text by Stoica and Moses [5].

7.1 Fundamentals of Matrix Algebra
Before beginning our development of AOA (spectral) estimation methods, it is important to review some matrix algebra basics. We will denote all vectors as lowercase with a bar. An example is the array vector $\bar{a}$. We will denote all matrices as uppercase also with a bar such as $\bar{A}$.

7.1.1 Vector Basics

Column Vector The vector $\bar{a}$ can be denoted as a column vector or as a row vector. If $\bar{a}$ is a column vector or a single column matrix, it can be described as

$$\bar{a} = \begin{bmatrix} a_1 \\ a_2 \\ \vdots \\ a_M \end{bmatrix} \tag{7.1}$$

Row Vectors If $\bar{b}$ is a row vector or a single row matrix, it can be described as

$$\bar{b} = [b_1 \quad b_2 \quad \cdots \quad b_N] \tag{7.2}$$

Vector Transpose Any column vector can be changed into a row vector or any row vector can be changed into a column vector by the transpose operation such that

$$\bar{a}^T = [a_1 \quad a_2 \quad \cdots \quad a_M] \tag{7.3}$$

$$\bar{b}^T = \begin{bmatrix} b_1 \\ b_2 \\ \vdots \\ b_N \end{bmatrix} \tag{7.4}$$

Vector Hermitian Transpose The *Hermitian transpose* is the conjugate transpose of a vector denoted by the operator H. Thus the Hermitian transpose[1] of $\bar{a}$ above can be demonstrated as

$$\bar{a}^H = [a_1^* \quad a_2^* \quad \cdots \quad a_M^*] \tag{7.5}$$

$$\bar{b}^H = \begin{bmatrix} b_1^* \\ b_2^* \\ \vdots \\ b_N^* \end{bmatrix} \tag{7.6}$$

Vector Dot Product (Inner Product) The *dot product* of row vector with itself is traditionally given by

$$\bar{b} \cdot \bar{b}^T = [b_1 \quad b_2 \quad \cdots \quad b_N] \cdot \begin{bmatrix} b_1 \\ b_2 \\ \vdots \\ b_N \end{bmatrix} \tag{7.7}$$

$$= b_1^2 + b_2^2 + \cdots + b_N^2$$

---

[1]The Hermitian transpose is also designated by the symbol †.

Vandermonde Vector A *Vandermonde vector* is an M-element vector such that

$$\bar{a} = \begin{bmatrix} x^0 \\ x^1 \\ \vdots \\ x^{(M-1)} \end{bmatrix} \quad (7.8)$$

Thus, the array steering vector of Eq. (4.8) is a Vandermonde vector.

7.1.2 Matrix Basics

A matrix is an $M \times N$ collection of elements such that

$$\bar{A} = \begin{bmatrix} a_{11} & a_{12} & \cdots & a_{1N} \\ a_{21} & a_{22} & \cdots & a_{2N} \\ \vdots & \vdots & \ddots & \vdots \\ a_{M1} & a_{M2} & \cdots & a_{MN} \end{bmatrix} \quad (7.9)$$

where $M \times N$ is the size or *order* of the *matrix*.

Matrix Determinant The *determinant* of a square matrix can be defined by the Laplace expansion and is given by

$$|\bar{A}| = \begin{bmatrix} a_{11} & a_{12} & \cdots & a_{1M} \\ a_{21} & a_{22} & \cdots & a_{2M} \\ \vdots & \vdots & \ddots & \vdots \\ a_{M1} & a_{M2} & \cdots & a_{MM} \end{bmatrix}$$

$$= \sum_{j=1}^{M} a_{ij} cof(a_{ij}) \quad \text{for any row index } i$$

$$= \sum_{j=1}^{M} a_{ij} cof(a_{ij}) \quad \text{for any column index } j \quad (7.10)$$

where, $cof(a_{ij})$ is the cofactor of the element a_{ij} and is defined by $cof(a_{ij}) = (-1)^{i+j} M_{ij}$ and M_{ij} is the minor of a_{ij}. The minor is determinant of the matrix left after striking the *i*th row and the *j*th column. If any two rows or two columns of a matrix are identical, the determinant is zero. The determinant operation in MATLAB is performed by the command det(A).

Example 7.1 Find the determinant of $\bar{A} = \begin{bmatrix} 1 & 2 & 0 \\ 3 & 2 & 1 \\ 5 & 1 & -1 \end{bmatrix}$

Solution Using the first row indices $|\bar{A}| = \sum_{j=1}^{M} a_{1j} cof(a_{1j})$. Thus

$$|\bar{A}| = 1 \cdot (2 \cdot (-1) - 1 \cdot 1) - 2 \cdot (3 \cdot (-1) - 5 \cdot 1) + 0 \cdot (3 \cdot 1 - 5 \cdot 2)$$

$$= 13$$

The same result may be found by using the two MATLAB commands:

```
>> A = [1 2 0;3 2 1;5 1 -1];
>> det(A)
ans = 13
```

Matrix Addition Matrices can be *added* or *subtracted* by simply adding or subtracting the same elements of each. Thus

$$\bar{C} = \bar{A} \pm \bar{B} \Rightarrow c_{ij} = a_{ij} \pm b_{ij} \tag{7.11}$$

Matrix Multiplication *Matrix multiplication* can occur if the column index of the first matrix equals the row index of the second. Thus an $M \times N$ matrix can be multiplied by an $N \times L$ matrix yielding an $M \times L$ matrix. The multiplication is such that

$$\bar{C} = \bar{A} \cdot \bar{B} \Rightarrow c_{ij} = \sum_{k=1}^{N} a_{ik} b_{kj} \tag{7.12}$$

Example 7.2 Multiply the two matrices $\bar{A} = \begin{bmatrix} 1 & -2 \\ 3 & 4 \end{bmatrix}$ $\bar{B} = \begin{bmatrix} 7 & 3 \\ -1 & 5 \end{bmatrix}$

Solution $\bar{A} \cdot \bar{B} = \begin{bmatrix} 9 & -7 \\ 17 & 29 \end{bmatrix}$

This also may be accomplished in MATLAB with the following commands:

```
>> A = [1 -2;3 4];
>> B = [7 3;-1 5];
>> A*B
```

ans = $\begin{bmatrix} 9 & -7 \\ 17 & 29 \end{bmatrix}$

Identity Matrix The *identity matrix*, denoted by $\bar{I}$, is defined as an $M \times M$ matrix with ones along the diagonal and zeros for all other matrix elements such that

$$\bar{I} = \begin{bmatrix} 1 & 0 & \cdots & 0 \\ 0 & 1 & \cdots & 0 \\ \vdots & \vdots & \ddots & \vdots \\ 0 & 0 & \cdots & 1 \end{bmatrix} \tag{7.13}$$

The product of an identity matrix $\bar{I}$ with any square matrix $\bar{A}$ yields $\bar{A}$ such that $\bar{I} \cdot \bar{A} = \bar{A} \cdot \bar{I} = \bar{A}$. The identity matrix can be created in MATLAB with the command eye(M). This produces an $M \times M$ identity matrix.

Cartesian Basis Vectors The columns of the identity matrix $\bar{I}$, are called *Cartesian basis vectors*. The Cartesian basis vectors are denoted by $\bar{u}_1, \bar{u}_2, \cdots, \bar{u}_M$ such that $\bar{u}_1 = [1 \ 0 \ \cdots \ 0]^T$, $\bar{u}_2 = [0 \ 1 \ \cdots \ 0]^T, \cdots,$ $\bar{u}_M = [0 \ 0 \ \cdots \ 1]^T$. Thus, the identity matrix can be defined as $\bar{I} = [\bar{u}_1 \ \bar{u}_2 \ \cdots \ \bar{u}_M]$.

Trace of a Matrix The *trace* of a square matrix is the sum of the diagonal elements such that

$$Tr(\bar{A}) = \sum_{i=1}^{N} a_{ii} \tag{7.14}$$

The trace of a matrix in MATLAB is found by the command trace (A).

Matrix Transpose The *transpose* of a matrix is the interchange of the rows and columns denoted by $\bar{A}^T$. The transpose of the product of two matrices is the product of the transposes in reverse order such that $(\bar{A} \cdot \bar{B})^T = \bar{B}^T \cdot \bar{A}^T$. The transpose in MATLAB is performed with the command transpose(A) or A.'

Matrix Hermitian Transpose The *Hermitian transpose* is the transpose of the conjugate (or the conjugate transpose) of the matrix elements denoted by $\bar{A}^H$. The Hermitian transpose in MATLAB is performed by the operation ctranspose(A) or A'. The determinant of a matrix is the same as the determinant of its transpose, thus $|\bar{A}| = |\bar{A}^T|$. The determinant of a matrix is the conjugate of its Hermitian transpose, thus $|\bar{A}| = |\bar{A}^H|^*$. The Hermitian transpose of the product of two matrices is the product of the Hermitian transposes in reverse order such that $(\bar{A} \cdot \bar{B})^H = \bar{B}^H \cdot \bar{A}^H$. This is an important property which will be used later in the text.

Inverse of a Matrix The *inverse* of a matrix is defined such that $\bar{A} \cdot \bar{A}^{-1} = \bar{I}$, where $\bar{A}^{-1}$ is the inverse of $\bar{A}$. The matrix $\bar{A}$ has an inverse provided that $|\bar{A}| \neq 0$. We define the cofactor matrix such that $\bar{C} = cof(\bar{A}) = [(-1)^{i+j} |\bar{A}_{ij}|]$ and $\bar{A}_{ij}$ is the remaining matrix after striking row i and column j. The inverse of a matrix in MATLAB is given by inv(A). Mathematically, the matrix inverse is given by

$$\bar{A}^{-1} = \frac{\bar{C}^T}{|\bar{A}|} \tag{7.15}$$

Example 7.3 Find the inverse of the matrix $\bar{A} = \begin{bmatrix} 1 & 3 \\ -2 & 5 \end{bmatrix}$.

Solution First find the cofactor matrix $\bar{C}$ and the determinant of $\bar{A}$.

$$\bar{C} = \begin{bmatrix} 5 & 2 \\ -3 & 1 \end{bmatrix} \quad |\bar{A}| = 11$$

The inverse of $\bar{A}$ is then given as

$$\bar{A}^{-1} = \frac{\begin{bmatrix} 5 & 2 \\ -3 & 1 \end{bmatrix}^T}{11} = \begin{bmatrix} .4545 & -.2727 \\ .1818 & .0909 \end{bmatrix}$$

This problem can also be easily solved in MATLAB using the following commands:

```
>> A = [1 3;-2 5];
>> inv(A)

ans =    .4545   -.2727
         .1818    .0909
```

Eigenvalues and Eigenvectors of a Matrix The German word *eigen* means *appropriate* or *peculiar*. Thus, the *eigenvalues* and *eigenvectors* of a matrix are the appropriate or peculiar values that satisfy a homogeneous condition. The values of λ, which are eigenvalues of the $N \times N$ square matrix $\bar{A}$, must satisfy the following condition:

$$|\lambda \bar{I} - \bar{A}| = 0 \tag{7.16}$$

The determinant above is called the *characteristic determinant* producing a polynomial of order N in λ having N roots. In other words

$$|\lambda \bar{I} - \bar{A}| = (\lambda - \lambda_1)(\lambda - \lambda_2) \cdots (\lambda - \lambda_N) \tag{7.17}$$

Each of the eigenvalues ($\lambda_1, \lambda_2, \ldots, \lambda_N$) satisfies Eq. (7.16). We may now also define eigenvectors associated with the matrix $\bar{A}$. If the $N \times N$ matrix $\bar{A}$ has N unique eigenvalues λ_j, it has N eigenvectors $\bar{e}_j$ satisfying the following homogeneous equation:

$$(\lambda_j \bar{I} - \bar{A}) \bar{e}_j = 0 \tag{7.18}$$

In MATLAB, one can find the eigenvectors and eigenvalues of a matrix A by the command [EV, V] = eig(A). The eigenvectors are the columns of the matrix EV and the corresponding eigenvalues are the diagonal elements of the matrix V. The MATLAB command diag(V) creates a vector of the eigenvalues along the diagonal of V.

Example 7.4 Use MATLAB to find the eigenvectors and eigenvalues of the matrix $\begin{bmatrix} 1 & 2 \\ 3 & 5 \end{bmatrix}$.

Solution The following lines create the matrix and compute the eigenvectors and eigenvalues:

```
>> A = [1 2;3 5];
>> [EV,V] = eig(A);
>> EV

EV =
  -0.8646  -0.3613
   0.5025  -0.9325
```

```
>> diag(V)
ans =
   -0.1623
    6.1623
```

The first column of EV is the first eigenvector and the corresponding eigenvalue is $\lambda_1 = -.1623$. The second column of EV is the second eigenvector and the corresponding eigenvalue is $\lambda_2 = 6.1623$.

These simple vector and matrix procedures will assist us in using MATLAB to solve for AOA estimation algorithms that are described in Sec. 7.3.

7.2 Array Correlation Matrix

Many of the AOA algorithms rely on the array correlation matrix. In order to understand the array correlation matrix, let us begin with a description of the array, the received signal, and the additive noise. Figure 7.1 depicts a receive array with incident planewaves from various directions.

Figure 7.1 shows D signals arriving from D directions. They are received by an array of M elements with M potential weights. Each received signal $x_m(k)$ includes additive, zero mean, Gaussian noise. Time is represented by the kth time sample. Thus, the array output y can be given in the following form:

$$y(k) = \bar{w}^T \cdot \bar{x}(k) \tag{7.19}$$

where

$$\bar{x}(k) = [\bar{a}(\theta_1) \quad \bar{a}(\theta_2) \quad \cdots \quad \bar{a}(\theta_D)] \cdot \begin{bmatrix} s_1(k) \\ s_2(k) \\ \vdots \\ s_D(k) \end{bmatrix} + \bar{n}(k) \tag{7.20}$$

$$= \bar{A} \cdot \bar{s}(k) + \bar{n}(k)$$

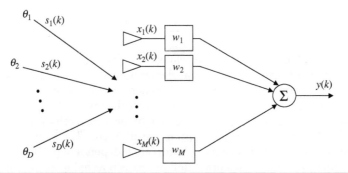

FIGURE 7.1 *M*-element array with arriving signals.

and

$\bar{w} = [w_1 \ w_2 \ \cdots \ w_M]^T$ = array weights
$\bar{s}(k)$ = vector of incident complex monochromatic signals at time k
$\bar{n}(k)$ = noise vector at each array element m, zero mean, variance σ_n^2
$\bar{a}(\theta_i)$ = M-element array steering vector for the θ_i direction of arrival
$\bar{A} = [\bar{a}(\theta_1) \ \bar{a}(\theta_2) \cdots \bar{a}(\theta_D)]$ $M \times D$ matrix of steering vectors $\bar{a}\theta_i$

Thus, each of the D-complex signals arrives at angles θ_i and is intercepted by the M antenna elements. It is initially assumed that the arriving signals are monochromatic and the number of arriving signals $D < M$. It is understood that the arriving signals are time varying, and thus our calculations are based on time snapshots of the incoming signal. Obviously if the transmitters are moving, the matrix of steering vectors is changing with time and the corresponding arrival angles are changing. Unless otherwise stated, the time dependence are suppressed in Eqs. (7.19) and (7.20). In order to simplify the notation, let us define the $M \times M$ array correlation matrix $\bar{R}_{xx}$ as

$$\bar{R}_{xx} = E[\bar{x} \cdot \bar{x}^H] = E[(\bar{A}\bar{s} + \bar{n})(\bar{s}^H \bar{A}^H + \bar{n}^H)]$$
$$= \bar{A}E[\bar{s} \cdot \bar{s}^H]\bar{A}^H + E[\bar{n} \cdot \bar{n}^H]$$
$$= \bar{A}\bar{R}_{ss}\bar{A}^H + \bar{R}_{nn} \qquad (7.21)$$

where $\bar{R}_{ss} = D \times D$ source correlation matrix
$\bar{R}_{nn} = \sigma_n^2 \bar{I}$ = $M \times M$ noise correlation matrix
$\bar{I} = N \times N$ identity matrix

The array correlation matrix $\bar{R}_{xx}$ and the source correlation matrix $\bar{R}_{ss}$ are found by the expected value of the respective absolute values squared (i.e., $\bar{R}_{xx} = E[\bar{x} \cdot \bar{x}^H]$ and $\bar{R}_{ss} = E[\bar{s} \cdot \bar{s}^H]$). If we do not know the exact statistics for the noise and signals, but we can assume that the process is ergodic, we can approximate the correlation by use of a time-averaged correlation. In that case the correlation matrices are defined by

$$\hat{R}_{xx} \approx \frac{1}{K}\sum_{k=1}^{K}\bar{x}(k)\bar{x}^H(k) \quad \hat{R}_{ss} \approx \frac{1}{K}\sum_{k=1}^{K}\bar{s}(k)\bar{s}^H(k) \quad \hat{R}_{nn} \approx \frac{1}{K}\sum_{k=1}^{K}\bar{n}(k)\bar{n}^H(k)$$

When the signals are uncorrelated, $\bar{R}_{ss}$, obviously has to be a diagonal matrix because off-diagonal elements have no correlation. When the signals are partly correlated, $\bar{R}_{ss}$ is nonsingular. When the signals are coherent, $\bar{R}_{ss}$ becomes singular because the rows are linear combinations of each other [5]. The matrix of steering vectors, $\bar{A}$, is an $M \times D$ matrix where all columns are different. Their structure is Vandermonde and hence the columns are independent [6, 7]. Often in the literature, the array correlation matrix is referred to as the covariance matrix. This is only true if the mean values of the signals and noise are zero. In that case, the covariance and the correlation matrices are identical. The arriving signal mean value must necessarily be zero because antennas cannot receive d.c. signals. The noise inherent in the receiver may or may not have zero mean depending on the source of the receiver noise.

There is much useful information to be discovered in the eigen-analysis of the array correlation matrix. Details of the eigenstructure are described in Godara [1] and are repeated here. Given M-array elements with D-narrowband signal sources and uncorrelated noise we can make some assumptions about the properties of the correlation matrix. First, $\bar{R}_{xx}$ is an $M \times M$ Hermitian matrix. A Hermitian matrix is equal to its complex conjugate transpose such that $\bar{R}_{xx} = \bar{R}_{xx}^H$. The array correlation matrix has M eigenvalues $(\lambda_1, \lambda_2, ..., \lambda_M)$ along with M-associated eigenvectors $\bar{E} = [\bar{e}_1 \bar{e}_2 \cdots \bar{e}_M]$. If the eigenvalues are sorted from smallest to largest, we can divide the matrix $\bar{E}$ into two subspaces such that $\bar{E} = [\bar{E}_N \bar{E}_S]$. The first subspace $\bar{E}_N$ is called the *noise subspace* and is composed of $M - D$ eigenvectors associated with the noise. For uncorrelated noise, the eigenvalues are given as $\lambda_1 = \lambda_2 = \cdots = \lambda_{M-D} = \sigma_n^2$. The second subspace $\bar{E}_S$ is called the *signal subspace* and is composed of D eigenvectors associated with the arriving signals. The noise subspace is an $M \times (M - D)$ matrix. The signal subspace is an $M \times D$ matrix.

The goal of AOA estimation techniques is to define a function that gives an indication of the angles of arrival based upon maxima versus angle. This function is traditionally called the *pseudospectrum* $P(\theta)$ and the units can be in energy or in watts (or at times energy or watts squared). There are several potential approaches to defining the pseudospectrum via beamforming, the array correlation matrix, eigenanalysis, linear prediction, minimum variance, maximum likelihood, minnorm, MUSIC, root-MUSIC, and many more approaches that are be addressed in this chapter. Both Stoica and Moses [5] and Van Trees [4] give an in-depth explanation of many of these possible approaches. We summarize some of the more popular pseudospectra solutions in the next section.

7.3 AOA Estimation Methods

7.3.1 Bartlett AOA Estimate

If the array is uniformly weighted, we can define the Bartlett AOA estimate [8] as

$$P_B(\theta) = \bar{a}^H(\theta)\bar{R}_{xx}\bar{a}(\theta) \qquad (7.22)$$

The Bartlett AOA estimate is the spatial version of an averaged periodogram and is a beamforming AOA estimate. Under the conditions where $\bar{s}$ represents uncorrelated monochromatic signals and there is no system noise, Eq. (7.22) is equivalent to the following long-hand expression:

$$P_B(\theta) = \left| \sum_{i=1}^{D} \sum_{m=1}^{M} e^{j(m-1)kd(\sin\theta - \sin\theta_i)} \right|^2 \quad (7.23)$$

The periodogram is thus equivalent to the spatial finite Fourier transform of all arriving signals. This is also equivalent to adding all beam-steered array factors for each AOA and finding the absolute value squared.

Example 7.5 Use MATLAB to plot the pseudospectrum using the Bartlett estimate for an $M = 6$-element array. With element spacing $d = \lambda/2$, uncorrelated, equal amplitude sources, (s_1, s_2), and $\sigma_n^2 = .1$, and the two different pairs of arrival angles given by $\pm 10°$ and $\pm 5°$, assume ergodicity.

Solution From the information given we can find the following:

$$\bar{s} = \begin{bmatrix} 1 \\ 1 \end{bmatrix} \quad \bar{a}(\theta) = [1 \quad e^{j\pi\sin\theta} \quad \cdots \quad e^{j5\pi\sin\theta}]^T$$

$$\bar{A} = [\bar{a}(\theta_1) \quad \bar{a}(\theta_2)] \quad \bar{R}_{ss} = \begin{bmatrix} 1 & 0 \\ 0 & 1 \end{bmatrix}$$

Applying Eq. (7.21) we can find $\bar{R}_{xx}$ for both sets of angles. Substituting $\bar{R}_{xx}$ into Eq. (7.22) and using MATLAB, we can plot the pseudospectrum as shown in Fig. 7.2a and b.

Recalling the half-power beamwidth of a linear array from Chap. 4, Eq. (4.21), we can estimate the beamwidth of this $M = 6$-element array to be $\approx 8.5°$. Thus, the two sources, which are 20° apart, are resolvable with the Bartlett approach. The two sources, which are 10° apart, are not resolvable. Herein lies one of the limitations of the Bartlett approach to AOA estimation: The ability to resolve angles is limited by the array half-power beamwidth. An increase in resolution requires a larger array. For large array lengths with $d = \lambda/2$ spacing, the AOA resolution is approximately $1/M$. Thus, $1/M$ is the AOA resolution limit of a periodogram and in the case above is an indicator of the resolution of the Bartlett method. It should be noted that when two emitters are separated by an angle wider than the array resolution, they can be resolved but a bias is introduced. This bias cause the peaks to deviate from the true AOA. This bias asymptotically decreases as the array length increases.

7.3.2 Capon AOA Estimate

The Capon AOA estimate [2, 4] is known as a *minimum variance distortionless response* (MVDR). It is also alternatively a maximum likelihood estimate of the power arriving from one direction while all other sources are considered as interference. Thus the goal is to maximize the signal-to-intereference ratio (SIR) while passing the signal of interest

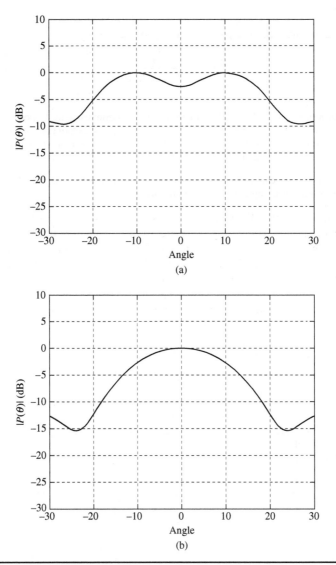

FIGURE 7.2 (a) Bartlett pseudospectrum for $\theta_1 = -10°$, $\theta_2 = 10°$. (b) Bartlett pseudospectrum for $\theta_1 = -5°$, $\theta_2 = 5°$.

undistorted in phase and amplitude. The source correlation matrix $\bar{R}_{ss}$ is assumed to be diagonal. This maximized SIR is accomplished with a set of array weights ($\bar{w} = [w_1 w_2 \cdots w_M]^T$) as shown in Fig. 7.1, where the array weights are given by

$$\bar{w} = \frac{\bar{R}_{xx}^{-1} \bar{a}(\theta)}{\bar{a}^H(\theta) \bar{R}_{xx}^{-1} \bar{a}(\theta)} \qquad (7.24)$$

where $\bar{R}_{xx}$ is the unweighted array correlation matrix.

Substituting the weights of Eq. (7.24) into the array of Fig. 7.1, we can then find that the pseudospectrum is given by

$$P_C(\theta) = \frac{1}{\bar{a}^H(\theta)\bar{R}_{xx}^{-1}\bar{a}(\theta)} \qquad (7.25)$$

Example 7.6 Use MATLAB to plot the pseudospectrum using the Capon estimate for an $M = 6$-element array. With element spacing $d = \lambda/2$, uncorrelated, equal amplitude sources, (s_1, s_2), and $\sigma_n^2 = .1$, and the pair of arrival angles given by $\pm 5°$, assume ergodicity.

Solution We can use the same array correlation matrix as was found in Example 7.5. Using MATLAB, we produce Fig. 7.3.

It is clear that the Capon AOA estimate has much greater resolution than the Bartlett AOA estimate. In the case where the competing sources are highly correlated, the Capon resolution can actually become worse. The derivation of the Capon (ML) weights was conditioned upon considering that all other sources are interferers. If the multiple signals can be considered as multipath signals, with Rayleigh amplitude and uniform phase, then the uncorrelated condition is met and the Capon estimate will work.

The advantage of the Bartlett and Capon estimation methods is that these are nonparametric solutions and one does not need an a priori knowledge of the specific statistical properties.

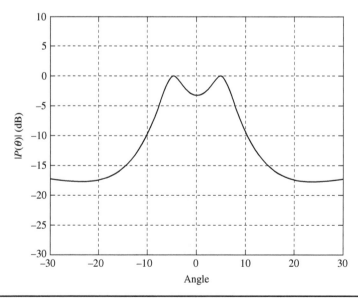

Figure 7.3 Capon (ML) pseudospectrum for $\theta_1 = -5°$, $\theta_2 = 5°$.

7.3.3 Linear Prediction AOA Estimate

The goal of the linear prediction method is to minimize the prediction error between the output of the mth sensor and the actual output [3, 9]. Our goal is to find the weights that minimize the mean-squared prediction error. In a similar vein as Eq. (7.24), the solution for the array weights is given as

$$\bar{w}_m = \frac{\bar{R}_{xx}^{-1}\bar{u}_m}{\bar{u}_m^T \bar{R}_{xx}^{-1}\bar{u}_m} \qquad (7.26)$$

where $\bar{u}_m$ is the *Cartesian basis vector*, which is the mth column of the $M \times M$ identity matrix.

Upon substitution of these array weights into the calculation of the pseudospectrum, it can be shown that

$$P_{LP_m}(\theta) = \frac{\bar{u}_m^T \bar{R}_{xx}^{-1} \bar{u}_m}{|\bar{u}_m^T \bar{R}_{xx}^{-1} \bar{a}(\theta)|^2} \qquad (7.27)$$

The particular choice for which mth element output for prediction is random. Although the choice made can dramatically affect the final resolution. If the array center element is chosen, the linear combination of the remaining sensor elements might provide a better estimate because the other array elements are spaced about the phase center of the array [3]. This would suggest that odd array lengths might provide better results than even arrays because the center element is precisely at the array phase center.

This linear prediction technique is sometimes referred to as an *autoregressive* method [4]. It has been argued that the spectral peaks using linear prediction are proportional to the square of the signal power [3]. This is true in Example 7.7.

Example 7.7 Use MATLAB to plot the pseudospectrum using the linear predictive estimate for an $M = 6$ element array. With element spacing $d = \lambda/2$, uncorrelated, equal amplitude sources, (s_1, s_2), and $\sigma_n^2 = .1$, and the pair of arrival angles given by $\pm 5°$, choose the 3rd element of the array as the reference element such that the Cartesian basis vector is $\bar{u}_3 = [0\ 0\ 1\ 0\ 0\ 0]^T$. Assume ergodicity.

Solution The pseudospectra is given as $P_{LP_3}(\theta) = \frac{\bar{u}_3^T \bar{R}_{xx}^{-1} \bar{u}_3}{|\bar{u}_3^T \bar{R}_{xx}^{-1}\bar{a}(\theta)|^2}$ and is plotted in Fig. 7.4.

It is very obvious that under these conditions, the linear predictive method provides superior performance over both the Bartlett estimate and the Capon estimate. The efficacy of the performance is dependent on the array element chosen and the subsequent $\bar{u}_n$ vector. When one selects the arrival signals to have different amplitudes, the linear predictive spectral peaks reflect the relative strengths of the incoming signals. Thus, the linear predictive method not only provides AOA information, but it also provides signal strength information.

Chapter Seven

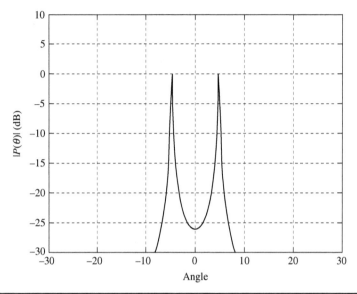

Figure 7.4 Linear predictive pseudospectrum for $\theta_1 = -5°$, $\theta_2 = 5°$.

7.3.4 Maximum Entropy AOA Estimate

The maximum entropy method is attributed to Burg [10, 11]. A further explanation of the maximum entropy approach is given in [1, 12]. The goal is to find a pseudospectrum that maximizes the entropy function subject to constraints. The details of the Burg derivation can be found in the references discussed previously. The pseudospectrum is given by

$$P_{ME_j}(\theta) = \frac{1}{\bar{a}(\theta)^H \bar{c}_j \bar{c}_j^H \bar{a}(\theta)} \qquad (7.28)$$

where $\bar{c}_j$ is the jth column of the inverse array correlation matrix ($\bar{R}_{xx}^{-1}$).

Example 7.8 Use MATLAB to plot the pseudospectrum using the maximum entropy AOA estimate for an $M = 6$-element array, element spacing $d = \lambda/2$, uncorrelated, equal amplitude sources, (s_1, s_2), and $\sigma_n^2 = .1$, and the pair of arrival angles given by $\pm 5°$. Choose the 3rd column ($\bar{c}_3$) of the array correlation matrix to satisfy Eq. (7.28). Assume ergodicity.

Solution The pseudospectrum is given as plotted in Fig. 7.5.

It should be noted that the maximum entropy method, when we select the $\bar{c}_3$ column from $\bar{R}_{xx}^{-1}$, gives the same pseudospectra as the linear predictive method. The choice of $\bar{c}_j$ can dramatically effect the resolution achieved. The center columns of the inverse array correlation matrix tend to give better results under the conditions assumed in this chapter.

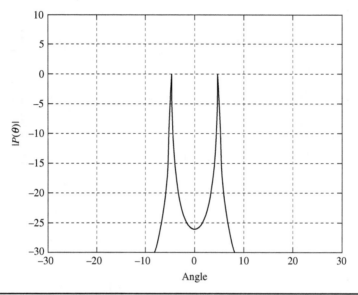

FIGURE 7.5 Maximum entropy pseudospectrum for $\theta_1 = -5°$, $\theta_2 = 5°$.

7.3.5 Pisarenko Harmonic Decomposition AOA Estimate

The *Pisarenko harmonic decomposition* (PHD) AOA estimate is named after the Russian mathematician who devised this minimum mean-squared error approach [13, 14]. The goal is to minimize the mean-squared error of the array output under the constraint that the norm of the weight vector be equal to unity. The eigenvector that minimizes the mean-squared error corresponds to the smallest eigenvalue. For an $M = 6$-element array, with two arriving signals, there will be two eigenvectors associated with the signal and four eigenvectors associated with the noise. The corresponding PHD pseudospectrum is given by

$$P_{PHD}(\theta) = \frac{1}{\left|\bar{a}^H(\theta)\bar{e}_1\right|^2} \qquad (7.29)$$

where $\bar{e}_1$ is the eigenvector associated with the smallest eigenvalue λ_1.

Example 7.9 Use MATLAB to plot the pseudospectrum using the Pisarenko harmonic decomposition estimate for an $M = 6$-element array, element spacing $d = \lambda/2$, uncorrelated, equal amplitude sources, (s_1, s_2), and $\sigma_n^2 = .1$, and the pair of arrival angles given by $\pm 5°$. Choose the first noise eigenvector to produce the pseudospectrum.

Solution After finding the array correlation matrix, we can use the eig( ) command in MATLAB to find the eigenvectors and corresponding eigenvalues. The eigenvalues are given by $\lambda_1 = \lambda_2 = \lambda_3 = \lambda_4 = \sigma_n^2 = .1$, $\lambda_5 = 2.95$, $\lambda_6 = 9.25$.

178 Chapter Seven

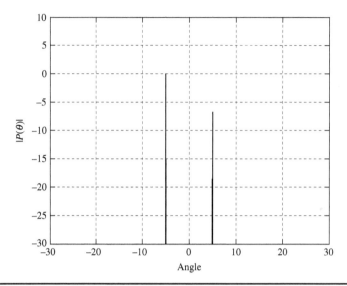

FIGURE 7.6 Pisarenko harmonic decomposition pseudospectrum for $\theta_1 = -5°$, $\theta_2 = 5°$.

The first eigenvector associated with λ_1 is given by

$$\bar{e}_1 = \begin{bmatrix} -0.143 \\ -0.195 \\ 0.065 \\ 0.198 \\ 0.612 \\ -0.723 \end{bmatrix}$$

Substituting this eigenvector in Eq. (7.29), we can plot Fig. 7.6.

The Pisarenko peaks are not an indication of the signal amplitudes. These peaks are the roots of the polynomial in the denominator of Eq. (7.29). It is clear that for this example, the Pisarenko solution has the best resolution.

7.3.6 Min-Norm AOA Estimate

The Minimum-Norm method was developed by Reddi [15] and Kumaresan and Tufts [16]. This method is also lucidly explained by Ermolaev and Gershman [17]. The Min-Norm method is only relevant for *uniform linear arrays* (ULA). The Min-Norm algorithm optimizes the weight vector by solving the optimization problem where

$$\min_{\bar{w}} \bar{w}^H \bar{w} \qquad \bar{E}_S^H \bar{w} = 0 \qquad \bar{w}^H \bar{u}_1 = 1 \qquad (7.30)$$

where $\bar{w}$ = array weights
$\bar{E}_S$ = subspace of D signal eigenvectors = $[\bar{e}_{M-D+1} \ \bar{e}_{M-D+2} \cdots \bar{e}_M]$
M = number of array elements

D = number of arriving signals
$\bar{u}_1$ = Cartesian basis vector (first column of the $M \times M$ identity matrix)
    = $[100 \ldots 0]^T$

The solution to the optimization yields the Min-Norm pseudospectrum

$$P_{MN}(\theta) = \frac{(\bar{u}_1^T \bar{E}_N \bar{E}_N^H \bar{u}_1)^2}{|\bar{a}(\theta)^H \bar{E}_N \bar{E}_N^H \bar{u}_1|^2} \qquad (7.31)$$

where $\bar{E}_N$ = subspace of $M - D$ noise eigenvectors = $[\bar{e}_1 \ \bar{e}_2 \ \cdots \ \bar{e}_{M-D}]$
$\bar{a}(\theta)$ = array steering vector

Because the numerator term in Eq. (7.31) is a constant, we can normalize the pseudospectrum such that

$$P_{MN}(\theta) = \frac{1}{|\bar{a}(\theta)^H \bar{E}_N \bar{E}_N^H \bar{u}_1|^2} \qquad (7.32)$$

Example 7.10 Use MATLAB to plot the pseudo-spectrum using the Min-Norm AOA estimate for an $M = 6$-element array, with element spacing $d = \lambda/2$, uncorrelated, equal amplitude sources, (s_1, s_2), and $\sigma_n^2 = .1$, and the pair of arrival angles given by $\pm 5°$. Use all noise eigenvectors to construct the noise subspace $\bar{E}_N$.

Solution After finding the array correlation matrix, we can use the eig( ) command in MATLAB to find the eigenvectors and corresponding eigenvalues. The eigenvalues are broken up into two groups. There are eigenvectors associated with the noise eigenvalues given by $\lambda_1 = \lambda_2 = \lambda_3 = \lambda_4 = \sigma_n^2 = .1$. There are eigenvectors associated with the signal eigenvalues $\lambda_5 = 2.95$ and $\lambda_6 = 9.25$. The subspace created by the $M - D = 4$ noise eigenvectors is given as

$$\bar{E}_N = \begin{bmatrix} -0.14 & -0.56 & -0.21 & 0.27 \\ -0.2 & 0.23 & 0.22 & -0.75 \\ 0.065 & 0.43 & 0.49 & 0.58 \\ 0.2 & 0.35 & -0.78 & 0.035 \\ 0.61 & -0.51 & 0.22 & -0.15 \\ -0.72 & -0.25 & 0 & 0.083 \end{bmatrix}$$

Applying this information to Eq. (7.32), we can plot the angular spectra in Fig. 7.7.

It should be noted that the pseudospectrum from the Min-Norm method is almost identical to the PHD pseudospectrum. The Min-Norm method combines all noise eigenvectors whereas the PHD method only uses the first noise eigenvector.

7.3.7 MUSIC AOA Estimate

MUSIC is an acronym that stands for MUltiple SIgnal Classification. This approach was first posed by Schmidt [18] and is a popular high resolution eigenstructure method. MUSIC promises to provide unbiased

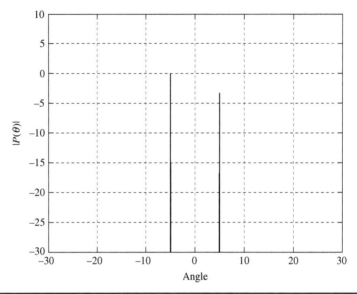

Figure 7.7 Min-Norm pseudospectrum for $\theta_1 = -5°$, $\theta_2 = 5°$.

estimates of the number of signals, the angles of arrival, and the strengths of the waveforms. MUSIC makes the assumption that the noise in each channel is uncorrelated making the noise correlation matrix diagonal. The incident signals may be somewhat correlated creating a nondiagonal signal correlation matrix. However, under high signal correlation the traditional MUSIC algorithm breaks down and other methods must be implemented to correct this weakness. These methods are discussed later in this chapter.

One must know in advance the number of incoming signals or one must search the eigenvalues to determine the number of incoming signals. If the number of signals is D, the number of signal eigenvalues and eigenvectors is D, and the number of noise eigenvalues and eigenvectors is $M - D$ (M is the number of array elements). Because MUSIC exploits the noise eigenvector subspace, it is sometimes referred to as a *subspace method*.

As before, we calculate the array correlation matrix assuming uncorrelated noise with equal variances.

$$\bar{R}_{xx} = \bar{A}\bar{R}_{ss}\bar{A}^H + \sigma_n^2 \bar{I} \tag{7.33}$$

We next find the eigenvalues and eigenvectors for $\bar{R}_{xx}$. We then produce D eigenvectors associated with the signals and $M - D$ eigenvectors associated with the noise. We choose the eigenvectors associated with the smallest eigenvalues. For uncorrelated signals, the smallest eigenvalues are equal to the variance of the noise. We can then construct ns such that the $M \times (M - D)$ dimensional subspace spanned by the noise eigenvectors such that

$$\bar{E}_N = [\bar{e}_1 \quad \bar{e}_2 \quad \cdots \quad \bar{e}_{M-D}] \tag{7.34}$$

The noise subspace eigenvectors are orthogonal to the array steering vectors at the angles of arrival $\theta_1, \theta_2, \ldots, \theta_D$. Because of this orthogonality condition, one can show that the Euclidean distance $d^2 = \bar{a}(\theta)^H \bar{E}_N \bar{E}_N^H \bar{a}(\theta) = 0$ for each and every arrival angle $\theta_1, \theta_2, \ldots, \theta_D$. Placing this distance expression in the denominator creates sharp peaks at the angles of arrival. The MUSIC pseudospectrum is now given as

$$P_{MU}(\theta) = \frac{1}{|\bar{a}(\theta)^H \bar{E}_N \bar{E}_N^H \bar{a}(\theta)|} \tag{7.35}$$

Example 7.11 Use MATLAB to plot the pseudospectrum using the MUSIC AOA estimate for an $M = 6$-element array. With element spacing $d = \lambda/2$, uncorrelated, equal amplitude sources, (s_1, s_2), and $\sigma_n^2 = .1$, and the pair of arrival angles given by $\pm 5°$. Use all noise eigenvectors to construct the noise subspace $\bar{E}_N$.

Solution After finding the array correlation matrix, we can use the eig( ) command in MATLAB to find the eigenvectors and corresponding eigenvalues. The eigenvalues are given by $\lambda_1 = \lambda_2 = \lambda_3 = \lambda_4 = \sigma_n^2 = .1$, $\lambda_5 = 2.95$, and $\lambda_6 = 9.25$. The eigenvalues and eigenvectors can be sorted in MATLAB from the least to the greatest by the following commands:

```
[V,Dia] = eig(Rxx);
[Y,Index] = sort(diag(Dia));
EN = V(:,Index(1:M-D));
```

The subspace created by the $M - D = 4$ noise eigenvectors again is given as

$$\bar{E}_N = \begin{bmatrix} -0.14 & -0.56 & -0.21 & 0.27 \\ -0.2 & 0.23 & 0.22 & -0.75 \\ 0.065 & 0.43 & 0.49 & 0.58 \\ 0.2 & 0.35 & -0.78 & 0.035 \\ 0.61 & -0.51 & 0.22 & -0.15 \\ -0.72 & -0.25 & 0 & 0.083 \end{bmatrix}$$

Applying this information to Eq. (7.35), we can plot the angular spectra in Fig. 7.8.

Under the conditions stated for the Pisarenko harmonic decomposition, the Min-Norm method, and the MUSIC method, the solutions all have similar resolution. It should be understood that in all examples discussed earlier, it was assumed that the array correlation matrix was of the form given in Eq. (7.33), that the noise variance for all elements was identical, and that the different signals were completely uncorrelated. In the case where the source correlation matrix is not diagonal, or the noise variances vary, the plots can change dramatically and the resolution will diminish.

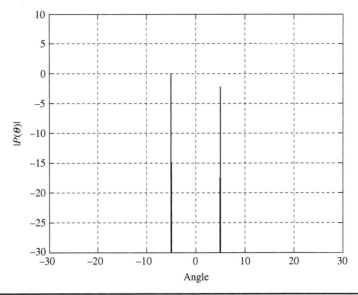

FIGURE 7.8 MUSIC pseudospectrum for $\theta_1 = -5°$, $\theta_2 = 5°$.

In the more practical application we must collect several time samples of the received signal plus noise, assume ergodicity, and estimate the correlation matrices via time averaging. We can repeat Eq. (7.33) without assuming that we know the signal statistics

$$\hat{R}_{xx} = E[\bar{x}(k) \cdot \bar{x}^H(k)] \approx \frac{1}{K}\sum_{k=1}^{K} \bar{x}(k) \cdot \bar{x}^H(k) \qquad (7.36)$$
$$\approx \bar{A}\hat{R}_{ss}\bar{A}^H + \bar{A}\hat{R}_{sn} + \hat{R}_{ns}\bar{A}^H + \hat{R}_{nn}$$

where

$$\hat{R}_{ss} = \frac{1}{K}\sum_{k=1}^{K} \bar{s}(k)\bar{s}^H(k) \qquad \hat{R}_{sn} = \frac{1}{K}\sum_{k=1}^{K} \bar{s}(k)\bar{n}^H(k)$$

$$\hat{R}_{ns} = \frac{1}{K}\sum_{k=1}^{K} \bar{n}(k)\bar{s}^H(k) \qquad \hat{R}_{nn} = \frac{1}{K}\sum_{k=1}^{K} \bar{n}(k)\bar{n}^H(k)$$

Example 7.12 Use MATLAB to plot the pseudospectrum using the MUSIC AOA estimate for an $M = 6$-element array. With element spacing $d = \lambda/2$, the pair of arrival angles are given by $\pm 5°$. Assume binary *Walsh-like* signals of amplitude 1, but with only K finite signal samples. Assume Gaussian distributed noise of $\sigma_n^2 = .1$ but with only K finite noise samples. Also, assume the process is ergodic and collect $K = 100$ time samples ($k = 1, 2, ..., K$) of the signal such that $s = \text{sign}(\text{randn}(M, K))$ and the noise such that $n = \text{sqrt(sig2)} * \text{randn}(M, K)$ (sig2 = σ_n^2). Calculate all correlation matrices via time averaging as defined in Eq. (7.36). This can be accomplished in MATLAB by the commands Rss = $s*s'/K$,

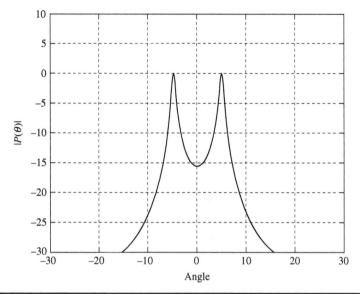

FIGURE 7.9 MUSIC pseudospectrum using time averages for $\theta_1 = -5°$, $\theta_2 = 5°$.

Rns = $n*s'/K$, Rsn = $n*n'/K$, and Rnn = $s*n'/K$. Assume a pair of arrival angles given by ±5°. Use all noise eigenvectors to construct the noise subspace $\bar{E}_N$ and find the pseudospectrum. (*Important:* MATLAB will not order the eigenvalues from least to greatest, so one must sort them before selecting the appropriate noise eigenvectors. The sorting method was shown in the previous example. The noise subspace is then given by EN = E(:,index (1:M − D)). The MATLAB code for this example demonstrates the sorting process.)

Solution We can generate the 100 time samples of the noise and signal as indicated earlier. After finding the array correlation matrix $\hat{R}_{xx}$, we can use the eig( ) command in MATLAB to find the eigenvectors and corresponding eigenvalues. The eigenvalues are given by $\lambda_1 = .08$, $\lambda_2 = .09$, $\lambda_3 = .12$, $\lambda_4 = .13$, $\lambda_5 = 2.97$, and $\lambda_6 = 9$.

Applying this information to Eq. (7.35), we can plot the angular spectra in Fig. 7.9.

It is clear from the last example that the resolution of the MUSIC algorithm begins to diminish as we have to estimate the correlation matrices by time averages so that we have $\hat{R}_{xx} = \bar{A}\hat{R}_{ss}\bar{A}^H + \bar{A}\hat{R}_{sn} + \hat{R}_{ns}\bar{A}^H + \hat{R}_{nn}$.

7.3.8 Root-MUSIC AOA Estimate

The MUSIC algorithm in general can apply to any arbitrary array regardless of the position of the array elements. *Root-MUSIC* implies that the MUSIC algorithm is reduced to finding roots of a polynomial as opposed to merely plotting the pseudospectrum or searching for peaks in the pseudospectrum. Barabell [12] simplified the MUSIC algorithm for the case where the antenna is a ULA. Recalling that the MUSIC pseudospectrum is given by

$$P_{MU}(\theta) = \frac{1}{|\bar{a}(\theta)^H \bar{E}_N \bar{E}_N^H \bar{a}(\theta)|} \tag{7.37}$$

One can simplify the denominator expression by defining the matrix $\bar{C} = E_N \bar{E}_N^H$, which is Hermitian. This leads to the root-MUSIC expression

$$P_{RMU}(\theta) = \frac{1}{|\bar{a}(\theta)^H \bar{C} \bar{a}(\theta)|} \tag{7.38}$$

If we have a ULA, the mth element of the array steering vector is given by

$$a_m(\theta) = e^{jkd(m-1)\sin\theta} \qquad m = 1, 2, \ldots, M \tag{7.39}$$

The denominator argument in Eq. (7.38) can be written as

$$\bar{a}(\theta)^H \bar{C} \bar{a}(\theta) = \sum_{m=1}^{M} \sum_{n=1}^{M} e^{-jkd(m-1)\sin\theta} C_{mn} e^{jkd(n-1)\sin\theta}$$

$$= \sum_{\ell=-M+1}^{M-1} c_\ell e^{-jkd\ell \sin\theta} \tag{7.40}$$

where c_ℓ is the sum of the diagonal elements of $\bar{C}$ along the ℓth diagonal such that

$$c_\ell = \sum_{n-m=\ell} C_{mn} \tag{7.41}$$

It should be noted that the matrix $\bar{C}$ has off-diagonal sums such that $c_0 > |c_\ell|$ for $\ell \neq 0$. Thus the sum of off-diagonal elements is always less than that of the main diagonal elements. In addition, $c_\ell = c_{-\ell}^*$. For a 6×6 matrix we have 11 diagonals ranging from diagonal numbers $\ell = -5, -4, \ldots, 0, \ldots, 4, 5$. The lower left diagonal is represented by $\ell = -5$ whereas the upper right diagonal is represented by $\ell = 5$. The c_ℓ coefficients are calculated by $c_{-5} = C_{61}, c_{-4} = C_{51} + C_{62}, c_{-3} = C_{41} + C_{52} + C_{63}$, and so on.

We can simplify Eq. (7.40) to be in the form of a polynomial whose coefficients are c_ℓ. Thus

$$D(z) = \sum_{\ell=-M+1}^{M-1} c_\ell z^\ell \tag{7.42}$$

where $z = e^{-jkd\sin\theta}$.

The roots of $D(z)$ that lie closest to the unit circle correspond to the poles of the MUSIC pseudospectrum. Thus, this technique is called *root-MUSIC*. The polynomial of Eq. (7.42) is of order $2(M-1)$ and thus has roots of $z_1, z_2, \ldots, z_{2(M-1)}$. Each root can be complex and using polar notation can be written as

$$z_i = |z_i| e^{j \arg(z_i)} \quad i = 1, 2, \ldots, 2(M-1) \tag{7.43}$$

where $\arg(z_i)$ is the phase angle of z_i.

Exact zeros in $D(z)$ exist when the root magnitudes $|z_i| = 1$. One can calculate the AOA by comparing $e^{j \arg(z_i)}$ to $e^{jkd \sin \theta_i}$ to get

$$\theta_i = -\sin^{-1}\left(\frac{1}{kd} \arg(z_i)\right) \tag{7.44}$$

Example 7.13 Repeat Example 7.12 by changing the noise variance to be $\sigma_n^2 = .3$. Change the angles of arrival to $\theta_1 = -4°$ and $\theta_2 = 8°$. Let the array be reduced to having only $M = 4$ elements. Approximate the correlation matrices by time averaging over $K = 300$ data points. Superimpose the plots of the pseudospectrum and the roots from root-MUSIC and compare.

Solution One can modify the MATLAB program with the new variables so that the 4-element array produces a 4×4 matrix $\bar{C}$ as defined previously. Thus $\bar{C}$ is given by

$$\bar{C} = \begin{bmatrix} .305 & -.388+.033i & -.092+.028i & .216-.064i \\ -.388-.033i & .6862 & -.225+.019i & -.11+.011i \\ -.092-.028i & -.225-.019i & .6732 & -.396+.046i \\ .216+.046i & -.11-.011i & -.396-.046i & .335 \end{bmatrix}$$

The root-MUSIC polynomial coefficients are given by the sums along the $2M-1$ diagonals. Thus

$$c = .216 + .065i, -.203 - .039i, -1.01 - .099i, 2.0,$$
$$-1.01 + .099i, -.203 + .039i, .216 - .065i$$

We can use the root command in MATLAB to find the roots and then solve for the magnitude and angles of the $2(M-1) = 6$ roots. We can plot the location of all 6 roots showing which roots are closest to the unit circle as shown in Fig. 7.10. It is clear that only the four on the right side of the y axis are nearest to the unit circle and are close to the expected angles of arrival.

We can choose the four roots closest to the unit circle and replot them along with the MUSIC pseudospectrum in Fig. 7.11.

The roots found with root-MUSIC earlier do not exactly reflect the actual location of the angles of arrival of $\theta_1 = -4°$ and $\theta_2 = 8°$, but they indicate two angles of arrival. The roots themselves show the existence of an angle of arrival at near $8°$, which is not obvious from the plot of the MUSIC pseudospectrum. The error in locating the correct root locations owes to the fact that the incoming signals are partially correlated, that we approximated the correlation matrix by time averaging, and that the S/N ratio is relatively low. One must exert care in exercising the use of root-MUSIC by knowing the assumptions and conditions under which the calculations are made.

186 Chapter Seven

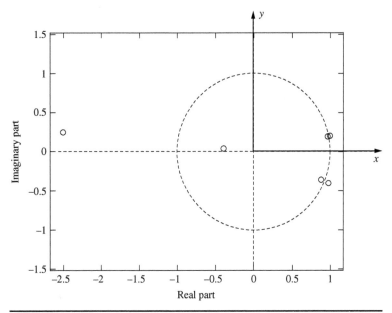

FIGURE 7.10 All 6 roots in Cartesian coordinates.

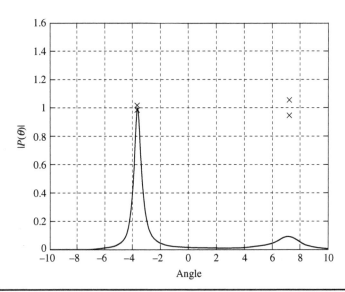

FIGURE 7.11 MUSIC pseudospectrum and roots found with root-MUSIC for $\theta_1 = -4°$ and $\theta_2 = 8°$.

It is interesting to note that the polynomial $D(z)$ is a self-reciprocal polynomial such that $D(z) = D*(z)$. The roots of the polynomial $D(z)$ are in reciprocal pairs meaning that $z_1 = \frac{1}{z_2^*}, z_3 = \frac{1}{z_4^*}, ..., z_{2M-3} = \frac{1}{z_{2M-3}^*}$.

Because of the self-reciprocal symmetry of $D(z)$ ([4]), one can factor $D(z)$ using the Fejér method such that

$$D(z) = p(z)\, p^*(1/z^*) \tag{7.45}$$

Under these conditions it is sufficient to solve for the roots of the $p(z)$ polynomial of degree $M-1$. The roots of $p(z)$ are on or within the unit circle while the roots of $p^*(1/z^*)$ or on or without the unit circle.

One method has been proposed by Ren and Willis [19] to reduce the order of the polynomial $D(z)$ and thus reduce the computational burden of finding the roots.

The polynomial rooting method can also be applied to the Capon algorithm where we substitute $\bar{C} = \hat{R}_{rr}^{-1}$ for $\bar{C} = \bar{E}_N \bar{E}_N^H$. However, because the accuracy of the Capon estimation algorithm is much less than the MUSIC approach, the root finding also suffers a loss in accuracy.

The same principles applied to root-MUSIC can also be applied to the Min-Norm method to create a root-Min-Norm solution. We can repeat Eq. (7.32)

$$P_{RMN}(\theta) = \frac{1}{\left| \bar{a}(\theta)^H \bar{C} \bar{u}_1 \right|^2} \tag{7.46}$$

where $\bar{u}_1$ = Cartesian basis vector (first column of the $M \times M$ identity matrix)
$= [1 0 0 \ldots 0]^T$
$\bar{C} = \bar{E}_N \bar{E}_N^H$ = a $M \times M$ Hermitian matrix
$\bar{E}_N$ = subspace of $M - D$ noise eigenvectors
$\bar{a}(\theta)$ = array steering vector

The product of the Cartesian basis vector and the Hermitian matrix results in creating a column vector composed of the first row of the matrix $\bar{C}$. The column vector based on the first column of $\bar{C}$ becomes $\bar{c}_1 = [C_{11} C_{12} \cdots C_{1M}]^T$, where the subscript 1 indicates the first column. We can substitute this into Eq. (7.46).

$$P_{RMN}(\theta) = \frac{1}{\left| \bar{a}(\theta)^H \bar{c}_1 \right|^2} = \frac{1}{\bar{a}(\theta)^H \bar{c}_1 \bar{c}_1^H \bar{a}(\theta)} \tag{7.47}$$

In a similar fashion to Eq. (7.42), we can create a polynomial from the denominator of Eq. (7.47) given by

$$D(z) = \sum_{\ell=-M+1}^{M-1} c_\ell z^\ell \tag{7.48}$$

188 Chapter Seven

The coefficients c_ℓ are again the sums of the $2M - 1$ matrix diagonals of $\bar{c}_1 \bar{c}_1^H$.

Example 7.14 Apply the root-MUSIC method to the Min-Norm method, where $\sigma_n^2 = .3$, $\theta_1 = -2°$, $\theta_2 = 4°$, and $M = 4$. Approximate the correlation matrices by again time averaging over $K = 300$ data points as was done in Example 7.12. Superimpose the plots of the Min-Norm pseudospectrum and the roots from root-Min-Norm and compare.

Solution The first column of the $\bar{C}$ matrix is given by

$$\bar{c}_1 = \begin{bmatrix} .19 \\ -.33 + .02i \\ -.06 + .04i \\ .2 - .05i \end{bmatrix}$$

We can calculate the matrix $\bar{c}_1 \bar{c}_1^H$ and find the polynomial coefficients by summing along the diagonals.

$$c = .04 - .01i, -.08 + .02i, -.06 - .01i, .1937, -.06 \\ + .01i, -.08 - .02i, .04 + .01i$$

We can use the root command in MATLAB to find the roots and then solve for the magnitude and angles of the $2(M-1) = 6$ roots. We can plot the location of all 6 roots showing which roots are closest to the unit circle as shown in Fig. 7.12. We can also superimpose the closest roots onto the plot of the Min-Norm pseudospectrum as shown in Fig. 7.13.

The Min-Norm pseudospectrum has much sharper resolution than MUSIC, but there is no indication of the AOA at $-2°$. However,

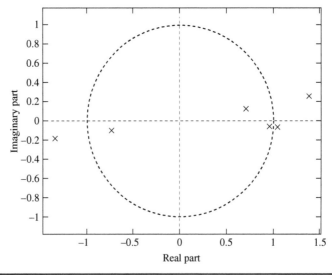

Figure 7.12 All 6 roots in Cartesian coordinates.

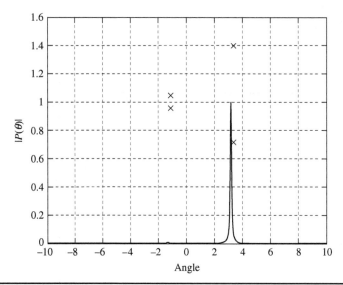

FIGURE 7.13 Min-Norm pseudospectrum and roots found with root-Min-Norm for $\theta_1 = -2°$ and $\theta_2 = 4°$.

the root-Min-Norm algorithm gives a fair indication of the location of both angles of arrival.

7.3.9 ESPRIT AOA Estimate

ESPRIT stands for *Estimation of Signal Parameters via Rotational Invariance Techniques* and was first proposed by Roy and Kailath [20] in 1989. Useful summaries of this technique are given by both Godara [1] and Liberti and Rappaport [21]. The goal of the ESPRIT technique is to exploit the rotational invariance in the signal subspace that is created by two arrays with a translational invariance structure. ESPRIT inherently assumes narrowband signals so that one knows the translational phase relationship between the multiple arrays to be used. As with MUSIC, ESPRIT assumes that there are $D < M$ narrow-band sources centered at the center frequency f_0. These signal sources are assumed to be of a sufficient range so that the incident propagating field is approximately planar. The sources can be either random or deterministic and the noise is assumed to be random with zero-mean. ESPRIT assumes multiple identical arrays called *doublets*. These can be separate arrays or can be composed of subarrays of one larger array. It is important that these arrays are displaced translationally but not rotationally. An example is shown in Fig. 7.14, where a 4-element linear array is composed of two identical 3-element subarrays or two doublets. These two subarrays are translationally displaced by the distance d. Let us label these arrays as array 1 and array 2.

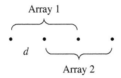

FIGURE 7.14 Doublet composed of two identical displaced arrays.

The signals induced on each of the arrays are given by

$$\bar{x}_1(k) = [\bar{a}_1(\theta_1) \ \bar{a}_1(\theta_2) \ \cdots \ \bar{a}_1(\theta_D)] \cdot \begin{bmatrix} s_1(k) \\ s_2(k) \\ \vdots \\ s_D(k) \end{bmatrix} + \bar{n}_1(k)$$

$$= \bar{A}_1 \cdot \bar{s}(k) + \bar{n}_1(k) \qquad (7.49)$$

and

$$\bar{x}_2(k) = \bar{A}_2 \cdot \bar{s}(k) + \bar{n}_2(k)$$

$$= \bar{A}_1 \cdot \bar{\Phi} \cdot \bar{s}(k) + \bar{n}_2(k) \qquad (7.50)$$

where $\bar{\Phi} = diag\{e^{jkd\sin\theta_1}, e^{jkd\sin\theta_2}, \ldots, e^{jkd\sin\theta_D}\}$
    $= $ a $D \times D$ diagonal unitary matrix with phase shifts between the doublets for each AOA
$\bar{A}_i = $ Vandermonde matrix of steering vectors for subarrays $i = 1, 2$

The complete received signal considering the contributions of both subarrays is given as

$$\bar{x}(k) = \begin{bmatrix} \bar{x}_1(k) \\ \bar{x}_2(k) \end{bmatrix} = \begin{bmatrix} \bar{A}_1 \\ \bar{A}_1 \cdot \bar{\Phi} \end{bmatrix} \cdot \bar{s}(k) + \begin{bmatrix} \bar{n}_1(k) \\ \bar{n}_2(k) \end{bmatrix} \qquad (7.51)$$

We can now calculate the correlation matrix for either the complete array or for the two subarrays. The correlation matrix for the complete array is given by

$$\bar{R}_{xx} = E[\bar{x} \cdot \bar{x}^H] = \bar{A}\bar{R}_{ss}\bar{A}^H + \sigma_n^2 \bar{I} \qquad (7.52)$$

whereas the correlation matrices for the two subarrays are given by

$$\bar{R}_{11} = E[\bar{x}_1 \cdot \bar{x}_1^H] = \bar{A}\bar{R}_{ss}\bar{A}^H + \sigma_n^2 \bar{I} \qquad (7.53)$$

and

$$\bar{R}_{22} = E[\bar{x}_2 \cdot \bar{x}_2^H] = \bar{A}\bar{\Phi}\bar{R}_{ss}\bar{\Phi}^H \bar{A}^H + \sigma_n^2 \bar{I} \qquad (7.54)$$

Each of the full rank correlation matrices given in Eqs. (7.53) and (7.54) has a set of eigenvectors corresponding to the D signals present. Creating the signal subspace for the two subarrays results in the two matrices $\bar{E}_1$ and $\bar{E}_2$. Creating the signal subspace for the entire array results in one signal subspace given by $\bar{E}_x$. Because of the invariance structure of the array, $\bar{E}_x$ can be decomposed into the subspaces $\bar{E}_1$ and $\bar{E}_2$.

Both $\bar{E}_1$ and $\bar{E}_2$ are $M \times D$ matrices whose columns are composed of the D eigenvectors corresponding to the largest eigenvalues of $\bar{R}_{11}$ and $\bar{R}_{22}$. Because the arrays are translationally related, the subspaces of eigenvectors are related by a unique nonsingular transformation matrix $\bar{\Psi}$ such that

$$\bar{E}_1 \bar{\Psi} = \bar{E}_2 \tag{7.55}$$

There must also exist a unique nonsingular transformation matrix $\bar{T}$ such that

$$\bar{E}_1 = \bar{A}\bar{T} \tag{7.56}$$

and

$$\bar{E}_2 = \bar{A}\bar{\Phi}\bar{T} \tag{7.57}$$

By substituting Eqs. (7.55) and (7.56) into Eq. (7.57) and assuming that $\bar{A}$ is of full rank, we can derive the relationship

$$\bar{T}\bar{\Psi}\bar{T}^{-1} = \bar{\Phi} \tag{7.58}$$

Thus, the eigenvalues of $\bar{\Psi}$ must be equal to the diagonal elements of $\bar{\Phi}$ such that $\lambda_1 = e^{jkd\sin\theta_1}$, $\lambda_2 = e^{jkd\sin\theta_2}$, ..., $\lambda_D = e^{jkd\sin\theta_D}$, and the columns of $\bar{T}$ must be the eigenvectors of $\bar{\Psi}$. $\bar{\Psi}$ is a rotation operator that maps the signal subspace $\bar{E}_1$ into the signal subspace $\bar{E}_2$. One is now left with the problem of estimating the subspace rotation operator $\bar{\Psi}$ and consequently finding the eigenvalues of $\bar{\Psi}$.

If we are restricted to a finite number of measurements and we also assume that the subspaces $\bar{E}_1$ and $\bar{E}_2$ are equally noisy, we can estimate the rotation operator $\bar{\Psi}$ using the *total least-squares* (TLS) criterion. Details of the TLS criterion can be found in van Huffel and Vandewalle [22]. This procedure is outlined as follows. (See Roy and Kailath [20].)

- Estimate the array correlation matrices $\bar{R}_{11}$, $\bar{R}_{22}$ from the data samples.
- Knowing the array correlation matrices for both subarrays, one can estimate the total number of sources by the number of large eigenvalues in either $\bar{R}_{11}$ or $\bar{R}_{22}$.

- Calculate the signal subspaces $\bar{E}_1$ and $\bar{E}_2$ based upon the signal eigenvectors of $\bar{R}_{11}$ and $\bar{R}_{22}$. For ULA, one can alternatively construct the signal subspaces from the entire array signal subspace $\bar{E}_x$. $\bar{E}_x$ is an $M \times D$ matrix composed of the signal eigenvectors. $\bar{E}_1$ can be constructed by selecting the first $M/2+1$ rows ($(M+1)/2+1$ for odd arrays) of $\bar{E}_x$. $\bar{E}_2$ can be constructed by selecting the last $M/2+1$ rows ($(M+1)/2+1$ for odd arrays) of $\bar{E}_x$.

- Next form a $2D \times 2D$ matrix using the signal subspaces such that

$$\bar{C} = \begin{bmatrix} \bar{E}_1^H \\ \bar{E}_2^H \end{bmatrix} [\bar{E}_1 \ \bar{E}_2] = \bar{E}_C \bar{\Lambda} \bar{E}_C^H \quad (7.59)$$

where the matrix $\bar{E}_C$ is from the *eigenvalue decomposition* (EVD) of $\bar{C}$ such that $\lambda_1 \geq \lambda_2 \geq \cdots \geq \lambda_{2D}$ and $\bar{\Lambda} = diag\{\lambda_1, \lambda_2, ..., \lambda_{2D}\}$

- Partition $\bar{E}_C$ into four $D \times D$ submatrices such that

$$\bar{E}_C = \begin{bmatrix} \bar{E}_{11} & \bar{E}_{12} \\ \bar{E}_{21} & \bar{E}_{22} \end{bmatrix} \quad (7.60)$$

- Estimate the rotation operator $\bar{\Psi}$ by

$$\bar{\Psi} = -\bar{E}_{12} \bar{E}_{22}^{-1} \quad (7.61)$$

- Calculate the eigenvalues of $\bar{\Psi}, \lambda_1, \lambda_2, ..., \lambda_D$.
- Now estimate the angles of arrival, given that $\lambda_i = |\lambda_i| e^{j \arg(\lambda_i)}$

$$\theta_i = \sin^{-1}\left(\frac{\arg(\lambda_i)}{kd}\right) \quad i = 1, 2, ..., D \quad (7.62)$$

If so desired, one can estimate the matrix of steering vectors from the signal subspace $\bar{E}_s$ and the eigenvectors of $\bar{\Psi}$ given by $\bar{E}_\psi$ such that $\hat{A} = \bar{E}_s \bar{E}_\psi$.

Example 7.15 Use the ESPRIT algorithm to predict the angles of arrival for an $M = 4$-element array where noise variance $\sigma_n^2 = .1$. Approximate the correlation matrices by again time averaging over $K = 300$ data points as was done in Example 7.12. The angles of arrival are $\theta_1 = -5°, \theta_2 = 10°$.

Solution The signal subspace for the entire ideal ULA array correlation matrix is given by

$$\bar{E}_x = \begin{bmatrix} .78 & .41+.1i \\ .12-.02i & .56 \\ -.22+.07i & .54-.05i \\ -.51+.25i & .46-.1i \end{bmatrix}$$

The two subarray signal subspaces can now be found by taking the first three rows of $\bar{E}_x$ to define $\bar{E}_1$ and the last three rows of $\bar{E}_x$ to define $\bar{E}_2$

$$\bar{E}_1 = \begin{bmatrix} .78 & .41+.1i \\ .12-.02i & .56 \\ -.22+.07i & .54-.05i \end{bmatrix} \quad \bar{E}_2 = \begin{bmatrix} .12-.02i & .56 \\ -.22+.07i & .54-.05i \\ -.51+.25i & .46-.1i \end{bmatrix}$$

Constructing the matrix of signal subspaces, we get

$$\bar{C} = \begin{bmatrix} \bar{E}_1^H \\ \bar{E}_2^H \end{bmatrix} [\bar{E}_1 \ \bar{E}_2] = \begin{bmatrix} .67 & .26+.06i & .2-.03i & .39 \\ .26-.06i & .78 & -.37+.12i & .78-.11i \\ .2+.03i & -.37-.12i & .4 & -.32-.08i \\ .39 & .78+.11i & -.32+.08i & .82 \end{bmatrix}$$

Performing the eigendecomposition, we can construct the matrix $\bar{E}_C$ such that

$$\bar{E}_C = \begin{bmatrix} \bar{E}_{11} & \bar{E}_{12} \\ \bar{E}_{21} & \bar{E}_{22} \end{bmatrix} = \begin{bmatrix} .31 & .8 & -.44+.1i & -22+.11i \\ .63-.09i & -.16 & .45 & -.61-.03i \\ -.26-.05i & .57+.1i & .75 & .15-.11i \\ .66 & .01+.02i & .07-.17i & .73 \end{bmatrix}$$

We can now calculate the rotation operator $\bar{\Psi} = -\bar{E}_{12}\bar{E}_{22}^{-1}$ given the rotation matrix

$$\bar{\Psi} = -\bar{E}_{12}\bar{E}_{22}^{-1} = \begin{bmatrix} .58-.079i & .2-.05i \\ -.67+.23i & .94-.11i \end{bmatrix}$$

Next we can calculate the eigenvalues of $\bar{\Psi}$ and solve for the angles of arrival using

$$\theta_1 = \sin^{-1}\left(\frac{\arg(\lambda_1)}{kd}\right) = -4.82°$$

$$\theta_2 = \sin^{-1}\left(\frac{\arg(\lambda_2)}{kd}\right) = 9.85°$$

7.4 References

1. Godara, L., "Application of Antenna Arrays to Mobile Communications, Part II: Beam-Forming and Direction-of-Arrival Considerations," *Proceedings of the IEEE*, Vol. 85, No. 8, pp. 1195–1245, Aug. 1997.
2. Capon, J., "High-Resolution Frequency-Wavenumber Spectrum Analysis," *Proceedings of the IEEE*, Vol. 57, No. 8, pp. 1408–1418, Aug. 1969.
3. Johnson, D., "The Application of Spectral Estimation Methods to Bearing Estimation Probems," *Proceedings of the IEEE*, Vol. 70, No. 9, pp. 1018–1028, Sept. 1982.
4. Van Trees, H., Optimum Array Processing: Part IV of Detection, Estimation, and Modulation Theory, Wiley Interscience, New York, 2002.
5. Stoica, P., and R. Moses, *Introduction to Spectral Analysis*, Prentice Hall, New York, 1997.
6. Shan, T. J., M. Wax, and T. Kailath, "Spatial Smoothing for Direction-of-Arrival Estimation of Coherent Signals," *IEEE Transactions on Acoustics, Speech, and Signal Processing*, Vol. ASSP-33, No. 4, pp. 806–811, Aug. 1985.
7. Minasyan, G., "Application of High Resolution Methods to Underwater Data Processing," Ph.D. Dissertation, N.N. Andreyev Acoustics Institute, Moscow, Oct. 1994 (In Russian).

8. Bartlett, M., *An Introduction to Stochastic Processes with Special References to Methods and Applications*, Cambridge University Press, New York, 1961.
9. Makhoul, J., "Linear Prediction: A Tutorial Review," *Proceedings of IEEE*, Vol. 63, pp. 561–580, 1975.
10. Burg, J. P., "Maximum Entropy Spectrum Analysis," Ph.D. dissertation, Dept. of Geophysics, Stanford University, Stanford CA, 1975.
11. Burg, J. P., "The Relationship between Maximum Entropy Spectra and Maximum Likelihood Spectra," *Geophysics*, Vol. 37, pp. 375–376, April 1972.
12. Barabell, A., "Improving the Resolution of Eigenstructure-Based Direction-Finding Algorithms," *Proceedings of ICASSP*, Boston, MA, pp. 336–339, 1983.
13. Pisarenko, V. F., "The Retrieval of Harmonics from a Covariance Function," *Geophysical Journal of the Royal Astronomical Society*, Vol. 33, pp. 347–366, 1973.
14. Johnson, D., and D. Dudgeon, *Array Signal Processing Concepts and Techniques*, Prentice Hall Signal Processing Series, New York, 1993.
15. Reddi, S. S., "Multiple Source Location—A Digital Approach," *IEEE Transactions on AES*, Vol. 15, No. 1, Jan. 1979.
16. Kumaresan, R., and D. Tufts, "Estimating the Angles of Arrival of Multiple Plane Waves," *IEEE Transactions on AES*, Vol. AES-19, pp. 134–139, 1983.
17. Ermolaev, V., and A. Gershman, "Fast Algorithm for Minimum-Norm Direction-of-Arrival Estimation," *IEEE Transactions on Signal Processing*, Vol. 42, No. 9, Sept. 1994.
18. Schmidt, R., "Multiple Emitter Location and Signal Parameter Estimation," *IEEE Transactions on Antenna Propogation*, Vol. AP-34, No. 2, pp. 276–280, March 1986.
19. Ren, Q., and A. Willis, "Fast Root-MUSIC Algorithm," *IEE Electronics Letters*, Vol. 33, No. 6, pp. 450–451, March 1997.
20. Roy, R., and T. Kailath, "ESPRIT—Estimation of Signal Parameters via Rotational Invariance Techniques," *IEEE Transactions on ASSP*, Vol. 37, No. 7, pp. 984–995, July 1989.
21. Liberti, J., and T. Rappaport, *Smart Antennas for Wireless Communications*, Prentice Hall, New York, 1999.
22. van Huffel, S., and J. Vandewalle, "The Total Least Squares Problem: Computational Aspects and Analysis," SIAM, Philadelphia, PA, 1991.

7.5 Problems

7.1 For the two matrices $\bar{A} = \begin{bmatrix} 1 & 2 \\ 3 & 4 \end{bmatrix}$ $\bar{B} = \begin{bmatrix} 1 & 1 \\ 5 & 2 \end{bmatrix}$

(a) What is $Trace(\bar{A})$ and $Trace(\bar{B})$?
(b) Show that $(\bar{A} \cdot \bar{B})^T = \bar{B}^T \cdot \bar{A}^T$.

7.2 For the two matrices $\bar{A} = \begin{bmatrix} 1 & 2j \\ 2j & 2 \end{bmatrix}$ $\bar{B} = \begin{bmatrix} j & 1 \\ 1 & 2j \end{bmatrix}$, show that $(\bar{A} \cdot \bar{B})^H = \bar{B}^H \cdot \bar{A}^H$.

7.3 For $\bar{A} = \begin{bmatrix} 1 & 2 \\ 3 & 4 \end{bmatrix}$

(a) Find the $\bar{A}^{-1}$ by hand.
(b) Find the $\bar{A}^{-1}$ using MATLAB.

7.4 For $\bar{A} = \begin{bmatrix} 1 & 2 \\ 3 & 1 \end{bmatrix}$

(a) Solve for all eigenvalues by hand using Eq. (7.16).
(b) Solve for all eigenvectors by hand using Eq. (7.17).
(c) Solve for all eigenvectors and eigenvalues using MATLAB.

7.5 Repeat Prob. 7.4 for $\bar{A} = \begin{bmatrix} 2 & 4 \\ .25 & 2 \end{bmatrix}$.

7.6 For $\bar{A} = \begin{bmatrix} 1 & 2 & 3 \\ 4 & 5 & 6 \\ 7 & 8 & 9 \end{bmatrix}$

(a) What are all eigenvalues and eigenvectors using MATLAB?
(b) Which eigenvalues are associated with which eigenvectors?
(c) Use Eq. (7.17), and by hand, prove that the first eigenvalue and the first eigenvector satisfy this equation.

7.7 An $N = 3$-element array exists with element spacing $d = \lambda/2$. One arriving signal is defined as $s_1(k) = .1, .2, .3; (k=1, 2, 3)$ arrives at the angle $\theta_1 = 0°$. The other signal is defined as $s_2(k) = .3, .4, .5; (k=1, 2, 3)$ arrives at the angle $\theta_2 = 30°$. The noise has a standard deviation $\sigma = .1$. Use Eqs. (7.15), (7.16), and (7.17). Use MATLAB.

(a) What are the array steering vectors $\bar{a}(\theta_1), \bar{a}(\theta_2)$?
(b) What is the matrix of steering vectors $\bar{A}$?
(c) What are the correlation matrices $\bar{R}_{ss}, \bar{R}_{nn}, \bar{R}_{xx}$?
(d) What are the eigenvalues and eigenvectors of $\bar{R}_{xx}$?

7.8 Repeat parts c and d of Prob. 7.7 using estimates on the covariance matrices ($\hat{R}_{ss}, \hat{R}_{nn}, \hat{R}_{xx}$). Allow the noise to be defined in MATLAB as n = 0.1*randn(3, 3);. Thus the noise is Gaussian distributed, but we only can work with three time samples. Are the eigenvalues and eigenvectors similar to those found in Prob. 7.7. Why or why not?

7.9 Plot the normalized Bartlett pseudospectrum $P_B(\theta)$ using MATLAB, for the case where $M = 7, d = \lambda/2, \sigma_n^2 = .2$, and $\theta_1 = 3°, \theta_2 = -3°$. Set the vertical scale to be −30 to 5 dB and horizontal scale to be −15° to 15°. Allow the signals s_1 and s_2 to be uncorrelated resulting in $\bar{R}_{ss} = \begin{bmatrix} 1 & 0 \\ 0 & 1 \end{bmatrix}$.

7.10 Plot the normalized Capon pseudospectrum $P_C(\theta)$ using MATLAB, for the case where $M = 7, d = \lambda/2, \sigma_n^2 = .2$, and $\theta_1 = 3°, \theta_2 = -3°$. Set the vertical scale to be −30 to 5 dB and horizontal scale to be −15° to 15°. Allow the signals s_1 and s_2 to be uncorrelated resulting in $\bar{R}_{ss} = \begin{bmatrix} 1 & 0 \\ 0 & 1 \end{bmatrix}$.

7.11 Plot the normalized linear prediction pseudospectrum $P_{LP4}(\theta)$ using MATLAB, for the case where $M = 7, d = \lambda/2, \sigma_n^2 = .2$, and $\theta_1 = 3°, \theta_2 = -3°$. Set the vertical scale to be −30 to 5 dB and horizontal scale to be −15° to 15°. Allow the signals s_1 and s_2 to be uncorrelated resulting in $\bar{R}_{ss} = \begin{bmatrix} 1 & 0 \\ 0 & 1 \end{bmatrix}$.

7.12 Plot the normalized maximum entropy pseudospectrum $P_{ME4}(\theta)$ using MATLAB, for the case where $M = 7, d = \lambda/2, \sigma_n^2 = .2$, and $\theta_1 = 3°, \theta_2 = -3°$. Set the vertical scale to be −30 to 5 dB and horizontal scale to be −15° to 15°. Allow the signals s_1 and s_2 to be uncorrelated resulting in $\bar{R}_{ss} = \begin{bmatrix} 1 & 0 \\ 0 & 1 \end{bmatrix}$.

7.13 For the Pisarenko harmonic decomposition pseudospectrum $P_{PHD}(\theta)$, where $M = 7, d = \lambda/2, \sigma_n^2 = .2$, and $\theta_1 = 3°, \theta_2 = -3°$. Set the vertical scale to be −30 to 5 dB and horizontal scale to be −15° to 15°. Allow the signals s_1 and s_2 to be uncorrelated resulting in $\bar{R}_{ss} = \begin{bmatrix} 1 & 0 \\ 0 & 1 \end{bmatrix}$.

(a) What is the smallest eigenvalue?
(b) What is the eigenvector associated with that eigenvalue?
(c) Plot the normalized pseudospectrum using MATLAB.

7.14 For the Min-Norm decomposition pseudospectrum $P_{MN}(\theta)$, where $M=7$, $d = \lambda/2$, $\sigma_n^2 = .2$, and $\theta_1 = 3°$, $\theta_2 = -3°$, set the vertical scale to be -30 to 5 dB and horizontal scale to be $-15°$ to $15°$. Allow the signals s_1 and s_2 to be uncorrelated resulting in $\bar{R}_{ss} = \begin{bmatrix} 1 & 0 \\ 0 & 1 \end{bmatrix}$.

(a) What is the signal subspace $\bar{E}_S$?
(b) What is the noise subspace $\bar{E}_N$?
(c) Plot the normalized pseudospectrum using MATLAB.

7.15 For the MUSIC decomposition pseudospectrum $P_{MUSIC}(\theta)$, where $M=7$, $d = \lambda/2$, $\sigma_n^2 = .2$, and $\theta_1 = 3°$, $\theta_2 = -3°$, set the vertical scale to be -30 to 5 dB and horizontal scale to be $-15°$ to $15°$. Allow the signals s_1 and s_2 to be uncorrelated resulting in $\bar{R}_{ss} = \begin{bmatrix} 1 & 0 \\ 0 & 1 \end{bmatrix}$.

(a) What is the signal subspace $\bar{E}_S$?
(b) What is the noise subspace $\bar{E}_N$?
(c) Plot the normalized pseudospectrum using MATLAB.

7.16 Repeat Example 7.12, using $M = 7$, $d = \lambda/2$, $\sigma_n^2 = .2$, and $\theta_1 = 3°$, $\theta_2 = -3°$, set the vertical scale to be -30 to 5 dB and horizontal scale to be $-15°$ to $15°$.

7.17 For an $M = 3$-element array with $d = \lambda/2$, $\sigma_n^2 = .2$, and $\theta_1 = 3°$, $\theta_2 = -3°$ and given that the signals s_1 and s_2 are uncorrelated, apply the root-MUSIC method.

(a) What is the matrix $\bar{C}$?
(b) What are the coefficients c_ℓ using Eq. (7.47)?
(c) What are the roots z_i using Eq. (7.49)?
(d) What are the angles θ_i using Eq. (7.50)?

7.18 For an $M = 3$-element array with $d = \lambda/2$, $\sigma_n^2 = .2$, and $\theta_1 = 3°$, $\theta_2 = -3°$ and given that the signals s_1 and s_2 are uncorrelated apply the root-Min-Norm method.

(a) What is the matrix $\bar{C}$?
(b) What is the first column $\bar{c}_1$?
(c) What are the coefficients c_ℓ using Eq. (7.47)?
(d) What are the roots z_i using Eq. (7.49)?
(e) What are the angles θ_i using Eq. (7.50)?

7.19 For an $M = 4$-element array with $d = \lambda/2$, $\sigma_n^2 = .2$, and $\theta_1 = 3°$, $\theta_2 = -3°$ and given that the signals s_1 and s_2 are uncorrelated, approximate the correlation matrices by time averaging over $K = 300$ data points as is done in Example 7.12. Apply the ESPRIT method.

(a) What are the correlation matrices $\bar{R}_{11}$, $\bar{R}_{22}$?
(b) Find the signal subspaces $\bar{E}_1$, $\bar{E}_2$.
(c) What is the matrix $\bar{E}_C$?
(d) What is the rotation operator $\bar{\Psi} = -\bar{E}_{12}\bar{E}_{22}^{-1}$?
(e) What are the eigenvalues of $\bar{\Psi}$?
(f) What are the predicted angles of arrival?

CHAPTER 8
Smart Antennas

8.1 Introduction

Traditional array antennas, where the main beam is steered to directions of interest, are called *phased arrays, beamsteered arrays*, or *scanned arrays*. The beam is steered via phase shifters and in the past these phase shifters were often implemented at RF frequencies. This general approach to phase shifting has been referred to as electronic beamsteering because of the attempt to change the phase of the current directly at each antenna element.

Modern beamsteered array antennas, where the pattern is shaped according to certain optimum criteria, are called *smart antennas*. Smart antennas have alternatively been called *digital beamformed* (DBF) arrays or *adaptive arrays* (when adaptive algorithms are employed). The term *smart* implies the use of signal processing in order to shape the beam pattern according to certain conditions. For an array to be *smart* implies sophistication beyond merely steering the beam to a direction of interest. Smart essentially means computer control of the antenna performance. Smart antennas hold the promise for improved radar systems, improved system capacities with mobile wireless, and improved wireless communications through the implementation of *space division multiple access* (SDMA).

Smart antenna patterns are controlled via algorithms based on certain criteria. These criteria could be maximizing the *signal-to interference ratio* (SIR), minimizing the variance, minimizing the *meansquare error* (MSE), steering toward a signal of interest, nulling the interfering signals, or tracking a moving emitter to name a few. The implementation of these algorithms can be performed *electronically* through analog devices but it is generally more easily performed using digital signal processing. This requires that the array outputs be digitized through the use of an A/D converter. This digitization can be performed at either IF or baseband frequencies. Since an antenna pattern (or beam) is formed by digital signal processing, this process is often referred to as *digital beamforming*. Figure 8.1 contrasts a traditional electronically steered array with a DBF array or smart antenna.

Chapter Eight

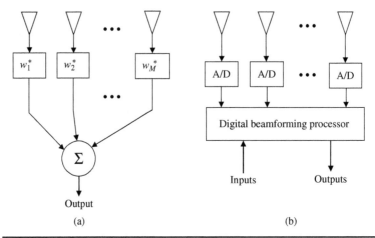

FIGURE 8.1 (a) Analog beamforming, (b) digital beamforming.

When the algorithms used are adaptive algorithms, this process is referred to as *adaptive beamforming*. Adaptive beamforming is a subcategory under the more general subject of digital beamforming. Digital beamforming has been applied to radar systems [1–5], sonar systems [6], and communications systems [7] to name a few. The chief advantage of digital beamforming is that phase shifting and array weighting can be performed on the digitized data rather than by being implemented in hardware. On receive, the beam is formed in the data processing rather than literally being forming in space. The digital beamforming method cannot be strictly called electronic steering since no effort is made to directly shift the phase of the antenna element currents. Rather, the *phase shifting* is computationally performed on the digitized signal. If the parameters of operation are changed or the detection criteria are modified, the beamforming can be changed by simply changing an algorithm rather than by replacing hardware.

Adaptive beamforming is generally the more useful and effective beamforming solution because the digital beamformer merely consists of an algorithm which dynamically *optimizes* the array pattern according to the changing electromagnetic environment.

Conventional array *static* processing systems are subject to degradation by various causes. The array SNR can be severely degraded by the presence of unwanted interfering signals, electronic countermeasures, clutter returns, reverberation returns (in acoustics), or multipath interference and fading. An *adaptive* array system consists of the antenna array elements terminated in an adaptive processor which is designed to specifically maximize certain criteria. As the emitters move or change, the adaptive array updates and compensates iteratively in order to track the changing environment. Many current modern radar systems still rely on older electronic scanning technologies. Recent efforts are being exerted to modify radar systems to include digital beamforming and adaptive beamforming techniques [4]. Though current modern mobile base stations tend to use older fixed beam technologies to satisfy SDMA, they also would benefit from the use of modern adaptive methods and thereby increase system capacities [8].

This chapter reviews the historical development of digital beamforming and adaptive arrays, explain some basic DBF principles, and covers the more popular DBF methods that include adaptive methods.

8.2 The Historical Development of *Smart Antennas*

"The development of adaptive arrays began in the late 1950s and has been more than four decades in the making." The word *adaptive array* was first coined by Van Atta [9], in 1959, to describe a *self-phased* array. Self-phased arrays merely reflected all incident signals back in the direction of arrival by using clever phasing schemes based upon phase conjugation. Because of the redirection of incident planewaves, these arrays were called *retrodirective arrays* (see an in-depth explanation of retrodirective arrays in Chap. 4, Sec. 4.9). Self-phased arrays are instantaneously adaptive arrays since they essentially reflect the incident signal in a similar fashion to the classic corner reflector.

Phase-locked loop (PLL) systems were incorporated into arrays in the 60s in an effort to construct better retrodirective arrays since it was assumed that retrodirection was the best approach [10]. PLLs still are used in single beam scanning systems [11].

Adaptive *sidelobe cancellation* (SLC) was first proposed by Howells [12, 13] in 1959. This technique allowed for interference nulling, thus raising the SIR. The Howells SLC was the first adaptive scheme which allowed for automatic interference nulling. By maximizing the generalized *signal-to-noise ratio* (SNR), Applebaum developed the algorithm governing adaptive interference cancellation [14, 15]. His algorithm became known as the *Howells-Applebaum algorithm*. At the same time, through the use of *least mean squares* (LMS), Widrow and others applied self-training to adaptive arrays [16, 17]. The Howells-Applebaum and *Widrow algorithms* both are steepest-descent/gradient-search methods. The Howells-Applebaum and Widrow algorithms both converge to the optimum Wiener solution. The convergence of these methods is dependent on the eigenvalue spread [18] such that larger spreads require longer convergence times. The convergence time constant is given by [19]

$$\tau_i = \frac{1}{2\mu\lambda_i} \qquad (8.1)$$

where μ = gradient step size
λ_i = i^{th} eigenvalue

The eigenvalues are derived from the correlation matrix where the largest eigenvalues correspond to the strongest signals and the smallest eigenvalues correspond to the weakest signals or noise. The larger eigenvalue spreads result in longer convergence times. In the case of SLC, the weakest interfering signals are cancelled last as can be seen from Eq. (8.1).

Because the convergence of SLC algorithm was slow for large eigenvalue spreads, Reed et al. developed the direct *sample matrix inversion* (SMI) technique in 1974 [20].

The next great complementary advance in adaptive arrays came with the application of spectral estimation methods to array processing. (Many of these spectral estimation methods are discussed at length in Chap. 7.) The spectral estimation methods, achieving higher angular resolutions, have come to be known as *supperresolution* algorithms [18]. Capon, in 1969, used a *maximum likelihood* (ML) method to solve for the *minimum variance distortionless response* (MVDR) of an array. His solution maximizes the SIR [21]. Additionally, the linear predictive method was used to minimize the mean-squared prediction error leading to optimum array weights. The array weights are dependent on the array correlation matrix and are given in Eq. (7.22) [22]. In 1972, Burg applied the maximum entropy method to spectral estimation and his technique was soon adapted to array signal processing [23, 24]. In 1973, Pisarenko developed the harmonic decomposition technique based on minimizing the MSE under the constraint that the norm of the weight vector be equal to unity [25]. The minimum-norm (Min-Norm) method was developed by Reddi [26] in 1979 and Kumaresan and Tufts [27] in 1983. The Min-Norm algorithm optimizes the weight vector by solving the optimization problem where the weight vector is orthogonal to the signal eigenvector subspace. This too is a spectral estimation problem applied to array signal processing. The now famous MUSIC algorithm was developed by Schmidt in 1986 [28]. MUSIC is a spectral estimation-based algorithm that exploits the orthogonality of the noise subspace with the array correlation matrix. The estimation of signal parameters via rotational iavariance (ESPRIT) technique was first proposed by Roy and Kailath [29] in 1989. The goal of ESPRIT is to exploit the rotational invariance in the signal subspace which is created by two arrays with a translational invariance structure.

This latter group of adaptive methods can be considered to be part of the array *superresolution* methods allowing the user to achieve higher resolutions than that allowed by the beamwidth of the array. The price of increased resolution comes at the cost of greater computational intensity.

Let us review some of the basics for fixed and adaptive algorithms.

8.3 Fixed Weight Beamforming Basics

8.3.1 Maximum Signal-to-Interference Ratio

One criterion that can be applied to enhancing the received signal and minimizing the interfering signals is based upon maximizing the SIR [7, 30]. Before we engage in the rigorous derivation of the SIR optimization, let us review a heuristic approach which was used in Chap. 4. It is intuitive that if we can cancel all interference by placing nulls at their angles of arrival, we will automatically maximize the SIR.

Repeating the basic development of Chap. 4, let us assume an $M = 3$-element array with one fixed known desired source and two fixed undesired interferers. All signals are assumed to operate at the same carrier frequency. Let us assume a 3-element array with the desired signal and interferers as shown in Fig. 8.2.

The array vector is given by

$$\bar{a} = [e^{-jkd\sin\theta} \quad 1 \quad e^{jkd\sin\theta}]^T \tag{8.2}$$

The, as yet to be determined, array weights for optimization are given by

$$\bar{w}^H = [w_1 \quad w_2 \quad w_3] \tag{8.3}$$

Therefore, the general total array output is given as

$$y = \bar{w}^H \cdot \bar{a} = w_1 e^{-jkd\sin\theta} + w_2 + w_3 e^{jkd\sin\theta} \tag{8.4}$$

The array output for the desired signal will be designated by y_s whereas the array output for the interfering or undesired signals will be designated by y_1 and y_2. Because there are three unknown weights, there must be three conditions satisfied.

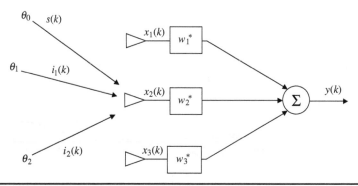

FIGURE 8.2 3-element array with desired and interfering signals.

Condition 1: $y_s = \overline{w}^H \cdot \overline{a}_0 = w_1 e^{-jkd\sin\theta_0} + w_2 + w_3 e^{jkd\sin\theta_0} = 1$

Condition 2: $y_1 = \overline{w}^H \cdot \overline{a}_1 = w_1 e^{-jkd\sin\theta_1} + w_2 + w_3 e^{jkd\sin\theta_1} = 0$

Condition 3: $y_2 = \overline{w}^H \cdot \overline{a}_2 = w_1 e^{-jkd\sin\theta_2} + w_2 + w_3 e^{jkd\sin\theta_2} = 0$

Condition 1 demands that $y_s = 1$ for the desired signal, thus allowing the desired signal to be received without modification. Conditions 2 and 3 reject the undesired interfering signals. These conditions can be recast in matrix form as

$$\overline{w}^H \cdot \overline{A} = \overline{u}_1^T \tag{8.5}$$

where $\overline{A} = [\overline{a}_0 \ \overline{a}_1 \ \overline{a}_2]$ = matrix of steering vectors
$\overline{u}_1 = [1 \ 0 \ \cdots \ 0]^T$ = Cartesian basis vector

One can invert the matrix to find the required complex weights w_1, w_2, and w_3 by using

$$\overline{w}^H = \overline{u}_1^T \cdot \overline{A}^{-1} \tag{8.6}$$

As an example, if the desired signal is arriving from $\theta_0 = 0°$ while $\theta_1 = -45°$ and $\theta_2 = 60°$, the necessary weights can be calculated to be

$$\begin{bmatrix} w_1^* \\ w_2^* \\ w_3^* \end{bmatrix} = \begin{bmatrix} .28 - .07i \\ .45 \\ .28 + .07i \end{bmatrix} \tag{8.7}$$

The array factor is shown plotted in Fig. 8.3.

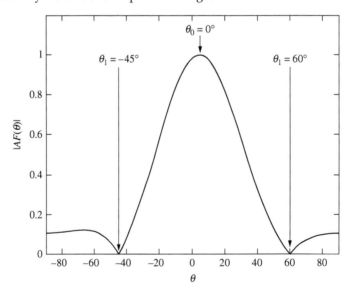

FIGURE 8.3 Sidelobe cancellation.

The Cartesian basis vector in Eq. (8.6) indicates that the array weights are taken from the first row of $\bar{A}^{-1}$.

The previous development is predicated on the fact that the desired signal and the total of the interfering signals make $\bar{A}$ an invertible square matrix. $\bar{A}$ must be an $M \times M$ matrix with M-array elements and M-arriving signals. In the case where the number of interferers is less than $M - 1$, Godara [19] has provided an equation which gives an estimate of the weights. However, his formulation requires noise be added in the system because the matrix inversion will be singular otherwise. Using the Godara method, we have

$$\bar{w}^H = \bar{u}_1^T \cdot \bar{A}^H (\bar{A} \cdot \bar{A}^H + \sigma_n^2 \bar{I})^{-1} \tag{8.8}$$

where $\bar{u}_1^T$ is the Cartesian basis vector whose length equals the total number of sources.

Example 8.1 For an $M = 5$-element array with $d = \lambda/2$, the desired signal arriving at $\theta = 0°$ and one interferer arrives at $-15°$ whereas the other interferer arrives at $+25°$. If the noise variance is $\sigma_n^2 = .001$ use the array weight estimation found in Eq. (8.8) to find the weights and plot.

Solution The problem is solved in MATLAB using sa_ex8_1.m. The matrix of steering vectors is given as

$$\bar{A} = [\bar{a}_0 \ \bar{a}_1 \ \bar{a}_2]$$

where

$$\bar{a}_0 = [1 \ 1 \ 1 \ 1 \ 1]^T$$
$$\bar{a}_n = [e^{-j2\pi \sin \theta_n} \ e^{-j\pi \sin \theta_n} \ 1 \ e^{j\pi \sin \theta_n} \ e^{j2\pi \sin \theta_n}]^T \quad n = 1, 2$$

Because only three sources are present, $\bar{u}_1 = [1 \ 0 \ 0]^T$. Substituting into Eq. (8.8)

$$\bar{w}^H = \bar{u}_1^T \cdot \bar{A}^H (\bar{A} \cdot \bar{A}^H + \sigma_n^2 \bar{I})^{-1} = \begin{bmatrix} .26 + .11i \\ .17 + .08i \\ .13 \\ .17 - .08i \\ .26 - .11i \end{bmatrix}^T$$

The plot of the array factor is shown in Fig. 8.4. The advantage of the Godara method is that the total number of sources can be less than the number of array elements.

This basic sidelobe canceling scheme works through an intuitive application of the array steering vector for the desired signal and interfering signals. However, by formally maximizing the SIR, we can derive the analytic solution for all arbitrary cases. We will closely follow the derivation given in both [7, 30].

The general nonadaptive conventional narrowband array is shown in Fig. 8.5.

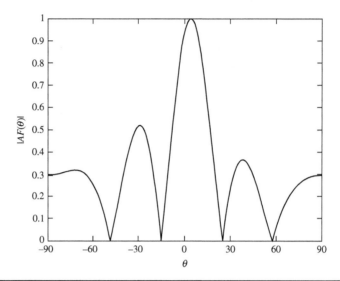

Figure 8.4 Array pattern with approximate nulls at −15° and 25°.

Figure 8.5 shows one desired signal arriving from the angle θ_0 and N interferers arriving from angles $\theta_1, \ldots, \theta_N$. The signal and the interferers are received by an array of M elements with M potential weights. Each received signal at element m also includes additive Gaussian noise. Time is represented by the kth time sample. Thus, the weighted array output y can be given in the following form:

$$y(k) = \bar{w}^H \cdot \bar{x}(k) \tag{8.9}$$

where

$$\bar{x}(k) = \bar{a}_0 s(k) + [\bar{a}_1 \quad \bar{a}_2 \quad \cdots \quad \bar{a}_N] \begin{bmatrix} i_1(k) \\ i_2(k) \\ \vdots \\ i_N(k) \end{bmatrix} + \bar{n}(k)$$

$$= \bar{x}_s(k) + \bar{x}_i(k) + \bar{n}(k) \tag{8.10}$$

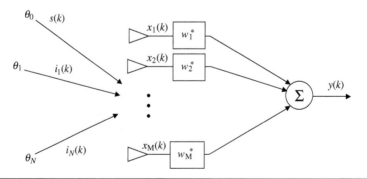

Figure 8.5 Traditional narrowband array.

with

$\bar{w} = [w_1 w_2 \cdots w_M]^T$ array weights
$\bar{x}_s(k)$ = desired signal vector
$\bar{x}_i(k)$ = interfering signals vector
$\bar{n}(k)$ = zero mean Gaussian noise for each channel
$\bar{a}_i$ = M-element array steering vector for the θ_i direction of arrival

We may rewrite Eq. (8.9), using the expanded notation in Eq. (8.10)

$$y(k) = \bar{w}^H \cdot [\bar{x}_s(k) + \bar{x}_i(k) + \bar{n}(k)] = \bar{w}^H \cdot [\bar{x}_s(k) + \bar{u}(k)] \qquad (8.11)$$

where

$$\bar{u}(k) = \bar{x}_i(k) + \bar{n}(k) = \text{undesired signal}$$

It is initially assumed that all arriving signals are monochromatic and the total number of arriving signals $N+1 \le M$. It is understood that the arriving signals are time varying, and thus our calculations are based on k-time snapshots of the incoming signal. Obviously, if the emitters are moving, the matrix of steering vectors is changing with time and the corresponding arrival angles are changing. Unless otherwise stated, the time dependence will be suppressed in Eqs. (8.9) through (8.11).

We can calculate the array correlation matrices for both the desired signal ($\bar{R}_{ss}$) and the undesired signals ($\bar{R}_{uu}$). The literature often calls these matrices the *array covariance* matrices. However, the covariance matrix is a mean removed correlation matrix. Since we do not generally know the statistical mean of the system noise or the front end detector output, it is best to label all $\bar{R}$ matrices as correlation matrices. If the process is ergodic and the time average is utilized, the correlation matrices can be defined with the time average notation as $\hat{R}_{ss}$ and $\hat{R}_{uu}$.

The weighted array output power for the desired signal is given by

$$\sigma_s^2 = E\left[|\bar{w}^H \cdot \bar{x}_s|^2\right] = \bar{w}^H \cdot \bar{R}_{ss} \cdot \bar{w} \qquad (8.12)$$

where

$$\bar{R}_{ss} = E[\bar{x}_s \bar{x}_s^H] = \text{signal correlation matrix}$$

The weighted array output power for the undesired signals is given by

$$\sigma_u^2 = E\left[|\bar{w}^H \cdot \bar{u}|^2\right] = \bar{w}^H \cdot \bar{R}_{uu} \cdot \bar{w} \qquad (8.13)$$

where it can be shown that

$$\bar{R}_{uu} = \bar{R}_{ii} + \bar{R}_{nn} \qquad (8.14)$$

with

$$\bar{R}_{ii} = \text{correlation matrix for interferers}$$

$$\bar{R}_{nn} = \text{correlation matrix for noise}$$

The SIR is defined as the ratio of the desired signal power divided by the undesired signal power.

$$SIR = \frac{\sigma_s^2}{\sigma_u^2} = \frac{\bar{w}^H \cdot \bar{R}_{ss} \cdot \bar{w}}{\bar{w}^H \cdot \bar{R}_{uu} \cdot \bar{w}} \quad (8.15)$$

The SIR can be maximized in Eq. (8.15) by taking the derivative with respect to $\bar{w}$ and setting the result equal to zero. This optimization procedure is outlined in Harrington [31]. Rearranging terms, we can derive the following relationship

$$\bar{R}_{ss} \cdot \bar{w} = SIR \cdot \bar{R}_{uu} \cdot \bar{w} \quad (8.16)$$

or

$$\bar{R}_{uu}^{-1} \bar{R}_{uu} \cdot \bar{w} = SIR \cdot \bar{w} \quad (8.17)$$

Equation (8.17) is an eigenvector equation with SIR being the eigenvalues. The maximum SIR (SIR_{max}) is equal to the largest eigenvalue λ_{max} for the Hermitian matrix $\bar{R}_{uu}^{-1} \bar{R}_{ss}$. The eigenvector associated with the largest eigenvalue is the optimum weight vector $\bar{w}_{opt}$. Thus

$$\bar{R}_{uu}^{-1} \bar{R}_{ss} \cdot \bar{w}_{SIR} = \lambda_{max} \cdot \bar{w}_{opt} = SIR_{max} \cdot \bar{w}_{SIR} \quad (8.18)$$

Because the correlation matrix is defined as $\bar{R}_{ss} = E[|s|^2] \bar{a}_0 \cdot \bar{a}_0^H$, we can pose the weight vector in terms of the optimum Wiener solution.

$$\bar{w}_{SIR} = \beta \cdot \bar{R}_{uu}^{-1} \cdot \bar{a}_0 \quad (8.19)$$

where

$$\beta = \frac{E[|s|^2]}{SIR_{max}} \bar{a}_0^H \cdot \bar{w}_{SIR} \quad (8.20)$$

Although Eq. (8.19) casts the weight vector in the optimum Wiener solution form, the weight vector is already known as the eigenvector found in Eq. (8.18).

Example 8.2 The $M = 3$-element array with spacing $d = .5\lambda$ has a noise variance $\sigma_n^2 = .001$, a desired received signal arriving at $\theta_0 = 30°$, and two interferers arriving at angles $\theta_1 = -30°$ and $\theta_2 = 45°$. Assume that the signal and interferer amplitudes are constant. Use MATLAB to calculate SIR_{max}, the normalized weights from Eq. (8.18), and plot the resulting pattern.

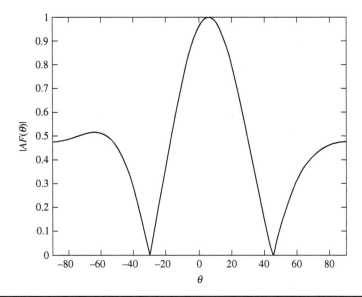

FIGURE 8.6 Maximum SIR pattern.

Solution Based on the incident angles of arrival for the desired signal and interferers along with the array vector $\bar{a}$, we can find the correlation matrices of the signal and undesired signals as

$$\bar{R}_{ss} = \begin{bmatrix} 1 & i & -1 \\ -i & 1 & i \\ -1 & -i & 1 \end{bmatrix}$$

$$\bar{R}_{uu} = \begin{bmatrix} 2.001 & -.61-.20i & -1.27-.96i \\ -.61+.20i & 2.001 & -.61-.20i \\ -1.27+.96i & -.61+.20i & 2.001 \end{bmatrix}$$

The largest eigenvalue for Eq. (8.18) is given in MATLAB as

$$SIR_{max} = \lambda_{max} = 679$$

The array weights are arbitrarily normalized by the center weight value. Thus

$$\bar{w}_{SIR} = \begin{bmatrix} 1.48+.5i \\ 1 \\ 1.48-.5i \end{bmatrix}$$

The derived pattern is shown plotted in Fig. 8.6.

8.3.2 Minimum Mean-Square Error

An alternative means for optimizing the array weights is found by minimizing the MSE. Figure 8.5 must be modified in such a way as to minimize the error while iterating the array weights. The modified array configuration is shown in Fig. 8.7.

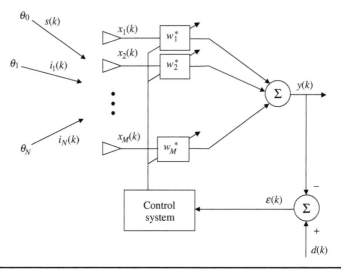

FIGURE 8.7 MSE adaptive system.

The signal $d(k)$ is the reference signal. Preferably the reference signal is either identical to the desired signal $s(k)$ or it is highly correlated with $s(k)$ and uncorrelated with the interfering signals $i_n(k)$. If $s(k)$ is not distinctly different from the interfering signals, the minimum mean-square technique will not work properly. The signal $\varepsilon(k)$ is the error signal such that

$$\varepsilon(k) = d(k) - \bar{w}^H \bar{x}(k) \tag{8.21}$$

Through some simple algebra, it can be shown that

$$|\varepsilon(k)|^2 = |d(k)|^2 - 2d(k)\bar{w}^H \bar{x}(k) + \bar{w}^H \bar{x}(k)\bar{x}^H(k)\bar{w} \tag{8.22}$$

For the purposes of simplification, we will suppress the time dependence notation k. Taking the expected value of both sides and simplifying the expression we get

$$E[|\varepsilon|^2] = E[|d|^2] - 2\bar{w}^H \bar{r} + \bar{w}^H \bar{R}_{xx} \bar{w} \tag{8.23}$$

where the following correlations are defined:

$$\bar{r} = E[d^* \cdot \bar{x}] = E[d^* \cdot (\bar{x}_s + \bar{x}_i + \bar{n})] \tag{8.24}$$

$$\bar{R}_{xx} = E[\bar{x}\bar{x}^H] = \bar{R}_{ss} + \bar{R}_{uu} \tag{8.25}$$

$$\bar{R}_{ss} = E[\bar{x}_s \bar{x}_s^H] \tag{8.26}$$

$$\bar{R}_{uu} = \bar{R}_{ii} + \bar{R}_{nn} \tag{8.27}$$

The expression in Eq. (8.23) is a quadratic function of the weight vector $\bar{w}$. This function is sometimes called the *performance surface* or *cost function* $J(\bar{w})$ and forms a quadratic surface in M-dimensional space. Because the optimum weights provide the minimum MSE, the extremum is the minimum of this function. A trivial example is given for a 2-element array that produces a two-dimensional surface as depicted in Fig. 8.8. As the desired angle of arrival (AOA) changes with time, the quadratic surface minimum changes with time in the weight plane.

In general, for an arbitrary number of weights, we can find the minimum value by taking the gradient of the MSE with respect to the weight vectors and equating it to zero. Thus the Wiener-Hopf equation is given as

$$\nabla_{\bar{w}}(E[J(\bar{w})]) = 2\bar{R}_{xx}\bar{w} - 2\bar{r} = 0 \tag{8.28}$$

Simple algebra can be applied to Eq. (8.28) to yield the optimum Wiener solution given as

$$\bar{w}_{MSE} = \bar{R}_{xx}^{-1}\bar{r} \tag{8.29}$$

If we allow the reference signal d to be equal to the desired signal s, and if s is uncorrelated with all interferers, we may simplify the correlation $\bar{r}$. Using Eqs. (8.10) and (8.24) the simplified correlation $\bar{r}$ is

$$\bar{r} = E[s*\cdot\bar{x}] = S\cdot\bar{a}_0 \tag{8.30}$$

where

$$S = E[|s|^2]$$

The optimum weights can then be identified as

$$\bar{w}_{MSE} = S\bar{R}_{xx}^{-1}\bar{a}_0 \tag{8.31}$$

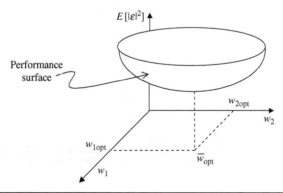

FIGURE 8.8 Quadratic surface for MSE.

Example 8.3 The $M = 5$-element array with spacing $d = .5\lambda$ has a received signal energy $S = 1$ arriving at $\theta_0 = 20°$, and two interferers arriving at angles $\theta_1 = -20°$ and $\theta_2 = 40°$, with noise variance $\sigma_n^2 = .001$. Use MATLAB to calculate the optimum weights and plot the resulting pattern.

Solution MATLAB code sa_ex8_3.m is used. The array vectors for the desired signal and the two interferers are given by

$$\bar{a}_0 = \begin{bmatrix} -.55 - .84i \\ .48 - .88i \\ 1.0 \\ .48 + .88i \\ -.55 + .84i \end{bmatrix} \quad \bar{a}_1 = \begin{bmatrix} -.55 + .84i \\ .48 + .88i \\ 1.0 \\ .48 - .88i \\ -.55 - .84i \end{bmatrix} \quad \bar{a}_2 = \begin{bmatrix} -.62 + .78i \\ -.43 - .90i \\ 1.0 \\ -.43 + .90i \\ -.62 - .78i \end{bmatrix}$$

The correlation matrix for the array is given by

$$\bar{R}_{xx} = \begin{bmatrix} 3.0 & .51 - .90i & -1.71 + .78i & -1.01 + .22i & -1.02 - .97i \\ .51 + .90i & 3.0 & .51 - .90i & -1.71 + .78i & -1.01 + .22i \\ -1.71 - .78i & .51 + .90i & 3.0 & .51 - .90i & -1.71 + .78i \\ -1.01 - .22i & -1.71 - .78i & .51 + .90i & 3.0 & .51 - .90i \\ -1.02 + .97i & -1.01 - .22i & -1.71 - .78i & .51 + .90i & 3.0 \end{bmatrix}$$

The weights can now be calculated by using Eq. (8.31) to get

$$\bar{w}_{MSE} = \begin{bmatrix} -.11 - .21i \\ .18 - .08i \\ .21 \\ .18 + .08i \\ -.11 + .21i \end{bmatrix}$$

Applying the weights to the array vector, we can plot the resulting minimum MSE pattern Fig. 8.9.

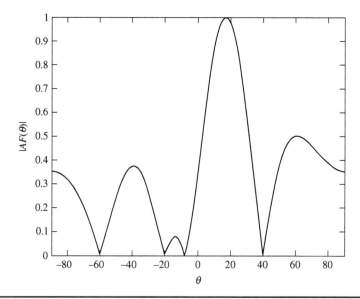

FIGURE 8.9 Minimum MSE pattern for 5-element array.

8.3.3 Maximum Likelihood

The maximum likelihood (ML) method is predicated on the assumption that we have an unknown desired signal $\bar{x}_s$ and that the unwanted signal $\bar{n}$ has a zero mean Gaussian distribution. The goal of this method is to define a likelihood function that can give us an estimate on the desired signal. Details of the ML approach are found in excellent treatments by Van Trees [32, 33]. This fundamental solution falls under the general category of estimation theory. Referring to Fig. 8.10, it should be noted that no feedback is given to the antenna elements. The input signal vector is given by

$$\bar{x} = \bar{a}_0 s + \bar{n} = \bar{x}_s + \bar{n} \tag{8.32}$$

The overall distribution is assumed to be Gaussian, but the mean is controlled by the desired signal $\bar{x}_s$. The probability density function can be described as the joint probability density $p(\bar{x} \mid \bar{x}_s)$. This density can be viewed as the likelihood function (Haykin [34] and Monzingo and Miller [30]) that can be used to estimate the parameter $\bar{x}_s$. The *probability density* can be described as

$$p(\bar{x} \mid \bar{x}_s) = \frac{1}{\sqrt{2\pi\sigma_n^2}} e^{-((\bar{x}-\bar{a}_0 s)^H \bar{R}_{nn}^{-1}(\bar{x}-\bar{a}_0 s))} \tag{8.33}$$

where σ_n = noise standard deviation
$\bar{R}_{nn} = \sigma_n^2 \bar{I}$ = noise correlation matrix

Because the parameter of interest is in the exponent, it is easier to work with the negative of the logarithm of the density function. We shall call this the *log-likelihood function*. Thus, we can define the log-likelihood function as

$$L[\bar{x}] = -\ln[p(\bar{x} \mid \bar{x}_s)] = C(\bar{x} - \bar{a}_0 s)^H \bar{R}_{nn}^{-1}(\bar{x} - \bar{a}_0 s) \tag{8.34}$$

where C = constant
$\bar{R}_{nn} = E[\bar{n}\bar{n}^H]$

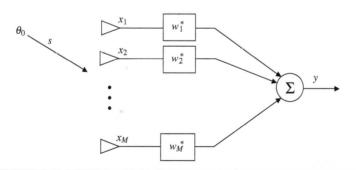

FIGURE 8.10 Traditional array.

Let us define our estimate of the desired signal, called $\hat{s}$, that maximizes the log-likelihood function. The maximum of $L[\bar{x}]$ is found by taking the partial derivative with respect to s and setting it equal to zero.

Thus

$$\frac{\partial L[\bar{x}]}{\partial s} = 0 = -2\bar{a}_0^H \bar{R}_{nn}^{-1} \bar{x} + 2\hat{s}\bar{a}_0^H \bar{R}_{nn}^{-1} \bar{a}_0 \tag{8.35}$$

Solving for $\hat{s}$ we get

$$\hat{s} = \frac{\bar{a}_0^H \bar{R}_{nn}^{-1}}{\bar{a}_0^H \bar{R}_{nn}^{-1} \bar{a}_0} \bar{x} = \bar{w}_{ML}^H \bar{x} \tag{8.36}$$

Thus

$$\bar{w}_{ML} = \frac{\bar{R}_{nn}^{-1} \bar{a}_0}{\bar{a}_0^H \bar{R}_{nn}^{-1} \bar{a}_0} \tag{8.37}$$

Example 8.4 The $M = 5$-element array with spacing $d = .5\lambda$ has a received signal arriving at the angle $\theta_0 = 30°$, with noise variance $\sigma_n^2 = .001$. Use MATLAB to calculate the optimum weights and plot the resulting pattern.

Solution Because we are assuming that the noise is zero-mean Gaussian noise, the noise correlation matrix is the identity matrix given as

$$\bar{R}_{nn} = \sigma_n^2 \begin{bmatrix} 1 & 0 & 0 & 0 & 0 \\ 0 & 1 & 0 & 0 & 0 \\ 0 & 0 & 1 & 0 & 0 \\ 0 & 0 & 0 & 1 & 0 \\ 0 & 0 & 0 & 0 & 1 \end{bmatrix}$$

The array steering vector is given by

$$\bar{a}_0 = \begin{bmatrix} -1 \\ -1i \\ 1 \\ 1i \\ -1 \end{bmatrix}$$

The calculated array weights are given as

$$\bar{w}_{ML} = \frac{\bar{R}_{nn}^{-1} \bar{a}_0}{\bar{a}_0^H \bar{R}_{nn}^{-1} \bar{a}_0} = .2\bar{a}_0$$

The subsequent array factor plot is shown in Fig. 8.11.

8.3.4 Minimum Variance

The *minimum variance solution* [7] is sometimes called the minimum variance distortionless response (MVDR) [34] or the *minimum noise variance performance measure* [30]. The term *distortionless* is applied when it is desired that the received signal is undistorted after the application of the array weights. The goal of the MV method is to minimize the array output noise variance. It is assumed that the

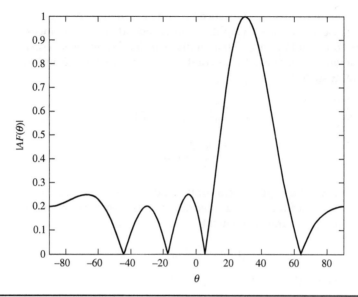

FIGURE 8.11 Maximum likelihood pattern for 5-element array.

desired and unwanted signals have zero mean. We may again use the array configuration of Fig. 8.10. The weighted array output is given by

$$y = \bar{w}^H \bar{x} = \bar{w}^H \bar{a}_0 s + \bar{w}^H \bar{u} \tag{8.38}$$

In order to ensure a distortionless response, we must also add the constraint that

$$\bar{w}^H \bar{a}_0 = 1 \tag{8.39}$$

Applying the constraint to Eq. (8.38), the array output is given as

$$y = s + \bar{w}^H \bar{u} \tag{8.40}$$

In addition, if the unwanted signal has zero mean, the expected value of the array output is given by

$$E[y] = s \tag{8.41}$$

We may now calculate the variance of y such that

$$\sigma^2_{MV} = E[|\bar{w}^H \bar{x}|^2] = E[|s + \bar{w}^H \bar{u}|^2]$$
$$= \bar{w}^H \bar{R}_{uu} \bar{w} \tag{8.42}$$

where

$$\bar{R}_{uu} = \bar{R}_{ii} + \bar{R}_{nn}$$

We can minimize this variance by using the method of Lagrange [35]. Because all the array weights are interdependent, we can incorporate the constraint in Eq. (8.39) to define a *modified performance criterion*, or *cost function*, that is a linear combination of the variance and the constraint such that

$$J(\bar{w}) = \frac{\sigma^2_{MV}}{2} + \lambda(1 - \bar{w}^H \bar{a}_0)$$

$$= \frac{\bar{w}^H \bar{R}_{uu} \bar{w}}{2} + \lambda(1 - \bar{w}^H \bar{a}_0) \tag{8.43}$$

where λ is the Lagrange multiplier and $J(\bar{w})$ is the cost function.

The cost function is a quadratic function and can be minimized by setting the gradient equal to zero. Thus

$$\nabla_{\bar{w}} J(\bar{w}) = \bar{R}_{uu} \bar{w}_{MV} - \lambda \bar{a}_0 = 0 \tag{8.44}$$

Solving for the weights, we conclude

$$\bar{w}_{MV} = \lambda \bar{R}_{uu}^{-1} \bar{a}_0 \tag{8.45}$$

In order to solve for the Lagrange multiplier (λ), we can substitute Eq. (8.39) into Eq. (8.45). Thus

$$\lambda = \frac{1}{\bar{a}_0^H \bar{R}_{uu}^{-1} \bar{a}_0} \tag{8.46}$$

Upon substituting Eq. (8.46) into Eq. (8.45), we arrive at the minimum variance optimum weights

$$\bar{w}_{MV} = \frac{\bar{R}_{uu}^{-1} \bar{a}_0}{\bar{a}_0^H \bar{R}_{uu}^{-1} \bar{a}_0} \tag{8.47}$$

It should be noted that the minimum variance solution is identical in form to the ML solution. The only difference is that the ML approach requires that all unwanted signals combined are zero mean and have a Gaussian distribution. However, with the minimum variance approach, the unwanted signal can include interferers arriving at unwanted angles as well as the noise. Thus, the minimum variance solution is more general in its application.

Example 8.5 The $M = 5$-element array with spacing $d = .5\lambda$ has a received signal arriving at the angle $\theta_0 = 30°$, one interferer arriving at $-10°$, and noise with a variance $\sigma_n^2 = .001$, Use MATLAB to calculate the optimum weights and plot the resulting pattern.

Solution It is a simple matter to slightly modify the MATLAB code from Example 8.3 to include the interferer at $-10°$. The MATLAB code used is sa_ ex8_ 5.m. The unwanted signal correlation matrix is given by

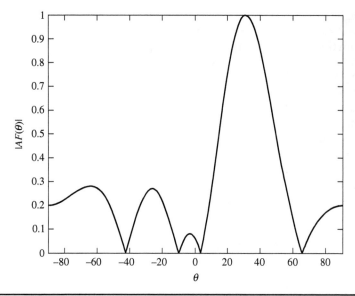

FIGURE 8.12 Minimum variance pattern for 5-element array.

$$\bar{R}_{uu} = \begin{bmatrix} 1.0 & .85+.52i & .46+.89i & -.07+.99i & -.57+.82i \\ .85-.52i & 1.0 & .85+.52i & .46+.89i & -.07+.99i \\ .46-.89i & .85-.52i & 1.0 & .85+.52i & .46+.89i \\ -.07-.99i & .46-.89i & .85-.52i & 1.0 & .85+.52i \\ -.57-.82i & -.07-.99i & .46-.89i & .85-.52i & 1.0 \end{bmatrix}$$

The desired array vector is

$$\bar{a}(\theta_0) = \begin{bmatrix} -1 \\ -1i \\ 1 \\ 1i \\ -1 \end{bmatrix}$$

Using Eq. (8.40), the calculated minimum variance weights are given as

$$\bar{w} = \begin{bmatrix} -.19+.04i \\ .03-.19i \\ .25 \\ .03+.19i \\ -.19-.04i \end{bmatrix}$$

The subsequent array factor plot, with cancellation for interference, is shown in Fig. 8.12. It should be noted that the minimum variance solution allows for canceling interferers and a null has been placed near −10°.

8.4 Adaptive Beamforming

The fixed beamforming approaches, mentioned in Sec. 8.3, which included the maximum SIR, the ML method, and the MV method were assumed to apply to fixed arrival angle emitters. If the arrival angles don't change with time, the optimum array weights won't need to be adjusted. However, if the desired arrival angles change with time, it is necessary to devise an optimization scheme that operates *on-the-fly* so as to keep recalculating the optimum array weights. The receiver signal processing algorithm then must allow for the continuous adaptation to an ever-changing electromagnetic environment. The adaptive algorithm takes the fixed beamforming process one step further and allows for the calculation of continuously updated weights. The adaptation process must satisfy a specified optimization criterion. Several examples of popular optimization techniques include LMS, SMI, *recursive least squares* (RLS), the *constant modulus algorithm* (CMA), *conjugate gradient*, and waveform diverse algorithms. We discuss and explain each of these techniques in the following sections.

8.4.1 Least Mean Squares

The least mean squares (LMS) algorithm is a gradient-based approach. Monzingo and Miller [30] give an excellent fundamental treatment of this approach. Gradient-based algorithms assume an established quadratic performance surface such as discussed in Sec. 8.3.2. When the performance surface is a quadratic function of the array weights, the performance surface $J(\bar{w})$ is in the shape of an elliptic paraboloid having one minimum. One of the best ways to establish the minimum is through the use of a gradient method. We can establish the performance surface (cost function) by again finding the MSE. The error, as indicated in Fig. 8.7, is

$$\varepsilon(k) = d(k) - \bar{w}^H(k)\bar{x}(k) \tag{8.48}$$

The squared error is given as

$$|\varepsilon(k)|^2 = |d(k) - \bar{w}^H(k)\bar{x}(k)|^2 \tag{8.49}$$

Momentarily, we will suppress the time dependence. As calculated in Sec. 8.3.2, the cost function is given as

$$J(\bar{w}) = D - 2\bar{w}^H \bar{r} + \bar{w}^H \bar{R}_{xx} \bar{w} \tag{8.50}$$

where

$$D = E[|d|^2]$$

We may use the gradient method to locate the minimum of Eq. (8.50). Thus

$$\nabla_{\bar{w}}(J(\bar{w})) = 2\bar{R}_{xx}\bar{w} - 2\bar{r} \tag{8.51}$$

The minimum occurs when the gradient is zero. Thus, the solution for the weights is the optimum Wiener solution as given by

$$\bar{w}_{opt} = \bar{R}_{xx}^{-1}\bar{r} \tag{8.52}$$

The solution in Eq. (8.52) is predicated on our knowledge of all signal statistics and thus in our calculation of the correlation matrix.

In general, we do not know the signal statistics and thus must resort to estimating the array correlation matrix ($\bar{R}_{xx}$) and the signal correlation vector ($\bar{r}$) over a range of snapshots or for each instant in time. The instantaneous estimates of these values are given as

$$\hat{R}_{xx}(k) \approx \bar{x}(k)\bar{x}^H(k) \tag{8.53}$$

and

$$\hat{r}(k) \approx d^*(k)\bar{x}(k) \tag{8.54}$$

We can use an iterative technique called the method of *steepest descent* to approximate the gradient of the cost function. The direction of steepest descent is in the opposite direction as the gradient vector. The method of steepest descent can be approximated in terms of the weights using the LMS method advocated by Widrow [16, 17]. The steepest descent iterative approximation is given as

$$\bar{w}(k+1) = \bar{w}(k) - \frac{1}{2}\mu \nabla_{\bar{w}}(J(\bar{w}(k))) \tag{8.55}$$

where, μ is the step-size parameter and $\nabla_{\bar{w}}$ is the gradient of the performance surface.

The gradient of the performance surface is given in Eq. (8.51). If we substitute the instantaneous correlation approximations, we have the LMS solution.

$$\begin{aligned}\bar{w}(k+1) &= \bar{w}(k) - \mu[\hat{R}_{xx}\bar{w} - \hat{r}] \\ &= \bar{w}(k) + \mu e^*(k)\bar{x}(k)\end{aligned} \tag{8.56}$$

where

$$e(k) = d(k) - \bar{w}^H(k)\bar{x}(k) = \text{error signal}$$

The convergence of the LMS algorithm in Eq. (8.56) is directly proportional to the *step-size parameter* μ. If the step size is too small, the convergence is slow and we will have the overdamped case. If the convergence is slower than the changing angles of arrival, it is possible that the adaptive array cannot acquire the signal of interest fast enough to track the changing signal. If the step size is too large, the LMS algorithm will overshoot the optimum weights of interest.

This is called the *underdamped case*. If attempted convergence is too fast, the weights will oscillate about the optimum weights but will not accurately track the solution desired. It is therefore imperative to choose a step size in a range that ensures convergence. It can be shown that stability is ensured provided that the following condition is met [30].

$$0 \leq \mu \leq \frac{1}{2\lambda_{max}} \tag{8.57}$$

where λ_{max} is the largest eigenvalue of $\hat{R}_{xx}$.

Because the correlation matrix is positive definite, all eigenvalues are positive. If all the interfering signals are noise and there is only one signal of interest, we can approximate the condition in Eq. (8.57) as

$$0 \leq \mu \leq \frac{1}{2\text{trace}[\hat{R}_{xx}]} \tag{8.58}$$

Example 8.6 An $M = 8$-element array with spacing $d = .5\lambda$ has a received signal arriving at the angle $\theta_0 = 30°$, an interferer at $\theta_1 = -60°$. Use MATLAB to write an LMS routine to solve for the desired weights. Assume that the desired received signal vector is defined by $\bar{x}_s(k) = \bar{a}_0 s(k)$, where $s(k) = \cos(2^*\text{pi}^*t(k)/T)$; with $T = 1$ ms and $t = (1:100)^*T/100$. Assume the interfering signal vector is defined by $\bar{x}_i(k) = \bar{a}_1 i(k)$, where $i(k) = \text{randn}(1,100)$. Both signals are nearly orthogonal over the time interval T. Let the desired signal $d(k) = s(k)$.

Use the LMS algorithm given in Eq. (8.56) to solve for the optimum array weights. Assume that the initial array weights are all zero. Allow for 100 iterations. Using MATLAB:
(a) Let step size $\mu = .02$.
(b) Calculate the eight array weights for 100 iterations.
(c) Plot the resulting weights magnitude versus iteration number.
(d) Plot the desired signal $s(k)$ and the array output $y(k)$.
(e) Plot the MSE $|e|^2$.
(f) Plot the array factor using the final weights calculated.

Solution The MATLAB code sa_ex8_6.m is used to solve the problem. The magnitude of the weights versus iteration number is shown in Fig. 8.13. Figure 8.14 shows how the array output acquires and tracks the desired signal after about 60 iterations. Figure 8.15 shows the resulting MSE that converges to near zero after 60 iterations. Figure 8.16 shows the final weighted array that has a peak at the desired direction of 30° and a null at the interfering direction of −60°.

8.4.2 Sample Matrix Inversion

One of the drawbacks of the LMS adaptive scheme is that the algorithm must go through many iterations before satisfactory convergence is achieved. If the signal characteristics are rapidly changing, the LMS adaptive algorithm may not allow tracking of the desired signal in a satisfactory manner. The rate of convergence of the weights is dictated by the eigenvalue spread of the array correlation matrix. In Example 8.6, the LMS algorithm did not converge until after 70 iterations.

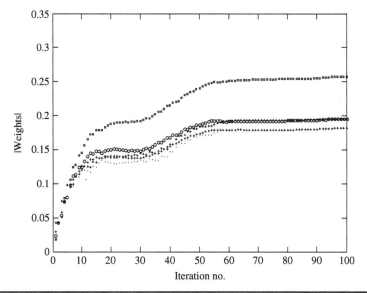

FIGURE 8.13 Magnitude of array weights.

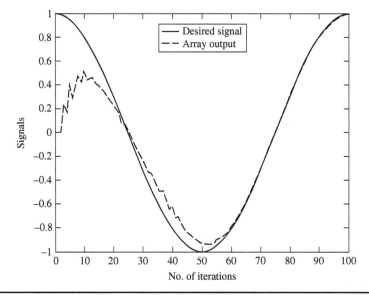

FIGURE 8.14 Acquisition and tracking of desired signal.

Seventy iterations corresponded to more than half of the period of the waveform of interest. One possible approach to circumventing the relatively slow convergence of the LMS scheme is by use of SMI method [7, 23, 36]. This method is also alternatively known

220 Chapter Eight

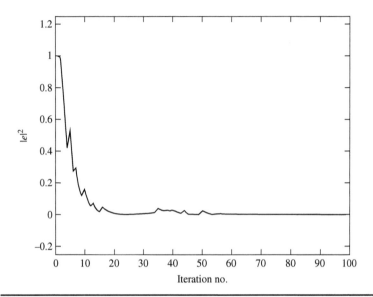

Figure 8.15 Mean square error.

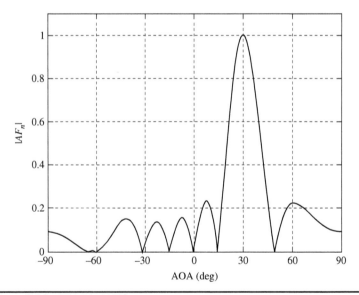

Figure 8.16 Weighted LMS array.

as *direct matrix inversion* (DMI) [30]. The *sample matrix* is a time average estimate of the covariance matrix using K-time samples. If the random process is ergodic in the covariance, the time average estimate will equal the actual covariance matrix.

Recalling the earlier discussion (Sec. 8.3.2) of the minimum MSE, the optimum array weights are given by the optimum Wiener solution as

$$\bar{w}_{opt} = \bar{R}_{xx}^{-1}\bar{r} \qquad (8.59)$$

where

$$\bar{r} = E[d^* \cdot \bar{x}]$$

$$\bar{R}_{xx} = E[\bar{x}\bar{x}^H]$$

As was shown with Eq. (7.32), we can estimate the correlation matrix by calculating the time average such that

$$\hat{R}_{xx} = \frac{1}{K}\sum_{k=1}^{K}\bar{x}(k)\bar{x}^H(k) \qquad (8.60)$$

where K is the observation interval.

The correlation vector $\bar{r}$ can be estimated by

$$\hat{r} = \frac{1}{K}\sum_{k=1}^{K}d^*(k)\bar{x}(k) \qquad (8.61)$$

Because we use a K-length block of data, this method is called a *block-adaptive approach*. We are thus adapting the weights block-by-block.

It is easy in MATLAB to calculate the array correlation matrix and the correlation vector by the following procedure. Define the matrix $\bar{X}_K(k)$ as the kth block of $\bar{x}$ vectors ranging over K-data snapshots. Thus

$$\bar{X}_K(k) = \begin{bmatrix} x_1(1+kK) & x_1(2+kK) & \cdots & x_1(K+kK) \\ x_2(1+kK) & x_2(1+kK) & & \vdots \\ \vdots & & \ddots & \\ x_M(1+kK) & \cdots & & x_M(K+kK) \end{bmatrix} \qquad (8.62)$$

where k is the block number and K is the block length. Thus, the estimate of the array correlation matrix is given by

$$\hat{R}_{xx}(k) = \frac{1}{K}\bar{X}_K(k)\bar{X}_K^H(k) \qquad (8.63)$$

In addition, the desired signal vector can be define by

$$\bar{d}(k) = [d(1+kK) \quad d(2+kK) \quad \cdots \quad d(K+kK)] \qquad (8.64)$$

Thus the estimate of the correlation vector is given by

$$\hat{r}(k) = \frac{1}{K}\bar{d}^*(k)\bar{X}_K(k) \qquad (8.65)$$

222 Chapter Eight

The SMI weights can then be calculated for the kth block of length K as

$$\bar{w}_{SMI}(k) = \hat{R}_{xx}^{-1}(k)\hat{r}(k)$$

$$= \left[\bar{X}_K(k)\bar{X}_K^H(k)\right]^{-1}\bar{d}^*(k)\bar{X}_K(k) \qquad (8.66)$$

Example 8.7 Let us compare the SMI solution to the LMS solution of Example 8.6. An $M = 8$-element array with spacing $d = .5\lambda$ has a received signal arriving at the angle $\theta_0 = 30°$, an interferer at $\theta_1 = -60°$. Use MATLAB to write an SMI routine to solve for the desired weights. Assume that the desired received signal vector is defined by $\bar{x}_s(k) = \bar{a}_0 s(k)$, where $s(k) = \cos(2*pi*t(k)/T)$ with $T = 1$ ms. Let the block length be $K = 30$. Time is defined as $t = (1:K)*T/K$. Assume the interfering signal vector is defined by $\bar{x}_i(k) = \bar{a}_1 i(k)$, where $i(k) = \text{randn}(1, K)$. Let the desired signal $d(k) = s(k)$. In order to keep the correlation matrix inverse from becoming singular, add noise to the system with variance $\sigma_n^2 = .01$.

Solution The MATLAB code sa_ex8_7.m is used to solve the problem. We may calculate the array input noise by use of the MATLAB command $n = \text{randn}(N,K)*\text{sqrt}(sig2)$, where sig2 = noise variance. After specifying the number of elements, the received signal angle, and the interfering angle, we calculate the received vector $\bar{x}(k)$. The correlation matrix is found using the simple MATLAB command Rxx = X*X'/K, the correlation vector is found using $r = X*S'/K$. The optimum Wiener weights are found and the resulting array pattern is seen in Fig. 8.17. The SMI pattern is similar to the LMS pattern and was generated with no iterations. The total number of snapshots K is less than the time to convergence for the LMS algorithm.

The SMI algorithm, although faster than the LMS algorithm, has several drawbacks. The correlation matrix may be ill conditioned resulting in errors or singularities when inverted. In addition, for

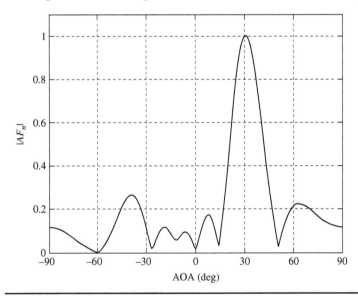

Figure 8.17 Weighted SMI array pattern.

large arrays, there is the challenge of inverting large matrices. To invert, the correlation matrix requires $N^3/2 + N^2$ complex multiplications [20]. The SMI update frequency will necessarily depend on signal frequencies and on channel fading conditions.

8.4.3 Recursive Least Squares

As was mentioned in the previous section, the SMI technique has several drawbacks. Even though the SMI method is faster than the LMS algorithm, the computational burden and potential singularities can cause problems. However, we can recursively calculate the required correlation matrix and the required correlation vector. Recall that in Eqs. (8.60) and (8.61) the estimate of the correlation matrix and vector was taken as the sum of the terms divided by the block length K. When we calculate the weights in Eq. (8.66), the division by K is cancelled by the product $\hat{R}_{xx}^{-1}(k)\bar{r}(k)$. Thus, we can rewrite the correlation matrix and the correlation vector omitting K as

$$\hat{R}_{xx}(k) = \sum_{i=1}^{k} \bar{x}(i)\bar{x}^H(i) \tag{8.67}$$

$$\hat{r}(k) = \sum_{i=1}^{k} d^*(i)\bar{x}(i) \tag{8.68}$$

where k is the block length and last time sample k, $\hat{R}_{xx}(k), \bar{r}(k)$ are the correlation estimates ending at time sample k.

Both summations (Eqs. (8.67) and (8.68)) use rectangular windows; thus they equally consider all previous time samples. Because the signal sources can change or slowly move with time, we might want to deemphasize the earliest data samples and emphasize the most recent ones. This can be accomplished by modifying Eqs. (8.67) and (8.68) such that we forget the earliest time samples. This is called a *weighted estimate*. Thus

$$\hat{R}_{xx}(k) = \sum_{i=1}^{k} \alpha^{k-i}\bar{x}(i)\bar{x}^H(i) \tag{8.69}$$

$$\hat{r}(k) = \sum_{i=1}^{k} \alpha^{k-i}d^*(i)\bar{x}(i) \tag{8.70}$$

where α is the *forgetting factor*.

The forgetting factor is also sometimes referred to as the *exponential weighting* factor [37]. α is a positive constant such that $0 \leq \alpha \leq 1$. When $\alpha = 1$, we restore the ordinary least squares algorithm. $\alpha = 1$ also indicates infinite memory. Let us break up the summation in Eqs. (8.69) and (8.70) into two terms: the summation for values up to $i = k - 1$ and last term for $i = k$.

$$\hat{R}_{xx}(k) = \alpha \sum_{i=1}^{k-1} \alpha^{k-1-i}\bar{x}(i)\bar{x}^H(i) + \bar{x}(k)\bar{x}^H(k)$$

$$= \alpha \hat{R}_{xx}(k-1) + \bar{x}(k)\bar{x}^H(k) \tag{8.71}$$

Chapter Eight

$$\hat{r}(k) = \alpha \sum_{i=1}^{k-1} \alpha^{k-1-i} d^*(i)\bar{x}(i) + d^*(k)\bar{x}(k)$$

$$= \alpha \hat{r}(k-1) + d^*(k)\bar{x}(k) \quad (8.72)$$

Thus, future values for the covariance estimate and the vector covariance estimate can be found using previous values.

Example 8.8 For an $M = 4$-element array, $d = \lambda/2$, one signal arrives at 45°, and $S(k) = \cos(2\pi(k-1)/(K-1))$. Calculate the array correlation for a block of length $K = 200$ using the standard SMI algorithm and the recursion algorithm with $\alpha = 1$. Plot the trace of the SMI correlation matrix for K data points and the trace of the recursion correlation matrix versus block length k, where $1 < k < K$.

Solution Using MATLAB code sa_ex8_8.m we can construct the array steering vector for the angle of arrival of 45°. After multiplying the steering vector by the signal $S(k)$, we can then find the correlation matrix to start the recursion relationship in Eq. (8.71). After K iterations, we can superimpose the traces of both correlation matrices. This is shown in Fig. 8.18.

It can be seen that the recursion formula oscillates for different block lengths and that it matches the SMI solution when $k = K$. The recursion formula always gives a correlation matrix estimate for any block length k but only matches SMI when the forgetting factor is 1. The advantage of the recursion approach is that one need not calculate the correlation for an entire block of length K. Rather, each update only requires one a block of length 1 and the previous correlation matrix.

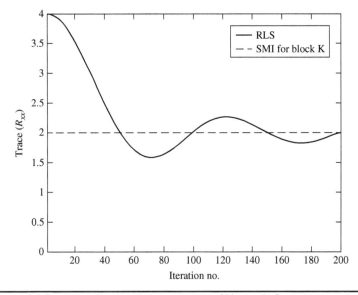

Figure 8.18 Trace of correlation matrix using SMI and RLS.

Not only can we recursively calculate the most recent correlation estimates, we can also use Eq. (8.71) to derive a recursion relationship for the inverse of the correlation matrix. The next steps follow the derivation in [37]. We can invoke the Sherman Morrison-Woodbury (SMW) theorem [38] to find the inverse of Eq. (8.71). Repeating the SMW theorem

$$(\bar{A} + \bar{z}\bar{z}^H)^{-1} = \bar{A}^{-1} - \frac{\bar{A}^{-1}\bar{z}\bar{z}^H\bar{A}^{-1}}{1 + \bar{z}^H\bar{A}^{-1}\bar{z}} \tag{8.73}$$

Applying Eq. (8.73) to Eq. (8.71), we have the following recursion formula:

$$\hat{R}_{xx}^{-1}(k) = \alpha^{-1}\hat{R}_{xx}^{-1}(k-1) - \frac{\alpha^{-2}\hat{R}_{xx}^{-1}(k-1)\bar{x}(k)\bar{x}^H(k)\hat{R}_{xx}^{-1}(k-1)}{1 + \alpha^{-1}\bar{x}^H(k)\hat{R}_{xx}^{-1}(k-1)\bar{x}(k)} \tag{8.74}$$

We can simplify Eq. (8.74) by defining the gain vector $\bar{g}(k)$

$$\bar{g}(k) = \frac{\alpha^{-1}\hat{R}_{xx}^{-1}(k-1)\bar{x}(k)}{1 + \alpha^{-1}\bar{x}^H(k)\hat{R}_{xx}^{-1}(k-1)\bar{x}(k)} \tag{8.75}$$

Thus

$$\hat{R}_{xx}^{-1}(k) = \alpha^{-1}\hat{R}_{xx}^{-1}(k-1) - \alpha^{-1}\bar{g}(k)\bar{x}^H(k)\hat{R}_{xx}^{-1}(k-1) \tag{8.76}$$

Equation (8.76) is known as the *Riccati equation* for the recursive least squares (RLS) method. We can rearrange Eq. (8.75) by multiplying the denominator times both sides of the equation to yield

$$\bar{g}(k) = \left[\alpha^{-1}\hat{R}_{xx}^{-1}(k-1) - \alpha^{-1}\bar{g}(k)\bar{x}^H(k)\hat{R}_{xx}^{-1}(k-1)\right]\bar{x}(k) \tag{8.77}$$

It is clear that the term inside of the brackets of Eq. (8.77) is equal to Eq. (8.76). Thus

$$\bar{g}(k) = \hat{R}_{xx}^{-1}(k)\bar{x}(k) \tag{8.78}$$

Now we can derive a recursion relationship to update the weight vectors. The optimum Wiener solution is repeated in terms of the iteration number k and we can substitute Eq. (8.72) yielding

$$\bar{w}(k) = \hat{R}_{xx}^{-1}(k)\hat{r}(k)$$
$$= \alpha\hat{R}_{xx}^{-1}(k)\hat{r}(k-1) + \hat{R}_{xx}^{-1}(k)\bar{x}(k)d^*(k) \tag{8.79}$$

We may now substitute Eq. (8.76) into the first correlation matrix inverse seen in Eq. (8.79).

$$\bar{w}(k) = \hat{R}_{xx}^{-1}(k-1)\hat{r}(k-1) - \bar{g}(k)\bar{x}^H(k)\hat{R}_{xx}^{-1}(k-1)\hat{r}(k-1) + \hat{R}_{xx}^{-1}(k)\bar{x}(k)d^*(k)$$
$$= \bar{w}(k-1) - \bar{g}(k)\bar{x}^H(k)\bar{w}(k-1) + \hat{R}_{xx}^{-1}(k)\bar{x}(k)d^*(k) \tag{8.80}$$

Finally we may substitute Eq. (8.78) into Eq. (8.80) to yield

$$\bar{w}(k) = \bar{w}(k-1) - \bar{g}(k)\bar{x}^H(k)\bar{w}(k-1) + \bar{g}(k)d^*(k)$$
$$= \bar{w}(k-1) + \bar{g}(k)[d^*(k) - \bar{x}^H(k)\bar{w}(k-1)] \qquad (8.81)$$

It should be noted that Eq. (8.81) is identical in form to Eq. (8.56).

Example 8.9 Use the RLS method to solve for the array weights and plot the resulting pattern. Let the array be an $M = 8$-element array with spacing $d = .5\lambda$ with a received signal arriving at the angle $\theta_0 = 30°$, an interferer at $\theta_1 = -60°$. Use MATLAB to write an RLS routine to solve for the desired weights. Use Eqs. (8.71), (8.78), and (8.81). Assume that the desired received signal vector is defined by $\bar{x}_s(k) = \bar{a}_0 s(k)$, where $s(k) = \cos(2^*\text{pi}^*\text{t}(k)/T)$; with $T = 1$ ms. Let there be $K = 50$ time samples such that $t = (0 : K-1)^*T/(K-1)$. Assume that the interfering signal vector is defined by $\bar{x}_i(k) = \bar{a}_1 i(k)$, where $i(k) = \sin(\text{pi}^*\text{t}(k)/T)$. Let the desired signal $d(k) = s(k)$. In order to keep the correlation matrix inverse from becoming singular, add noise to the system with variance $\sigma_n^2 = .01$. Begin with the assumption that all array weights are zero such that $\bar{w}(1) = [\ 0\ \ 0\ \ 0\ \ 0\ \ 0\ \ 0\ \ 0\ \ 0\ ]^T$. Set the forgetting factor $\alpha = .9$.

Solution MATLAB code sa_ex8_9.m is used to solve for the array weights and to plot the resulting pattern shown in Fig. 8.19.

The advantage of the RLS algorithm over SMI is that it is no longer necessary to invert a large correlation matrix. The recursive equations allow for easy updates of the inverse of the correlation matrix. The RLS algorithm also converges much more quickly than the LMS algorithm.

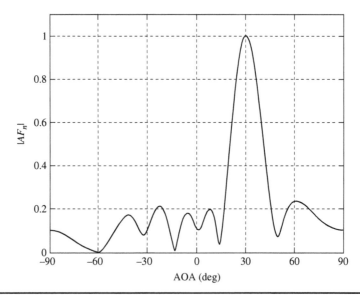

Figure 8.19 RLS array pattern.

8.4.4 Constant Modulus

Many adaptive beamforming algorithms are based on minimizing the error between a reference signal and the array output. The reference signal is typically a training sequence used to *train* the adaptive array or a desired signal based on an a priori knowledge of nature of the arriving signals. In the case where a reference signal is not available one must resort to an assortment of optimization techniques that are *blind* to the exact content of the incoming signals.

Many wireless communication and radar signals are frequency- or phase-modulated signals. Some examples of phase and frequency modulated signals are FM, PSK, FSK, QAM, and polyphase. This being the case, the amplitude of the signal should ideally be a constant. Thus the signal is said to have a constant magnitude or *modulus*. However, in fading channels, where multipath terms exist, the received signal is the composite of all multipath terms. Thus, the channel introduces an amplitude variation on the signal magnitude. Frequency selective channels by definition destroy the constant modulus property of the signal. If we know that the arriving signals of interest should have a constant modulus, we can devise algorithms that restore or *equalize* the amplitude of the original signal.

Godard [39] was the first to capitalize on the *constant modulus* (CM) property in order to create a family of blind equalization algorithms to be used in two-dimensional data communication systems. Specifically, Godard's algorithm applies to phase modulating waveforms. Godard used a cost function called a *dispersion function* of order p and, after minimization, the optimum weights are found. The Godard cost function is given by

$$J(k) = E[(|y(k)|^p - R_p)^q] \qquad (8.82)$$

where p is the positive integer and q is the positive integer $= 1$.

Godard showed that the gradient of the cost function is zero when R_p is defined by

$$R_p = \frac{E[|s(k)|^{2p}]}{E[|s(k)|^p]} \qquad (8.83)$$

where $s(k)$ is the zero-memory estimate of $y(k)$.

The resulting error signal is given by

$$e(k) = y(k)|y(k)|^{p-2}(R_p - |y(k)|^p) \qquad (8.84)$$

This error signal can replace the traditional error signal in the LMS algorithm to yield

$$\bar{w}(k+1) = \bar{w}(k) + \mu e^*(k)\bar{x}(k) \qquad (8.85)$$

The $p = 1$ case reduces the cost function to the form

$$J(k) = E[(|y(k)| - R_1)^2] \tag{8.86}$$

where

$$R_1 = \frac{E[|s(k)|^2]}{E[|s(k)|]} \tag{8.87}$$

If we scale the output estimate $s(k)$ to unity, we can write the error signal in Eq. (8.84) as

$$e(k) = \left(y(k) - \frac{y(k)}{|y(k)|}\right) \tag{8.88}$$

Thus the weight vector, in the $p = 1$ case, becomes

$$\bar{w}(k+1) = \bar{w}(k) + \mu\left(1 - \frac{1}{|y(k)|}\right)y^*(k)\bar{x}(k) \tag{8.89}$$

The $p = 2$ case reduces the cost function to the form

$$J(k) = E[(|y(k)|^2 - R_2)^2] \tag{8.90}$$

where

$$R_2 = \frac{E[|s(k)|^4]}{E[|s(k)|^2]} \tag{8.91}$$

If we scale the output estimate $s(k)$ to unity, we can write the error signal in Eq. (8.84) as

$$e(k) = y(k)(1 - |y(k)|^2) \tag{8.92}$$

Thus the weight vector, in the $p = 2$ case, becomes

$$\bar{w}(k+1) = \bar{w}(k) + \mu(1 - |y(k)|^2)y^*(k)\bar{x}(k) \tag{8.93}$$

The cases where $p = 1$ or 2 are referred to as *constant modulus algorithms* (CMA). The $p = 1$ case has been proven to converge much more rapidly than the $p = 2$ case [40]. A similar algorithm was developed by Treichler and Agee [41] and is identical to the Godard case for $p = 2$.

Example 8.10 Allow the same constant modulus signal to arrive at the receiver via a direct path and two additional multipaths and assume that the channel is frequency selective. Let the direct path arriving signal be defined as a *32-chip binary sequence* where the chip values are ±1 and are sampled four times per chip. The direct path signal arrives at 45°. The first multipath signal arrives at −30° but is 30 percent of the direct path in amplitude. The second multipath signal arrives at 0° but is 10 percent of the direct path in amplitude. Because of multipath, there will be slight time delays in the binary sequences causing dispersion.

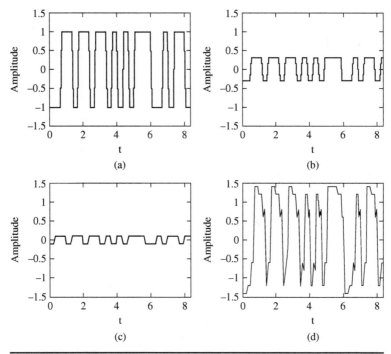

FIGURE 8.20 (a) Direct path, (b) path 2, (c) path 3, (d) combined signals.

This time delay can be implemented by zero padding the multipath signals to effect a time delay. We will use the $P = 1$ CMA algorithm to define the optimum weights. Choose $\mu = .5$, $M = 8$ elements, and $d = \lambda/2$. Define the initial weights $\bar{w}(1)$ to be zero. Plot the resulting pattern.

Solution The three received waveforms are shown in Fig. 8.20. The last block indicates the combined received waveforms. It can be seen that the combined received signal has an amplitude variation due to channel dispersion.

The array output is defined as $y(k) = \bar{w}^H(k)\bar{x}(k)$. The recursion relationship for the array weights is given in Eq. (8.89). MATLAB code sa_ex8_10.m is used to solve for the weights. The resulting pattern plot is given in Fig. 8.21. It should be noted that the CMA algorithm suppresses the multipath but does not cancel it.

8.4.5 Least Squares Constant Modulus

One severe disadvantage of the Godard CMA algorithm is the slow convergence time. The slow convergence limits the usefulness of the algorithm in dynamic environments where the signal must be captured quickly. This also limits the usefulness of CMA when channel conditions are rapidly changing. The previous CMA method is based on the method of steepest descent by taking the gradient of the cost function in Eq. (8.82). A faster algorithm was developed by Agee [42] using the method of nonlinear least-squares. The least-squares method is also known as the *Gauss method* based upon the work of

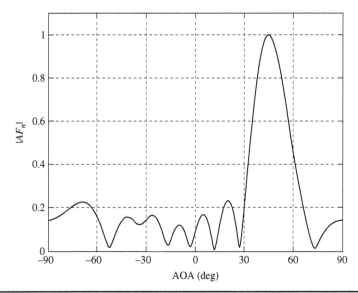

FIGURE 8.21 CMA pattern.

Gauss in 1795 [43]. This method is referred to as the *LS-CMA algorithm* [44] and is also known as an *autoregressive estimator* based on a least squares minimization [45].

The following derivation is taken directly from [42] and [44]. In the method of least squares, one defines a cost function which is the weighted sum of error squares or the total error energy. The energies are the energies of a finite sample set K. The cost function is defined by

$$C(\bar{w}) = \sum_{k=1}^{K} |\phi_k(\bar{w})|^2 = \|\bar{\Phi}(\bar{w})\|_2^2 \tag{8.94}$$

where $\phi_k(\bar{w}) = $ error at kth data sample
$\Phi(\bar{w}) = [\phi_1(\bar{w}) \quad \phi_1(\bar{w}) \quad \cdots \quad \phi_K(\bar{w})]^T$
$K = $ number of data samples in one block

Equation (8.94) has a partial Taylor-series expansion with a sum-of-squares form given as

$$C(\bar{w} + \bar{\Delta}) \approx \|\bar{\Phi}(\bar{w}) + \bar{J}^H(\bar{w})\bar{\Delta}\|_2^2 \tag{8.95}$$

where the complex Jacobian of $\bar{\Phi}(\bar{w})$ is defined as

$$\bar{J}\bar{w} = [\nabla \phi_1(\bar{w}) \quad \nabla \phi_2(\bar{w}) \quad \cdots \quad \nabla \phi_K(\bar{w})] \tag{8.96}$$

and $\bar{\Delta}$ is the offset that updates weights.

We wish to find the offset $\bar{\Delta}$ which minimizes the sum-of-squared errors. Taking the gradient of Eq. (8.95) and setting it equal to zero, we can find the optimum offset vector to be defined as

$$\bar{\Delta} = -\left[\bar{J}(\bar{w})\bar{J}^H(\bar{w})\right]^{-1}\bar{J}(\bar{w})\bar{\Phi}(\bar{w}) \qquad (8.97)$$

The new updated weight vector is then given by

$$\bar{w}(n+1) = \bar{w}(n) - \left[\bar{J}(\bar{w}(n))\bar{J}^H(\bar{w}(n))\right]^{-1}\bar{J}(\bar{w}(n))\bar{\Phi}(\bar{w}(n)) \qquad (8.98)$$

The new weight vector is the previous weight vector adjusted by the offset $\bar{\Delta}$. The number n is the iteration number not to be confused with the time sample k.

Let us now apply the least squares method to the CMA using the 1-2 cost function [8].

$$C(\bar{w}) = \sum_{k=1}^{K}|\phi_k(\bar{w})|^2 = \sum_{k=1}^{K}||y(k)|-1|^2 \qquad (8.99)$$

where $y(k) = \bar{w}^H \bar{x}(k) =$ array output at time k.

We may write ϕ_k as a vector such that

$$\bar{\phi}(\bar{w}) = \begin{bmatrix} |y(1)|-1 \\ |y(2)|-1 \\ \vdots \\ |y(K)|-1 \end{bmatrix} \qquad (8.100)$$

We may now define the Jacobian of the error vector $\bar{\phi}(\bar{w})$

$$\bar{J}(\bar{w}) = [\nabla(\phi_1(\bar{w})) \quad \nabla(\phi_2(\bar{w})) \quad \cdots \quad \nabla(\phi_K(\bar{w}))]$$

$$= \left[\bar{x}(1)\frac{y^*(1)}{|y(1)|} \quad \bar{x}(2)\frac{y^*(2)}{|y(2)|} \quad \cdots \quad \bar{x}(K)\frac{y^*(K)}{|y(K)|}\right] \qquad (8.101)$$

$$= \bar{X}\bar{Y}_{CM}$$

where

$$\bar{X} = [\bar{x}(1) \quad \bar{x}(2) \quad \cdots \quad \bar{x}(K)] \qquad (8.102)$$

and

$$\bar{Y}_{CM} = \begin{bmatrix} \frac{y^*(1)}{|y(1)|} & 0 & \cdots & 0 \\ 0 & \frac{y^*(2)}{|y(2)|} & & 0 \\ \vdots & & \ddots & \vdots \\ 0 & 0 & \cdots & \frac{y^*(K)}{|y(K)|} \end{bmatrix} \qquad (8.103)$$

Multiplying the Jacobian times its Hermitian transpose, we get

$$\bar{J}(\bar{w})\bar{J}^H(\bar{w}) = \bar{X}\bar{Y}_{CM}\bar{Y}_{CM}^H\bar{X}^H = \bar{X}\bar{X}^H \quad (8.104)$$

The product of the Jacobian times the energy matrix is given by

$$\bar{J}(\bar{w})\bar{\Phi}(\bar{w}) = \bar{X}\bar{Y}_{CM}\begin{bmatrix} |y(1)|-1 \\ |y(2)|-1 \\ \vdots \\ |y(K)|-1 \end{bmatrix} = \bar{X}\begin{bmatrix} y^*(1) - \frac{y^*(1)}{|y(1)|} \\ y^*(2) - \frac{y^*(2)}{|y(2)|} \\ \vdots \\ y^*(K) - \frac{y^*(K)}{|y(K)|} \end{bmatrix} = \bar{X}(\bar{y} - \bar{r})^* \quad (8.105)$$

where

$$\bar{y} = [y(1) \quad y(2) \quad \cdots \quad y(K)]^T \quad (8.106)$$

and

$$\bar{r} = \left[\frac{y(1)}{|y(1)|} \quad \frac{y(2)}{|y(2)|} \quad \cdots \quad \frac{y(K)}{|y(K)|} \right]^T = L(\bar{y}) \quad (8.107)$$

where $L(\bar{y})$ is a hard-limiter acting on $\bar{y}$.
Substituting Eqs. (8.103) and (8.105) into Eq. (8.98), we get

$$\begin{aligned}\bar{w}(n+1) &= \bar{w}(n) - \left[\bar{X}\bar{X}^H\right]^{-1}\bar{X}(\bar{y}(n) - \bar{r}(n))^* \\ &= \bar{w}(n) - \left[\bar{X}\bar{X}^H\right]^{-1}\bar{X}\bar{X}^H\bar{w}(n) + \left[\bar{X}\bar{X}^H\right]^{-1}\bar{X}\bar{r}^{*(n)} \\ &= \left[\bar{X}\bar{X}^H\right]^{-1}\bar{X}\bar{r}^{*(n)}\end{aligned} \quad (8.108)$$

where

$$\bar{r}*(n) = \left[\frac{\bar{w}^H(n)\bar{x}(1)}{|\bar{w}^H(n)\bar{x}(1)|} \quad \frac{\bar{w}^H(n)\bar{x}(2)}{|\bar{w}^H(n)\bar{x}(2)|} \quad \cdots \quad \frac{\bar{w}^H(n)\bar{x}(K)}{|\bar{w}^H(n)\bar{x}(K)|} \right]^H \quad (8.109)$$

While only one block of data is used to implement the LS-CMA, the algorithm iterates through n values until convergence. Initial weights $\bar{w}(1)$ are chosen, the complex-limited output data vector $\bar{r}^*(1)$ is calculated, then the next weight $\bar{w}(2)$ is calculated, and the iteration continues until satisfactory convergence is satisfied. This is called the *static* LS-CMA algorithm because only one static block, of length K, is used for the iteration process. The LS-CMA algorithm bears a striking resemblance to the SMI algorithm in Eq. (8.66).

Example 8.11 Allow the same constant modulus signal to arrive at the receiver via a direct path and one additional multipath and assume that the channel is frequency selective. Let the direct path arriving signal be defined as a 32-*bit binary chipping sequence*, where the chip values are ±1 and are sampled four times per chip. Let the block length $K = 132$. The direct path signal arrives at 45°. A multipath signal arrives at −30° but is 30 percent of the direct path in amplitude. Because of multipath, there will be slight time delays in the binary sequences

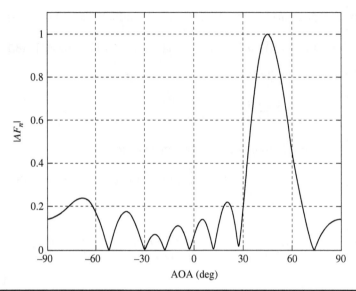

FIGURE 8.22 Static LS-CMA.

causing dispersion. Let the dispersion be implemented by zero padding the two signals. Zero pad the direct path by adding four zeros to the back of the signal. Zero pad the multipath signal with two zeros in front and two zeros in back. Use the LS-CMA algorithm to find the optimum weights. $M = 8$ elements and $d = \lambda/2$. Add zero mean Gaussian noise for each element in the array and let the noise variance be $\sigma_n^2 = .01$. Define the initial weights $\bar{w}(1)$ to be one. Iterate three times. Plot the resulting pattern.

Solution The MATLAB code sa_ex8_11.m can be utilized for this purpose. The CM waveforms are generated identically to Example 8.10. The resulting pattern is shown in Fig. 8.22.

It should be noted that LS-CMA algorithm does a better job of nulling the multipath terms than does the CMA algorithm from the previous example.

The chief advantage of the static LS-CMA is that it can converge up to 100 times faster than the conventional CMA algorithm. In fact, in this example, the weights effectively converged after only a few iterations.

The static LS-CMA algorithm computed the weights simply based upon a fixed block of sampled data. In order to maintain up-to-date adaptation in a dynamic signal environment, it is better to update the data blocks for each iteration. Thus a dynamic LS-CMA algorithm is more appropriate. The *dynamic* LS-CMA is a modification of the previous static version. Let us define a dynamic block of data as the array output before applying weights. For the nth iteration, the nth block of length K is given as

$$\bar{X}(n) = [\bar{x}(1+nK) \quad \bar{x}(2+nK) \quad \cdots \quad \bar{x}(K+nK)] \qquad (8.110)$$

The weighted array output, for the nth iteration is now defined as

$$\bar{y}(n) = [y(1+nK) \quad y(2+nK) \quad \cdots \quad y(K+nK)]^T = [\bar{w}^H(n)\bar{X}(n)]^T \quad (8.111)$$

The complex limited output data vector is given as

$$\bar{r}(n) = \left[\frac{y(1+nK)}{|y(1+nK)|} \quad \frac{y(2+nK)}{|y(2+nK)|} \quad \cdots \quad \frac{y(K+nK)}{|y(K+nK)|} \right]^T \quad (8.112)$$

Replacing Eq. (8.108) with the dynamic version, we have

$$\bar{w}(n+1) = [\bar{X}(n)\bar{X}^H(n)]^{-1}\bar{X}(n)\bar{r}^*(n) \quad (8.113)$$

where

$$\bar{r}^*(n) = \left[\frac{\bar{w}^H(n)\bar{x}(1+nK)}{|\bar{w}^H(n)\bar{x}(1+nK)|} \quad \frac{\bar{w}^H(n)\bar{x}(2+nK)}{|\bar{w}^H(n)\bar{x}(2+nK)|} \quad \cdots \quad \frac{\bar{w}^H(n)\bar{x}(K+nK)}{|\bar{w}^H(n)\bar{x}(K+nK)|} \right]^H$$

(8.114)

We can further simplify Eq. (8.113) by defining the array correlation matrix and the correlation vector as

$$\hat{R}_{xx}(n) = \frac{\bar{X}(n)\bar{X}^H(n)}{K} \quad (8.115)$$

and

$$\hat{\rho}_{xr}(n) = \frac{\bar{X}(n)\bar{r}^*(n)}{K} \quad (8.116)$$

The dynamic LS-CMA is now defined as

$$\bar{w}(n+1) = \hat{R}_{xx}^{-1}(n)\hat{\rho}_{xr}(n) \quad (8.117)$$

Example 8.12 Repeat Example 8.11 using the dynamic LS-CMA. Define the block length $K = 22$ data points. Allow the block to update for every iteration n. $M = 8$ elements and $d = \lambda/2$. Define the initial weights $\bar{w}(1)$ to be one. Solve for the weights after 6 iterations. Plot the resulting pattern.

Solution The MATLAB code sa_ex8_12.m is used for this purpose. The CM waveforms are generated identically to Example 8.10 except that the block of K-data points moves with the iteration number. The resulting pattern is shown in Fig. 8.23.

8.4.6 Conjugate Gradient Method

The problem with the steepest descent method has been the sensitivity of the convergence rates to the eigenvalue spread of the correlation matrix. Greater spreads result in slower convergences. The convergence rate can be accelerated by use of the *conjugate gradient method* (CGM). The goal of CGM is to iteratively search for

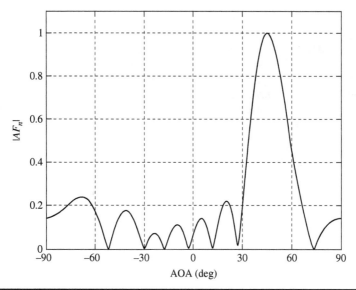

Figure 8.23 Dynamic LS-CMA.

the optimum solution by choosing conjugate (perpendicular) paths for each new iteration. Conjugacy in this context is intended to mean *orthogonal*. The method of CGM produces orthogonal search directions resulting in the fastest convergence. Figure 8.24 depicts a top view of a two-dimensional performance surface where the conjugate steps show convergence toward the optimum solution. Note that the path taken at iteration $n + 1$ is perpendicular to the path taken at the previous iteration n.

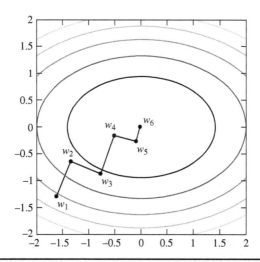

Figure 8.24 Convergence using conjugate directions.

The CGM method has its early roots in the desire to solve a system of linear equations. One of the earliest references to CGM is found in a 1952 document by Hestenes and Stiefel [46]. Additionally, early work was performed by Fletcher and Powell in 1963 [47] and Fletcher and Reeves in 1964 [48]. The CGM has also has been called the *accelerated gradient approach* (AG) by Monzingo and Miller [30]. The gradient is accelerated by virtue of choosing conjugate directions. The CGM method was modified for use in predicting array weights by Choi [49, 50]. A survey of this method has been written by Godara [51] and a concise summary of the method can be found in Sadiku [52]. The following summary is taken from [51, 52].

CGM is an iterative method whose goal is to minimize the quadratic cost function

$$J(\bar{w}) = \frac{1}{2}\bar{w}^H \bar{A}\bar{w} - \bar{d}^H \bar{w} \qquad (8.118)$$

where

$$\bar{A} = \begin{bmatrix} x_1(1) & x_2(1) & \cdots & x_M(1) \\ x_1(2) & x_2(2) & \cdots & x_M(2) \\ \vdots & \vdots & \ddots & \vdots \\ x_1(K) & x_2(K) & \cdots & x_M(K) \end{bmatrix} K \times M \text{ matrix of array snapshots}$$

K = number of snapshots
M = number of array elements
$\bar{w}$ = unknown weight vector
$\bar{d} = [d(1)\, d(2) \cdots d(K)]^T$ = desired signal vector of K snapshots

We may take the gradient of the cost function and set it to zero in order to find the minimum. It can be shown that

$$\nabla_{\bar{w}} J(\bar{w}) = \bar{A}\bar{w} - \bar{d} \qquad (8.119)$$

We may use the method of steepest descent in order to iterate to minimize Eq. (8.119). We wish to slide to the bottom of the quadratic cost function choosing the least number of iterations. We may start with an initial guess for the weights $\bar{w}(1)$ and find the residual $\bar{r}(1)$. The first residual value at the first guess is given as

$$\bar{r}(1) = -J'(\bar{w}(1)) = \bar{d} - \bar{A}\bar{w}(1) \qquad (8.120)$$

We can next choose a direction vector $\bar{D}$ which gives us the new conjugate direction to iterate toward the optimum weight. Thus

$$\bar{D}(1) = \bar{A}^H \bar{r}(1) \qquad (8.121)$$

The general weight update equation is given by

$$\bar{w}(n+1) = \bar{w}(n) - \mu(n)\bar{D}(n) \qquad (8.122)$$

where the step size is determined by

$$\mu(n) = \frac{\bar{r}^H(n)\bar{A}\bar{A}^H\bar{r}(n)}{\bar{D}^H(n)\bar{A}^H\bar{A}\bar{D}(n)} \tag{8.123}$$

We may now update the residual and the direction vector. We can premultiply Eq. (8.122) by $-\bar{A}$ and add $\bar{d}$ to derive the updates for the residuals.

$$\bar{r}(n+1) = \bar{r}(n) + \mu(n)\bar{A}\bar{D}(n) \tag{8.124}$$

The direction vector update is given by

$$\bar{D}(n+1) = \bar{A}^H\bar{r}(n+1) - \alpha(n)\bar{D}(n) \tag{8.125}$$

We can use a linear search to determine $\alpha(n)$ that minimizes $J(\bar{w}(n))$.

Thus

$$\alpha(n) = \frac{\bar{r}^H(n+1)\bar{A}\bar{A}^H\bar{r}(n+1)}{\bar{r}^H(n)\bar{A}\bar{A}^H\bar{r}(n)} \tag{8.126}$$

Thus, the procedure to use CGM is to find the residual and the corresponding weights and update until convergence is satisfied. It can be shown that the true solution can be found in no more than K iterations. This condition is known as *quadratic convergence*.

Example 8.13 For the $M = 8$-element array with elements a half-wavelength apart, find the optimum weights under the following conditions: The arriving signal of interest is $s = \cos(\pi k/K)$; arriving at an angle of 45°. One interfering signal is defined as $I_1 = \text{randn}(1,K)$; arriving at $-30°$. The other interfering signal is defined as $I_2 = \text{randn}(1,K)$; arriving at 0°. The noise has a variance of $\sigma^2 = .001$. Thus, $n = \sigma^*\text{randn}(1,K)$. Use the CGM method to find the optimum weights when using a block size $K = 20$. Plot the norm of the residual for all iterations. Plot the resulting pattern.

Solution The MATLAB code sa_ex8_13.m is used to solve for the optimum weights using CGM. The plot of the norm of the residuals is given in Fig. 8.25. It can be seen that the residual drops to very small levels after 14 iterations. The plot of the resulting pattern is shown in Fig. 8.26. It can be seen that two nulls are placed at the two angles of arrival of the interference.

It should be especially noted that MATLAB provides a conjugate gradient function that can solve for the optimum weights. The function is the least squares implementation of CG and is given by $w = \text{lsqr}(A,d)$; A is the matrix as defined earlier and d is the desired signal vector containing K-data samples. Using the parameters in Example 8.12, we can use the MATLAB function lsqr(A,d) and produce the array factor plot in Fig. 8.27. The MATLAB function produces nulls in the correct locations but the sidelobe levels are higher than produced by the code sa_ex8_12.m.

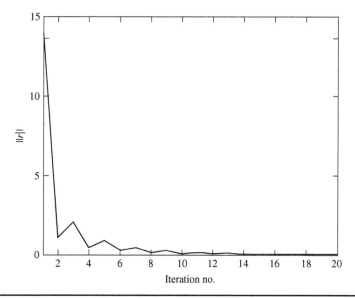

Figure 8.25 Norm of the residuals for each iteration.

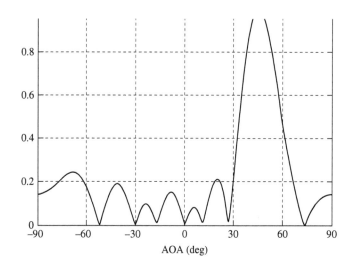

Figure 8.26 Array pattern using CGM.

8.4.7 Spreading Sequence Array Weights

A radically different approach to wireless beamforming has been proposed, which can augment or replace traditional approaches to DBF or adaptive arrays. This new approach does not perform electronic or

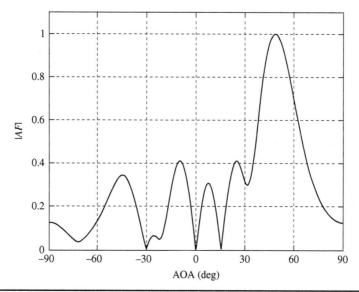

Figure 8.27 Array pattern resulting from use of the MATLAB CGM.

digital phase shifting. It does not rely on adaptive methods; however, receive beams are produced in directions of interest. An array correlation matrix is not computed, array weights are not adapted, an a priori knowledge of the signal environment is not required, and a desired or pilot signal is not used. Rather, a clever spreading technique is used that uniquely defines all received directions of interest. This algorithm can create virtual contiguous beams like a Butler matrix, or it can selectively look in any specific directions of interest.

This new DBF receiver is based upon a radical shift in traditional DBF methods. This *out of the box* approach blends the virtues of switched beam and adaptive array technologies while avoiding many of their respective weaknesses. This solution does not direct beams by any previously known method, although this approach falls under the general topic of *waveform diversity*. This new method provides the same spatial resolution as traditional arrays while simultaneously looking in all desired directions of interest.

The previous adaptive methods discussed rely on subspace methods, steepest descent methods, gradients, blind adaptive algorithms, signal coherence, constant moduli, and other known signal properties in order to provide feedback to control the array weights. There are many drawbacks to the previous methods including computational intensity, a required knowledge of signal statistics, signal independence, and slow convergence rates.

This wholly new approach is based on a patent by Elam [53]. In this new method, adaptation is not necessary and tracking is not performed, while excellent results are achieved. This approach can use any antenna configuration and, within the resolution of the antenna geometry, can simultaneously receive numerous signals of unequal strength at various angles. This novel approach can use any arbitrary two-dimensional array to produce an instantaneous pincushion (i.e., 3D) array.

The new technique essentially works by designing the array weights to be time-varying random phase functions. The weights act to modulate each array element output. Specifically, the array outputs are weighted or modulated with a set of statistically independent polyphase chipping sequences. A different and independent modulating waveform is used for each antenna output. The chips are the same as chips defined in the traditional communications sense. The phase modulation waveforms purposely shred the phase of each antenna output by chipping at rates much higher than the baseband frequency of the message signal. This shredding process breaks up the phase relationship between all array elements and thus purposely eliminates array phase coherence. This is opposite of the traditional goal to achieve phase coherence for a specific look angle. The receiver then can *see* all incoming signals simultaneously without the necessity of steering or adapting because the array elements, for the moment, become statistically independent from one another. The chipped incoming waveforms are processed in a quadrature receiver and are subsequently compared to similar chipped waveforms stored in memory. The memory waveforms are created based on expected angles of arrival. The theoretical spatial capacity is simply the angular space of interest divided by the antenna array beam solid angle.

8.4.8 Description of the New SDMA Receiver

The novel DBF receiver is depicted in Fig. 8.28. The novelty of this new approach is vested in the nature of the signals $\beta_n(t)$, the unique array signal memory, and correlation based detection.

The new SDMA digital beamformer can be used with any arbitrary N-element antenna array. It can be a linear array but should preferably be a two- or three-dimensional random array such that the antenna geometry and element phasing is unique for each incoming angle-of-arrival. Ideally the receive array should have an equal angular resolution over all receive angles of interest. For purposes of illustration, the array used in this discussion will be an N-element linear array.

The incoming signals arrive at angles θ_ℓ, where $\ell = 1, 2, \ldots, L$. Each different angle of arrival produces a unique array element output with a unique phase relationship between each element. These phase relationships will be used in conjunction with the modulations $\beta_n(t)$ to produce a unique summed signal $y^r(t)$.

Corresponding to the actual N-element antenna array is a second virtual array modeled in memory. The virtual array is modeled after

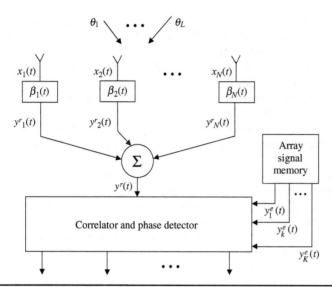

FIGURE 8.28 SDMA quadrature receiver.

the actual physical array used. The memory array has K possible virtual outputs for each expected direction $\theta_k (k = 1, 2, \ldots, K)$. The total number of expected directions K should be less than or equal to the number of antenna elements N. These memory signals are generated based upon a knowledge of the antenna array geometry and the calculated phase delays for each specific direction of interest. The expected directions are generally chosen by the user to be no closer than the angular resolution allowed by the array. Consideration can be given to the local topography and potential multipath directions in eliminating possible unwanted expected directions. All potential incoming directions may not be used if it is desired to block interfering signals from known directions.

The N-element array outputs and the N-antenna array memory outputs are both *phase modulated* (PM) by the same set of N pseudo-noise (*pn*) phase modulating sequences. The nth phase modulating sequence will be designated as $\beta_n(t)$. $\beta_n(t)$ is composed of M polyphase chips. Each chip is of length τ_c and the entire sequence is of length $T = M\tau_c$. The chip rate is chosen to be much greater than the Nyquist rate of the incoming baseband signal modulation. The purpose for this *over sampling* is so that the phase modulation of the incoming signal is nearly constant over the entire group of M chips. In general, the goal should be such that $T \leq 1/(4B_m)$, where B_m is the message signal bandwidth.

Each phase modulating waveform $\beta_n(t)$ is used to modulate or *tag* each array output with a unique marking or identifying waveform. $\beta_n(t)$ deliberately shreds or scrambles the phase of the signal at array element n. This shredding process, temporarily scrambles the phase relationship between all other array elements. The desired element phasing is restored in the correlator if the incoming signal correlates with one of the memory signals.

Examples of some usable *pn* modulating sequences are Gold, Kasami, Welti, Golay, or any other sequences with similar statistical and orthogonal properties [54–56]. These properties will aid in the identification of the exact directions of arrival of the L incoming signals. Figure 8.29 shows a typical example of the first two of the set of N bi-phase sequences applied to an N-element array. The total code length $TB_m = .25$.

Because the phase out of each antenna element is intentionally scrambled by the modulating sequences, the array pattern is randomized during each chip in the sequence. The momentarily scrambled pattern changes every τ_c seconds. As an example, the array patterns are plotted for the first four chips of an $N = 10$-element array as shown in Fig. 8.30. Since the phase relationship between all elements is scrambled, the array patterns for each new set of chips is random. In the limit, as the number of chips increases, the average array pattern, over all chips, becomes uniform.

Each baseband output of the receive array will have a complex voltage waveform whose phase will consist of each emitter's message signal $m_\ell(t)$ and the unique receive antenna element phase contributions.

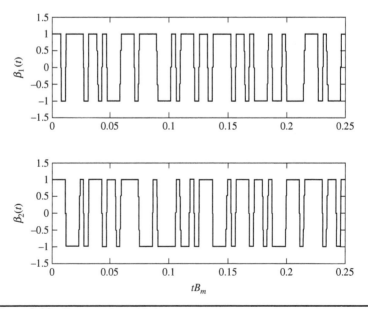

Figure 8.29 Two sample pn bi-phase sequences.

Smart Antennas 243

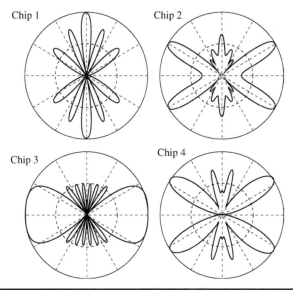

FIGURE 8.30 Scrambled patterns for the first four chips.

Ignoring the space loss and polarization mismatches, the received baseband array output is given in vector form as

$$\bar{x}^r(t) = \begin{bmatrix} 1 & \cdots & 1 \\ e^{jkd\sin(\theta_1)} & \cdots & e^{jkd\sin(\theta_L)} \\ \vdots & \cdots & \vdots \\ e^{j(n-1)kd\sin(\theta_1)} & \cdots & e^{j(n-1)kd\sin(\theta_L)} \end{bmatrix} \begin{bmatrix} e^{jm_1(t)} \\ \vdots \\ e^{jm_L(t)} \end{bmatrix}$$

$$= \overline{A^r} \cdot \bar{s}^r(t) \qquad (8.127)$$

where $m_\ell(t) = \ell$th emitter's phase modulation
d = array element spacing
k = wavenumber
θ_ℓ = angle of arrival of the ℓth incoming signal
$\bar{a}_\ell^r = [1, e^{jkd\sin(\theta_\ell)}, \cdots, e^{j(n-1)kd\sin(\theta_\ell)}]^T$
= steering vector for direction θ_ℓ
$\overline{A^r}$ = matrix of steering vectors for all angles of arrival θ_ℓ
$\bar{s}^r$ = vector of arriving signal baseband phasors

The received signals, for each array output, are phase modulated with the chipping sequences as described earlier. The chipping waveforms can be viewed as phase-only array weights. These array weights can be depicted as the vector $\bar{\beta}(t)$. The total weighted or chipped array output is called the *received signal vector* and is given by

$$y^r(t) = \bar{\beta}(t)^T \cdot \bar{x}^r(t) \qquad (8.128)$$

In a similar way, the array signal memory steering vectors are created based on M expected angles-of-arrival θ_m.

$$\overline{A^e} = \begin{bmatrix} 1 & \cdots & 1 \\ e^{jkd\sin(\theta_1)} & \cdots & e^{jkd\sin(\theta_K)} \\ \vdots & \cdots & \vdots \\ e^{j(n-1)kd\sin(\theta_1)} & \cdots & e^{j(n-1)kd\sin(\theta_K)} \end{bmatrix}$$

$$= [\overline{a}_1^e \ \cdots \ \overline{a}_K^e] \tag{8.129}$$

where $\overline{a}_K^e$ is the steering vector for expected direction θ_k and $\overline{A^e}$ is the matrix of steering vectors for expected direction θ_k.

The memory has K outputs, one for each expected direction θ_k. Each memory output, for the expected angle θ_k is given by

$$y_k^e(t) = \overline{B}(t)^T \cdot \overline{a}_k^e \tag{8.130}$$

The signal correlator is designed to correlate the actual received signal with the conjugate of the various expected direction memory signals. This is similar to matched filter detection. The best correlations occur when the actual AOA matches the expected AOA. The correlation can be used as a discriminant for detection. Because the arriving signals have a random arrival phase delay, a quadrature correlation receiver should be employed such that the random carrier phase does not affect the detection (Haykin [54]). The general complex correlation output, for the kth expected direction, is given as

$$R_k = \int_t^{t+T} y^r(t) \cdot y_k^{e*}(t) dt = |R_k| e^{j\phi k} \tag{8.131}$$

where R_k is the correlation magnitude at expected angle θ_k and θ_k is the correlation phase at expected angle θ_k.

The new SDMA receiver does not process the incoming signals with phase shifters or beam steering. We do not *look* for the emitter direction by steering but actually find the direction by correlation. The correlation magnitude $|R_k|$ is used as the discriminant to determine if a signal is present at the expected angle θ_k. If the discriminant exceeds a predetermined threshold, a signal is deemed present and the phase is calculated. Because it is assumed that the emitter PM is nearly constant over the code length $M\tau_c$, the correlator output phase angle is approximately the average of the emitter PM. Thus

$$\phi_k = \arg(R_k) \approx \tilde{m}_k \tag{8.132}$$

where

$$\tilde{m}_k = \frac{1}{T} \int_t^{t+T} m_k(t) dt = \text{average of emitter's modulation at angle } \theta_k$$

The average phase $\tilde{m}_k$ retrieved for each quarter cycle of the $m(t)$ can be used to reconstruct the user's phase modulation using an FIR filter.

Example Using Bi-phase Chipping

We can let all chipping waveforms, $\beta_n(t)$, be defined as *bi-phase pn sequences*. It is instructive to show the appearance of the signals at various stages in the receiver. Figure 8.31 shows the nth element received baseband modulation, the nth phase modulating waveform $\beta_n(t)$, the nth phase modulated output $y_n^r(t)$ and finally the N combined phase modulated outputs $y^r(t)$.

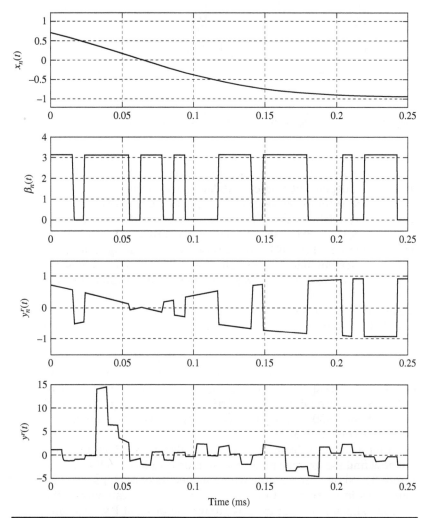

FIGURE 8.31 Display of waveforms throughout the receiver.

Chapter Eight

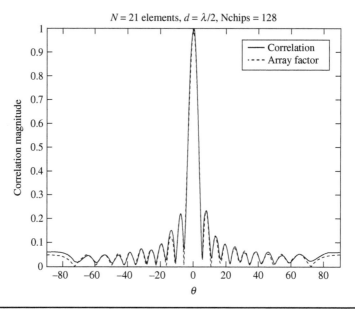

FIGURE 8.32 Comparing the new SDMA receiver to a traditional linear array.

The fourth waveform is obviously unique owing to the arrival phase angle of the incoming signal from direction θ_ℓ and the chipping by the N independent phase modulating waveforms. This specific chipped and summed waveform is unique to the direction of arrival and can be correlated with pre-calculated waveforms stored in memory.

Figure 8.32 compares the correlation magnitude in Eq. (8.131) for an incoming signal ($\theta = 0°$) superimposed with the array factor for a typical N-element linear array. The following receiver values are used: $N = 21$, $d = \lambda/2$, $M = 128$ chips. Figure 8.32 demonstrates that the new SDMA receiver achieves the same angular resolution as the conventional linear array provided that the number of chips M is adequately large. Shorter sequences can be used when the sequences are not only independent but also orthogonal.

When multiple signals arrive at the receive antenna array, they can be detected through the same process of correlating all incoming signals with the expected signals stored in memory. Figure 8.33 demonstrates the correlation for five arriving signals at equally spaced angles of $-60°$, $-30°$, $0°$, $30°$, $60°$. The chipping sequences are binary Welti codes of length 32.

After a signal is deemed present near the expected angle θ_k, we can now find the average phase modulation ϕ_k defined in Eq. (8.133). Figure 8.34 shows an example where sequences of M chips are used four times in succession to reconstruct a piecewise estimate of the message signal $m_k(t)$. Figure 8.34 shows the original PM $m_k(t)$ and the estimate calculated for each quarter cycle.

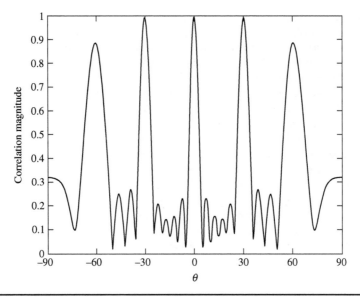

FIGURE 8.33 Correlation for five equally spaced arrival angles.

It can be seen that the predicted average phase for each quarter cycle can be fed into an FIR filter to reconstruct the original emitter modulation.

In summary, the new SDMA receiver uses spreading sequences, as array weights, to shred the phase out of each array element. This approach provides a radical and novel alternative to traditional fixed

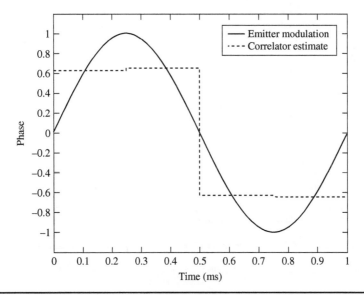

FIGURE 8.34 Emitter modulation with receiver estimate.

beam and adaptive array approaches. The spreading sequences are composed of polyphase codes that have chip rates much greater than the anticipated emitter baseband modulation. The shredding momentarily randomizes the phase relationship between each element output. This in turn causes the pattern to be randomized during each chip of the spreading sequence. Over the M chip duration of the spreading sequence, the average pattern for the array approaches that of an omnidirectional array. The shredded signal is reconstructed with a quadrature correlation receiver by correlating with memory signals. The memory signals are created by a similar shredding process but for expected angles of arrival. Correlator outputs are generated for each expected AOA. The correlation magnitude is used as a discriminant to decide if a signal is present at the expected angle θ_k. The correlation phase angle is a piecewise estimate of the original emitter's modulation.

This new approach has numerous advantages over traditional SDMA array antennas. The receiver is superior to a switched beam array because contiguous beams can be formed in any regions of interest without the need for hardware phase shifters. The beams are created by correlation. The receiver memory can simply be altered to redirect the beams to new regions of interest if so desired. The elimination of phase shifters can result in a significant cost savings.

The new receiver also has advantages over adaptive arrays because it does not require adaptation, it can process multiple angles of arrival simultaneously, and it is not limited by acquisition or tracking speeds. Interfering signals are minimized because the shredded waveforms at interferer angles do not correlate well with expected direction waveforms stored in memory.

Any arbitrary and/or random antenna array geometry can be incorporated into this new approach as long as the expected memory signals are based upon the array geometry. The array resolution achieved is consistent with the limitations of the array geometry.

8.5 References

1. Barton, P., "Digital Beamforming for Radar," *IEE Proceedings on Pt. F*, Vol. 127, pp. 266–277, Aug. 1980.
2. Brookner, E., "Trends in Array Radars for the 1980s and Beyond," IEEE Antenna and Propagation Society Newsletter, April 1984.
3. Steyskal, H., "Digital Beamforming Antennas—An Introduction," *Microwave Journal*, pp. 107–124, January 1987.
4. Skolnik, M., "System Aspects of Digital Beam Forming Ubiquitous Radar," Naval Research Lab, Report No. NRL/MR/5007—02-8625, June 28, 2002.
5. Skolnik, M., *Introduction to Radar Systems*, 3d ed. McGraw-Hill, New York, 2001.
6. Curtis, T., "Digital Beamforming for Sonar Systems," *IEE Proceedings on Pt. F*, Vol. 127, pp. 257–265, Aug. 1980.
7. Litva, J., and T. Kowk-Yeung Lo, *Digital Beamforming in Wireless Communications*, Artech House, 1996.
8. Liberti, J., and T. Rappaport, *Smart Antennas for Wireless Communications: IS-95 and Third Generation CDMA Applications*, Prentice Hall, New York, 1999.

9. Van Atta, L., "Electromagnetic Reflection," U.S. Patent 2908002, Oct. 6, 1959.
10. Margerum, D., "Self-Phased Arrays," in *Microwave Scanning Antennas*, Vol. 3, *Array Systems*, Ch. 5. ed. Hansen, R.C. Academic Press, New York, 1966.
11. York, R., and T. Itoh, "Injection and Phase-Locking Techniques for Beam Control," *IEEE Transactions on MTT*, Vol. 46, No. 11, pp. 1920–1920, Nov. 1998.
12. Howells, P., "Intermediate Frequency Sidelobe Canceller," U. S. Patent 3202990, Aug. 24, 1965.
13. Howells, P., "Explorations in Fixed and Adaptive Resolution at GE and SURC," *IEEE Transactions on Antenna and Propagation, Special Issue on Adaptive Antennas*, Vol. AP-24, No. 5, pp. 575–584, Sept. 1976.
14. Applebaum, S., "Adaptive Arrays," Syracuse University Research Corporation, Rep. SPL TR66-1, August 1966.
15. Applebaum, S., "Adaptive Arrays," *IEEE Transactions on Antenna and Propagation*, Vol. AP-24, No. 5, pp. 585–598, Sept. 1976.
16. Widrow, B., and M. Hoff, "Adaptive Switch Circuits," IRE Wescom, Convention Record, Part 4, pp. 96–104, 1960.
17. Widrow, B., P. Mantey, L. Griffiths, et al., "Adaptive Antenna Systems," *Proceedings of the IEEE*, Vol. 55, Dec. 1967.
18. Gabriel, W., "Adaptive Processing Antenna Systems," *IEEE Antenna and Propagation Newsletter*, pp. 5–11, Oct. 1983.
19. Godara, L., "Application of Antenna Arrays to Mobile Communications, Part II: Beam-Forming and Direction-of-Arrival Considerations," *Proceedings of the IEEE*, Vol. 85, No. 8, pp. 1195–1245, Aug. 1997.
20. Reed, I., J. Mallett, and L. Brennen, "Rapid Convergence Rate in Adaptive Arrays," *IEEE Transactions on Aerospace on Electronics Systems*, Vol. AES-10, pp. 853–863, Nov. 1974.
21. Capon, J., "High-Resolution Frequency-Wavenumber Spectrum Analysis," *Proceedings of the IEEE*, Vol. 57, No. 8, pp. 1408–1418, Aug. 1969.
22. Makhoul, J., "Linear Prediction: A Tutorial Review," *Proceedings of the IEEE*, Vol. 63, pp. 561–580, 1975.
23. Burg, J. P., "The Relationship between Maximum Entropy Spectra and Maximum Likelihood Spectra," *Geophysics*, Vol. 37, pp. 375–376, April 1972.
24. Burg, J. P., "Maximum Entropy Spectrum Analysis," Ph.D. Dissertation, Dept. of Geophysics, Stanford University, Stanford CA, 1975.
25. Pisarenko, V. F., "The Retrieval of Harmonics from a Covariance Function," *Geophysical Journal of the Royal Astronomical Society*, Vol. 33, pp. 347–366, 1973.
26. Reddi, S. S., "Multiple Source Location—A Digital Approach," *IEEE Transactions on AES*, Vol. 15, No. 1, Jan. 1979.
27. Kumaresan, R., and D. Tufts, "Estimating the Angles of Arrival of Multiple Plane Waves," *IEEE Transactions on AES*, Vol. AES-19, pp. 134–139, 1983.
28. Schmidt, R., "Multiple Emitter Location and Signal Parameter Estimation," *IEEE Transactions on Antenna Propagation*, Vol. AP-34, No. 2, pp. 276–280, March 1986.
29. Roy, R., and T. Kailath, "ESPRIT—Estimation of Signal Parameters via Rotational Invariance Techniques," *IEEE Transactions on ASSP*, Vol. 37, No. 7, pp. 984–995, July 1989.
30. Monzingo, R., and T. Miller, *Introduction to Adaptive Arrays*, Wiley Interscience, John Wiley & Sons, New York, 1980.
31. Harrington, R., *Field Computation by Moment Methods*, MacMillan, New York, Chap. 10, p. 191, 1968.
32. Van Trees, H., *Detection, Estimation, and Modulation Theory: Part I*, Wiley, New York, 1968.
33. Van Trees, H., *Optimum Array Processing, Part IV of Detection, Estimation, and Modulation Theory*, Wiley Interscience, New York, 2002.
34. Haykin, S., H. Justice, N. Owsley, et al., *Array Signal Processing*, Prentice Hall, New York, 1985.

35. Cohen, H., *Mathematics for Scientists and Engineers*, Prentice Hall, New York, 1992.
36. Godara, L., *Smart Antennas*, CRC Press, Boca Raton, FL, 2004.
37. Haykin, S., *Adaptive Filter Theory*, 4th ed., Prentice Hall, New York, 2002.
38. Golub, G. H., and C. H. Van Loan, *Matrix Computations*, The Johns Hopkins University Press, 3d ed., 1996.
39. Godard, D. N., "Self-Recovering Equalization and Carrier Tracking in Two-Dimensional Data Communication Systems," *IEEE Transactions on Communications*, Vol. Com-28, No. 11, pp. 1867–1875, Nov. 1980.
40. Larimore, M., and J. Treichler, "Convergence Behavior of the Constant Modulus Algorithm, Acoustics," *IEEE International Conference on ICASSP '83*, Vol. 8, pp. 13–16, April 1983.
41. Treichler, J., and B. Agee, "A New Approach to Multipath Correction of Constant Modulus Signals," *IEEE Transactions on Acoustics, Speech, and Signal Processing*, Vol. Assp-31, No. 2, pp. 459–472, April 1983.
42. Agee, B., "The Least-Squares CMA: A New Technique for Rapid Correction of Constant Modulus Signals," *IEEE International Conference on ICASSP '86*, Vol. 11, pp. 953–956, April 1986.
43. Sorenson, H., "Least-Squares Estimation: From Gauss to Kalman," *IEEE Spectrum*, Vol. 7, pp. 63–68, July 1970.
44. Rong, Z., "Simulation of Adaptive Array Algorithms for CDMA Systems," Master's Thesis MPRG-TR-96-31, Mobile & Portable Radio Research Group, Virginia Tech, Blacksburg, VA, Sept. 1996.
45. Stoica, P., and R. Moses, *Introduction to Spectral Analysis*, Prentice Hall, New York, 1997.
46. Hestenes, M., and E. Stiefel, "Method of Conjugate Gradients for Solving Linear Systems," *Journal of Research of the National Bureau of Standards*, Vol. 49, pp. 409–436, 1952.
47. Fletcher, R., and M. Powell, "A Rapidly Convergent Descent Method for Minimization," *Computer Journal*, Vol. 6, pp. 163–168, 1963.
48. Fletcher, R., and C. Reeves, "Function Minimization by Conjugate Gradients," *Computer Journal*, Vol. 7, pp. 149–154, 1964.
49. Choi, S., *Application of the Conjugate Gradient Method for Optimum Array Processing*, Book Series on PIER (Progress in Electromagnetics Research), Vol. 5, Elsevier, Amsterdam, 1991.
50. Choi, S., and T. Sarkar, Adaptive Antenna Array Utilizing the Conjugate Gradient Method for Multipath Mobile Communication, Signal Processing, Vol. 29, pp. 319–333, 1992.
51. Godara, L., *Smart Antennas*, CRC Press, Boca Raton, FL, 2004.
52. Sadiku, M., *Numerical Techniques in Electromagnetics*, 2d ed., CRC Press, Boca Raton, FL, 2001.
53. Elam, C., "Method and Apparatus for Space Division Multiple Access Receiver," Patent No. 6,823,021, Rights assigned to Greenwich Technology Associates, One Soundview way, Darien, CT.
54. Simon, H., *Communication Systems*, 2d ed., p. 580. Wiley, New York, 1983.
55. Ziemer, R. E., and R. L. Peterson, *Introduction to Digital Communication*, Prentice Hall, pp. 731–742, 2001.
56. Skolnik, M. *Radar Handbook*, 2d ed., McGraw-Hill, pp. 10.17–10.26, 1990.

8.6 Problems

8.1 For an $M = 5$-element array with $d = \lambda/2$, the desired signal arrives at $\theta = 20°$, one interferer arrives at $-20°$, while the other interferer arrives at $+45°$. The noise variance is $\sigma_n^2 = .001$. Use the Godara method outlined in Eq. (8.8).

(a) What are the array weights?

(b) Plot the magnitude of the weighted array pattern for $-90° < \theta < 90°$.

8.2 Maximum SIR method: Given an $M = 5$-element array with spacing $d = .5\lambda$ and noise variance $\sigma_n^2 = .001$, a desired received signal arriving at $\theta_0 = 20°$, and two inteferers arrive at angles $\theta_1 = -30°$ and $\theta_2 = -45°$, assume that the signal and interferer amplitudes are constant and that $\bar{R}_{ss} = \bar{R}_{ii} = \begin{bmatrix} 1 & \cdots & 0 \\ \vdots & \ddots & \vdots \\ 0 & \cdots & 1 \end{bmatrix}$.

(a) Use MATLAB to calculate SIR_{max}.
(b) What are the normalized weights?
(c) Plot the resulting pattern.

8.3 Minimum Mean Square Error (MMSE) method: For an $M = 2$-element array with elements a half-wavelength apart, find the optimum Wiener weights under the following conditions: The arriving signal of interest is $s(t) = e^{j\omega t}$ arriving at the arbitrary angle of θ_s. The noise is zero-mean Gaussian with arbitrary variance σ_n^2. Allow the desired signal $d(t) = s(t)$.

(a) What is the signal correlation matrix, $\bar{R}_{ss}$, in symbolic form?
(b) What is the array correlation matrix, $\bar{R}_{xx}$, in symbolic form?
(c) What is the correlation vector $\bar{r}$?
(d) What is the correlation matrix inverse, $\bar{R}_{xx}^{-1}$, in symbolic form?
(e) Symbolically derive the equation for the weights.
(f) What are the exact weights for $\theta_s = 30°$ and for $\sigma_n^2 = .1$?

8.4 MMSE method: Let the array be a 2-element array where the arriving signal is coming in at $\theta = 0°$. Let the desired signal equal the arriving signal with $s = 1$. Thus, $\bar{x} = \begin{bmatrix} 1 \\ e^{jkd \sin(\theta)} \end{bmatrix}$ Let $d = \lambda/2$.

(a) Derive the performance surface using Eq. (8.23) and plot for $-4 < w1, w2 < 4$.
(b) Derive the solutions for $w1$ and $w2$ by using $\frac{\partial}{\partial w1}$ and $\frac{\partial}{\partial w2}$.
(c) How is the derived solution consistent with the plot? (Recall that the $\nabla_w E[|\varepsilon^2|] = 0$ gives the *bottom* of the curve.)
(d) Derive the performance surface using Eq. (8.23) for the arrival angle of $\theta = 30°$.
(e) Derive the solutions for w_1 and w_{21} by using $\frac{\partial}{\partial w1}$ and $\frac{\partial}{\partial w2}$ for the angle given in part d.

8.5 MMSE method: Given an $M = 5$-element array with spacing $d = .5\lambda$, a received signal energy $S = 1$ arriving at $\theta_0 = 30°$, and two interferers arriving at angles $\theta_1 = -20°$ and $\theta_2 = 40°$, with noise variance $\sigma_n^2 = .001$.

Assume that the signal and interferer amplitudes are constant and that $\bar{R}_{ss} = \bar{R}_{ii} = \begin{bmatrix} 1 & 0 \\ 0 & 1 \end{bmatrix}$.

(a) Use MATLAB to calculate the optimum weights.
(b) Plot the resulting pattern.

8.6 Maximum likelihood (ML) method: Given an $M = 5$-element array with spacing $d = .5\lambda$ which has a received signal arriving at the angle $\theta_0 = 45°$, with noise variance $\sigma_n^2 = .01$.

(a) Use MATLAB to calculate the optimum weights.
(b) Plot the magnitude of the weighted array pattern for $-90° < \theta < 90°$.

252   Chapter Eight

8.7 Minimum variance (MV) method: Given an $M = 5$-element array with spacing $d = .5\lambda$ which has a received signal arriving at the angle $\theta_0 = 40°$, one interferer arriving at $-20°$, and noise with a variance $\sigma_n^2 = .001$.
  (a) Use MATLAB to calculate the optimum weights.
  (b) Plot the magnitude of the weighted array pattern for $-90° < \theta < 90°$.

8.8 Least mean squares (LMS) method: For an $M = 2$-element array with elements a half-wavelength apart, find the LMS weights using Eqs. (8.48) and (8.56) under the following conditions: $\mu = .5$, $\sigma_n^2 = 0$, the arriving signal is $s(t) = 1$, the angle of arrival is $\theta_s = 45°$, the desired signal $d(t) = s(t)$. Set the initial array weights such that $\bar{w}(1) = \begin{bmatrix} 0 \\ 0 \end{bmatrix}$. There are no other arriving signals.

  (a) By hand calculate the array weights for the next three iterations (i.e., $\bar{w}(2), \bar{w}(3), \bar{w}(4)$).
  (b) What is the error $|\varepsilon(k)|$ for $k = 2, 3, 4$?
  (c) Use MATLAB to calculate the weights and error for 20 iterations. Plot the absolute value of each weight versus iteration k on one plot and the absolute value of the error vs. iteration k on another plot.

8.9 LMS method: Given an $M = 8$-element array with spacing $d = .5\lambda$ which has a received signal arriving at the angle $\theta_0 = 40°$, an interferer at $\theta_1 = -20°$. Assume that the desired received signal vector is defined by $\bar{x}_s(k) = \bar{a}_0 s(k)$ where $s(k) = \sin(\text{pi}*t(k)/T)$; with $T = 1$ ms and $t = (1:100)*T/100$. Assume the interfering signal vector is defined by $\bar{x}_i(k) = \bar{a}_1 i(k)$, where $i(k) = \text{randn}(1, 100)$. Both signals are nearly orthogonal over the time interval T. Let the desired signal $d(k) = s(k)$. Assume that the initial array weights are all zero. Allow for 100 iterations. Let step size $\mu = .02$.
  (a) Calculate the 8 array weights for 100 iterations.
  (b) Plot the magnitude of the each weight versus iteration number on the same plot.
  (c) Plot the MSE $|e|^2$ versus iteration number.
  (d) What are the array weights at the 100th iteration?
  (e) Plot the magnitude of the weighted array pattern for $-90° < \theta < 90°$.

8.10 Sample matrix inversion (SMI) method: Use the SMI method to find the weights for an $M = 8$-element array with spacing $d = .5\lambda$. Let the received signal arrive at the angle $\theta_0 = 45°$. One interferer arrives at $\theta_1 = -45°$. Assume that the desired received signal vector is defined by $\bar{x}_s(k) = \bar{a}_0 s(k)$, where $s(k) = \sin(2 * \text{pi} * t(k)/T)$ with $T = 2$ ms. Let the block length be $K = 50$. Time is defined as $t = (1:K)*T/K$. Assume the interfering signal vector is defined by $\bar{x}_i(k) = \bar{a}_1 i(k)$, where $i(k) = \text{randn}(1, K) + \text{randn}(1, K) j$. Let the desired signal $d(k) = s(k)$. In order to keep the correlation matrix inverse from becoming singular, add noise to the system with variance $\sigma_n^2 = .01$. Define the noise in MATLAB as $n = \text{randn}(N, K) * \text{sqrt(sig2)}$.
  (a) Find the weights.
  (b) Plot the magnitude of the weighted array pattern for $-90° < \theta < 90°$.

8.11 Repeat Prob. 8.8 for $M = 5$, $d = .5\lambda$, $\theta_0 = 30°$, $\theta_1 = -20°$, and allow the received signal to be a phase modulation such that $s(k) = \exp(1 j*.5*\text{pi}* \sin(\text{pi}*t(k)/T))$.

8.12 Recursive least squares (RLS) method: Let the array be an $M = 7$-element array with spacing $d = .5\lambda$ with a received signal arriving at the angle $\theta_0 = 30°$, an interferer at $\theta_1 = -20°$, and the other interferer at $\theta_2 = -40°$. Use MATLAB to write an RLS routine to solve for the desired weights. Assume that the desired received signal vector is defined by $\bar{x}_s(k) = \bar{a}_0 s(k)$, where $s(k) = \exp(1j * \sin(\text{pi} * t(k)/T)$; with $T = 1$ ms. Let there be $K = 50$ time samples such that $t = (0 : K- 1) * T/(K - 1)$. Assume that the interfering signals $i1(k) = \exp(1j*\text{pi}*\text{randn}(1, K))$; $i2(k) = \exp(1j*\text{pi}*\text{randn}(1, K))$. MATLAB changes the random numbers each time so that $i1$ and $i2$ are different. Let the desired signal $d(k) = s(k)$. In order to keep the correlation matrix inverse from becoming singular, add noise to the system with variance $\sigma_n^2 = .01$. Initialize all array weights as zero. Set the forgetting factor $\alpha = .995$.
(a) Plot the magnitude of the first weight for all iterations.
(b) Plot the magnitude of the weighted array pattern for $-90° < \theta < 90°$.

8.13 Constant modulus algorithm (CMA) method: Allow the same constant modulus signal to arrive at the receiver via a direct path and two additional multipaths and assume that the channel is frequency selective. Let the direct path arriving signal be defined as a 32-chip binary sequence where the chip values are ±1 and are sampled four times per chip (see Example 8.10). The direct path signal arrives at 30°. The first multipath signal arrives at 0° but is 30 percent of the direct path in amplitude. The second multipath signal arrives at 20° but is 10 percent of the direct path in amplitude. Because of multipath, there will be slight time delays in the binary sequences causing dispersion. Let the dispersion be implemented by zero padding the signals. Zero pad the direct path by adding eight zeros to the back of the signal. Zero pad the first multipath signal with two zeros in front and two zeros in back. Zero pad the second multipath signal with four zeros up front. We will use the $P = 1$ CMA algorithm to define the optimum weights. Choose $\mu = .6$, $N = 6$ elements, and $d = \lambda/2$. Define the initial weights, $\bar{w}(1)$, to be zero. Plot the resulting pattern using MATLAB.

8.14 Least squares constant modulus algorithm (LS-CMA) method: Use the static version of the LS-CMA. Allow the same constant modulus signal to arrive at the receiver via a direct path and one additional multipath and assume that the channel is frequency selective. Let the direct path arriving signal be defined as a 32-chip binary sequence where the chip values are ±1 and are sampled four times per chip. Let the block length $K = 132$. The direct path signal arrives at 30°. A multipath signal arrives at −30° but is 50 percent of the direct path in amplitude. Because of multipath, there will be slight time delays in the binary sequences causing dispersion. Let the dispersion be implemented by zero padding the signals. Zero pad the direct path by adding four zeros to the back of the signal. Zero pad the multipath signal with two zeros in front and two zeros in back. Use the LS-CMA algorithm to find the optimum weights. $N = 9$ elements and $d = \lambda/2$. Add noise for each element in the array and let the noise variance be $\sigma_n^2 = .01$. Define the initial weights, $\bar{w}(1)$, to be one. Iterate three times. Plot the resulting pattern using MATLAB.

8.15 Repeat Prob. 8.14 but use the dynamic LS-CMA algorithm. Let the block length be $K = 22$. Allow the block to update for every iteration n.

Stop after 5 iterations using 6 (6 · K = 132) blocks. Plot the resulting pattern using MATLAB.

8.16 Conjugate gradient method (CGM): For the M = 9-element array with elements a half-wavelength apart, find the optimum weights under the following conditions: The arriving signal of interest is $s = \sin(\pi k/K)$; arriving at an angle of 30°. One interfering signal is defined as $I_1 = \text{randn}(1, K)$; arriving at −20°. The other interfering signal is defined as $I_2 = \text{randn}(1, K)$; arriving at −45°. The noise has a variance of $\sigma_n^2 = .001$. Thus, $n = \sigma*\text{randn}(1, K)$. Use the CGM to find the optimum weights when using a block size K = 20.
   (a) Plot the norm of the residual for all iterations.
   (b) Plot the resulting pattern using MATLAB.

CHAPTER 9
Direction Finding

Radio direction finding (DF) is older than radio itself, beginning with Heinrich Hertz' experiments with antenna directionality in 1888 [1]. In today's wireless world direction finding is considered part of geolocation services available to almost everyone, but at the turn of the 1900s the original "wireless world" was energized to seriously consider radio DF as a significant navigation aide by the tragic sinking of the *Titanic*. During World War I radio intercept and DF became important military assets, and by World War II secure radio communication and navigation became a continual race against radio intercept and direction finding. While Axis submarine wolf packs were using simple loop antennas to find convoy ships and attack them, networks of land-based intercept and DF sites using Adcock arrays were providing coordinated locations to direct Allied destroyers to sink the submarines. Axis aircraft sorties were guided by radio navigation beacons for bombing at night, whereas the British responded with navigation spoofing to protect London and Coventry. Modern DF systems owe a lot to these early pioneers.

9.1 Loop Antennas

9.1.1 Early Direction Finding with Loop Antennas

A number of direction-finding experiments were conducted by the U.S. Navy before 1915, most ending in failure, and all were received with disinterest by Navy admirals. Neither the National Electric Signaling Company nor Marconi Wireless Telegraph Company performed well.

The first success came with Dr. Frederick Kolster at Stone Radio & Telegraph Company, where he assisted in the first U.S. naval direction-finding experiments on the collier USS *Lebanon* in 1906. The antenna, a "delta loop" strung between the ship's stacks, required the ship to change course in order to measure changes in signal amplitude and determine signal direction. The U.S. Navy disregarded this method of radio DF as impractical. However, Dr. Kolster retained interest in direction finding

---

This chapter was written by Robert L. Kellogg, Technical Fellow, The Boeing Company.

and by 1915, now as an employee of the U.S. Bureau of Standards, discovered that a vertical coil of wire wound on a rectangular frame could easily be rotated to incoming signals allowing practical radio DF. After demonstrations of his antenna, Navy admirals changed their attitude toward radio DF and accepted it as a valuable navigation aid. Twenty Kolster "SE 74" loop DF systems were quickly installed on U.S. ships, marking the beginning of U.S. maritime radio direction-finding systems [2].

9.1.2 Loop Antenna Fundamentals

In Chap. 3 we found the solution for a horizontal loop antenna in the x-y plane with loop normal aligned to the z axis. For a constant loop current I_o the far field electric and magnetic field vector components are described in terms of first-order Bessel functions and a $1/r$ dissipation term with a phase exponential:

$$\bar{E}_\phi = \eta \left(\frac{ka}{2}\right) I_o \frac{e^{-jkr}}{r} J_1(ka\sin\theta)\hat{\phi} \tag{9.1}$$

$$\bar{H}_\vartheta = -\frac{E_\phi}{\eta}\hat{\theta} = -\left(\frac{ka}{2}\right) I_o \frac{e^{-jkr}}{r} J_1(ka\sin\theta)\hat{\theta} \tag{9.2}$$

The Poynting vector of radiation propagation is directly derived as the cross product of the electric and conjugate magnetic vectors:

$$S(\theta,\phi)\hat{r} = \frac{1}{2}(\bar{E}_\phi \times \bar{H}_\theta^*)$$

$$S(\theta,\phi)\hat{r} = \frac{1}{2}\left(\eta\left(\frac{ka}{2}\right) I_o \frac{e^{-jkr}}{r} J_1(ka\sin\theta)\hat{\phi} \times -\left(\frac{ka}{2}\right) I_o \frac{e^{+jkr}}{r} J_1(ka\sin\theta)\hat{\theta}\right)$$

$$S(\theta,\phi)\hat{r} = -\frac{\eta}{8} I_o^2 \frac{(ka)^2}{r^2} J_1^2(ka\sin\theta)\hat{r} \tag{9.3}$$

where θ, ϕ = polar (zenith) angle and azimuth angle from x axis, respectively
a = loop radius, m
k = wave number $2\pi/\lambda$ where λ is the signal wavelength, m
k can be written in terms of permeability μ and permittivity ε as $k = \frac{2\pi}{\lambda} = \frac{2\pi f}{c} = \frac{\omega}{c} = \omega\sqrt{\mu\varepsilon}$
r = far field distance from the center of the loop, m
η = complex impedance of the medium = $\sqrt{\frac{\mu}{\varepsilon}}$ (ohms) in free space $\eta = 120\pi = 377$ ohms
I_o = loop current, A
J_1 = Bessel function of the first kind

The antenna radiation intensity pattern $U(\theta, \phi)$ is formed using the range independent terms of the Poynting vector in absolute terms taken from Eq. (9.3):

$$U(\theta,\phi) = r^2|S(\theta,\phi)| = \frac{\eta}{8} I_o^2 (ka)^2 |J_1^2(ka\sin\theta)| \tag{9.4}$$

In the VLF, HF, and lower VHF bands where the loop radius is only a fraction of the wavelength ($a < .03\ \lambda$), the Bessel function can be approximated as a Taylor series expansion:

$$J_1(ka\sin\theta) = \frac{1}{2}(ka\sin\theta) - \frac{1}{16}(ka\sin\theta)^3 + \cdots \tag{9.5}$$

Using the Bessel approximation in the electric and magnetic vectors and radiation intensity pattern gives

$$\bar{E}_\phi = \eta\left(\frac{ka}{2}\right)^2 I_o \frac{e^{-jkr}}{r}\sin\theta\ \hat{\phi} \tag{9.6}$$

$$\bar{H}_\theta = -\left(\frac{ka}{2}\right)^2 I_o \frac{e^{-jkr}}{r}\sin\theta\ \hat{\theta} \tag{9.7}$$

$$U(\theta,\phi) = \frac{\eta}{32}I_o^2(ka)^4\left|\sin^2\theta\right| \tag{9.8}$$

The horizontal loop acts much like the vertical dipole, replacing the electric moment with a magnetic moment. The result is an electric field vector horizontally polarized (E_ϕ) and a magnetic field vector that is vertical (H_θ).

We make several observations about the radiation pattern of the horizontal loop: (a) it depends only on the zenith polar angle θ, which means that the radiation pattern is axially symmetric about the polar z axis; (b) the horizontal loop is only sensitive to field components that project onto a horizontal E-field ($E_h = E_\phi$) with vertically aligned magnetic field.

The development of Eqs. (9.1) and (9.2) and their approximations (9.6) through (9.8) were done using a constant circular loop current I_o. If the current remains constant, other small loop geometries can be evaluated, such as a square or octagonal loop. The result is that Eqs. (9.6) through (9.8) still obtain, but are scaled simply by the loop area. Hence an important property of small loops is that the loop shape does not matter, *only* the loop area, A_{loop}.

9.1.3 Vertical Loop Antennas

By turning a horizontal loop to the vertical position (with the loop current in the x-z plane and the loop normal pointing toward the y axis), we discover why the Kostler loop was so successful for radio direction finding.

An inspection of the radiation pattern (Fig. 9.1) shows that on the horizon ($\theta = 90°$) the radiation pattern appears to respond to a vertical polarized wave as a simple spatial dipole pattern with maxima at $\phi = 0°$ and $180°$ along the x axis and minima at $\phi = \pm 90°$ on the y axis.

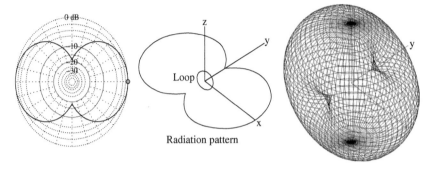

Figure 9.1 Radiation pattern of a vertical loop on x-z axis. (*Pattern created in EZNEC.*)

9.1.4 Vertical Loop Matched Polarization

To describe the vertical loop radiation pattern, we use an auxiliary angle ξ to measure the polar distance from the loop's normal now located along the y axis. This angle alone describes the loop response to a *matched* polarized wave. Maintaining the standard definition of polar zenith angle θ from the z axis and azimuth ϕ from the x axis, the Law of Cosines for a spherical triangle gives

$$\cos\xi = \sin\theta\sin\phi \qquad (9.9)$$

Using the trigonometric relation $\sin^2\xi = 1-\cos^2\xi$, the vertical loop radiation intensity pattern for a matched polarized wave becomes

$$U(\theta,\phi) = \beta|(1-\sin^2\theta\sin^2\phi)| \qquad (9.10)$$

where

$$\beta = \frac{\eta}{2}I_o^2 k^4 \left(\frac{A_{\text{loop}}}{4\pi}\right)^2 = \frac{\eta}{2}\frac{\pi^2}{\lambda^4}I_o^2 N^2 A_{\text{loop}}^2 \qquad (9.11)$$

The radiation pattern for matched polarization [Eq. (9.10) seen in Fig. 9.2] maintains maximum response at $\phi = 0°$ and $180°$, but as elevation angle increase, the minimum at $\phi = \pm 90°$, becomes shallower and shallower.

9.1.5 Vertical Loop with Polarized Signal

We would like to refer to the loop response in the presence of signal polarization. At $\theta = 90°$, we expect that a vertical loop only responds to the horizontal magnetic component. However, this is awkward because the majority of signal propagation refers to the orientation of the electric field component. Keeping this in mind, we let the incoming wave be described by a combination of vertical (E_v) and horizontal (E_h) electric field components *even though the loop responds to the magnetic component*.

Direction Finding 259

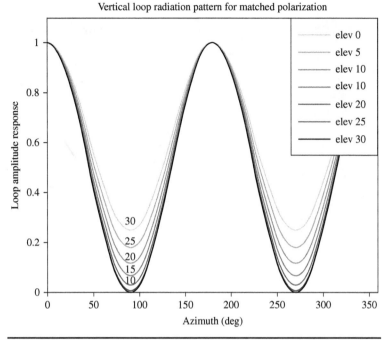

FIGURE 9.2 Vertical loop normalized radiation intensity pattern $U(\theta, \phi)$ for matched polarization where elevation is $90° - \theta$.

To determine how these components interact with the vertical loop, again use spherical trigonometry: this time to create the angle ψ between the polar arc α that extends from the signal direction origin at (θ, ϕ) to the y axis and the vertical arc $90° - \theta$ (elevation arc angle) that extends from the signal direction origin directly to the horizon. This last arc makes a $90°$ angle with the horizon as shown in Fig. 9.3. Using both the Law of Sines and Law of Cosines gives

$$\cos\Psi = \frac{\cos\phi}{\sin\xi} \quad \text{and} \quad \sin\Psi = \frac{\cos\theta \sin\phi}{\sin\xi} \qquad (9.12)$$

The vertical loop will respond to an electric field that is perpendicular to arc ξ. [This is the old E_ϕ of the horizontal loop described in Eq. (9.1).] Denoting this matched response as E_ψ in terms of the electric field vectors

$$E_\psi = E_v \cos\Psi - E_h \sin\Psi \qquad (9.13)$$

Therefore, the total loop response will be proportional to $\sin\xi$, the loop's polar angle

$$E_{\text{loop}} = E_\psi \sin\xi \qquad (9.14)$$

260 Chapter Nine

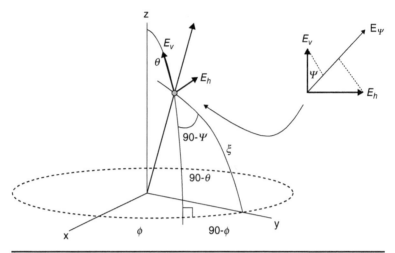

FIGURE 9.3 Spherical trigonometry relations for E_v and E_h components.

Expanding E_ψ and using trigonometric identities, we get

$$E_{\text{loop}} = E_v \cos\phi - E_h \sin\phi \cos\theta \qquad (9.15)$$

This is an important result showing that for an incoming vertical wave with no E_h component, the loop response is *independent* of the zenith angle θ. Put another way, the vertical loop response is independent of signal elevation angle for vertically polarized waves.

Equation (9.15) only accounts for linear polarization. A more general description of polarization must account for the phase offset between the vertical electric and magnetic field vectors (creating left or right circularly or elliptically polarized waves). Therefore, let the electric field be represented by the general term

$$E_{\text{loop}} = E_v \cos\phi - E_h \sin\phi \cos\theta \cdot e^{+j\Phi} \qquad (9.16)$$

where $\tan\Psi = \dfrac{E_h}{E_v}$ and ψ is the polarization angle from vertical and Φ is the polarization phase ($\pi/2$ = right circular received signals).

9.1.6 Cross-Loop Array and the Bellini-Tosi Radio Goniometer

While Kolster created the first practical loop DF antenna, other improvements were being made. Bellini and Tosi created the first workable, if unappreciated, DF system in 1907 [3, 4]. The B-T DF system used crossed-loop antennas, originally made in the shape of the triangle "delta loops" and later the Kolster vertical loop antennas. On ships the delta loops were very large, attached between ship masts or stack and run at angles to the ship deck. Cables carried the magnetically induced currents below deck into two small crossed loops with

Direction Finding

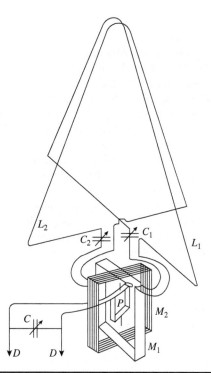

FIGURE 9.4 Schematic of Bellini-Tosi radio goniometer direction-finding system. (*From Radio Engineering Principles, Lauer & Brown, McGraw-Hill, 1920.*)

the same orientations as the above-deck loops. The below-deck B-T equipment used a rotating pickup coil, called a search coil or "radio goniometer" inside the two small sense loops to determine the direction of radio energy. The radio goniometer responded with maximum current output (D-D in Fig. 9.4) when aligned with the incoming signal direction.

In 1915, the Marconi Radio Company purchased the rights to the B-T system and used it widely on ships because the main loop antennas could remain fixed in the superstructure of the ship and a below-deck operator could easily turn the radio goniometer coil by hand to find maximum or minimum signal amplitude. The whole below-deck equipment was in one Marconi equipment box.

By 1915, British naval intelligence had been intercepting and direction-finding almost every German signal. The British not only determined the singal's line of bearing, but with good luck, obtained the German naval codes such that the Admiralty knew of the departure of each German U-boat as it left for patrol [5].

Originally operators listened to the signal as they rotated the goniometer to determine both minimum and maximum signal amplitude. Maximum signal was used to identify the radio signal and minimum was used for DF determination. After World War I when audion (amplifier) tubes became widely available, Robert Watson-Watt, the British radar pioneer, experimented with cathode-ray tubes (CRT) for DF display. In 1936 a French émigré Henri Busignies developed a naval HFDF system called the Model DAJ, using CRT display for the International Telephone and Telegraph Company. Deployment was postponed as issues of shipboard power required redesign. In the end another system won the honor: "The first shipboard HFDF to employ a CRT to display bearing information was called FH4, an experimental system installed on HMS Culver (Y-87), the former U.S. Coast Guard cutter *Mendota*, in October 1941. Unfortunately, the ship was torpedoed and sunk in January 1942" [6].

We begin to show how crossed loop perform DF by using Eq. (9.15) and create the ratio of voltage from a loop called B_x for the north-south loop (placed in the y-z plane and normal vector along the x axis) and B_y for the east-west loop (placed in the x-z plane and normal vector along the y axis). Keeping the east-west loop for azimuth ϕ reference and assuming linear polarization:

$$\frac{B_x}{B_y} = \frac{E_v \sin\phi - E_h \cos\theta \cos\phi}{E_v \cos\phi - E_h \cos\theta \sin\phi} = \frac{E_v \sin\phi - E_h \sin(elev)\cos\phi}{E_v \cos\phi - E_h \sin(elev)\sin\phi} \quad (9.17)$$

Equation (9.17) reduces a simple tangent of approximate signal direction $\alpha(+n\pi)$ when only vertically polarized signal waves are present.

$$\tan\alpha = \frac{B_x}{B_y} = \frac{\sin\phi}{\cos\phi} = \tan\phi \text{ (vertically polarized wave response)} \quad (9.18)$$

If a horizontal component is present, the additional term in Eq. (9.17) alters the B_x/B_y ratio and alters the estimate of signal direction if only the tangent approximation is used. (Generally the elevation angle of the incoming signal is unknown.) In the 1920s the alteration of apparent signal direction was known as the "night-effect." For example, Smith-Rose [7] graphed the changes in observed line of bearing from a transmitter at Saint Assise, France, to his Physical Laboratory at Teddington, England (Fig. 9.5).

Two research groups, one in the United States and one in Great Britain, proved the existence of the Kennelly-Heaviside propagation layer. In the U.S. Gregory Breit and Merle Antony Tuve used a pulsed transmitter (the forerunner of radar) to measure the time of echoes from the Kennelly-Heaviside layer. At the same time Edward Appleton at Ditton Park verified the existence of the free electrons 95 to 100 km above the earth's atmosphere [8], which Robert Watson-Watt, pioneer inventor of radar, dubbed the "ionosphere" in 1926 [9]. At kilohertz frequencies, the radio waves were refracted by the E-layer ionosphere. As can be seen in Smith-Rose's graph, the directional error

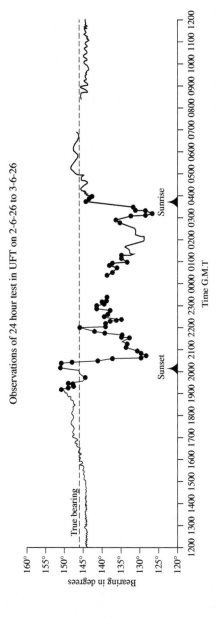

Figure 9.5 Observed signal bearing of Saint Assise taken at Teddington during the 24-hour period June 2 to 3, 1926 from a signal transmitted at 21 kHz (λ = 14.3 km). (*Smith-Rose.*)

is particularly large at sunrise and sunset when the ionosphere has electron gradients that create significant horizontal radio components to shift the apparent line of bearing by 15°.

How do we measure the ratio B_x/B_y and determine line of bearing? In today's digital world, start by making vectors of paired B_x and B_y measurements, written in the low transpose form as

$$\mathbf{B}_x = [B_x(1), B_x(2), ... B_x(K)]^T \qquad k = 1, 2, ... K$$
$$\mathbf{B}_y = [B_y(1), B_y(2), ... B_y(K)]^T \qquad (9.19)$$

and combine them into a measurement matrix

$$\mathbf{B} = [\mathbf{B}_x, \mathbf{B}_y] \qquad (9.20)$$

From the measurement matrix we create the covariance matrix $\mathbf{R}$. If we have tuned our radio DF system correctly, $\mathbf{R}$ will be predominantly signal energy.

$$\mathbf{R} = [\mathbf{B} \cdot \mathbf{B}^T] = \begin{bmatrix} <\mathbf{B}_x^2> & <\mathbf{B}_x\mathbf{B}_y> \\ <\mathbf{B}_x\mathbf{B}_y> & <\mathbf{B}_y^2> \end{bmatrix} \qquad (9.21)$$

where $<> = \sum_{k=1}^{K} ()_k$.

The covariance matrix has two principal axes, with the major axis aligned with B_y/B_x. The semimajor and semiminor axis lengths are equal to the square root of the eigenvalue solutions resulting from the covariance matrix determinant:

$$|\mathbf{R} - \lambda \mathbf{I}| = \left| \begin{bmatrix} <\mathbf{B}_x^2> - \lambda & <\mathbf{B}_x\mathbf{B}_y> \\ <\mathbf{B}_x\mathbf{B}_y> & <\mathbf{B}_y^2> - \lambda \end{bmatrix} \right| = 0 \qquad (9.22)$$

This in turn gives rise to the polynomial solution for the eigenvalues as

$$\lambda = \frac{<\mathbf{B}_x^2> + <\mathbf{B}_y^2>}{2} \pm \frac{[(<\mathbf{B}_x^2> + <\mathbf{B}_y^2>)^2 - 4(<\mathbf{B}_x^2><\mathbf{B}_y^2> - <\mathbf{B}_x\mathbf{B}_y>^2)]^{1/2}}{2}$$

(9.23)

and rotation angle (from the x axis) complemented to azimuth is given by

$$\alpha = \frac{\pi}{2} - \frac{1}{2}\tan^{-1}\left(\frac{2<\mathbf{B}_x\mathbf{B}_y>}{<\mathbf{B}_x^2> - <\mathbf{B}_y^2>}\right) \qquad (9.24)$$

When determining the bearing angle by comparing the loop voltage from the B_x and B_y antennas or using the covariance matrix approach above, the bearing is ambiguous, with the possibility of 180° in error. In most navigation applications the operator taking bearings of a shore beacon had sufficient information to determine which angle is true.

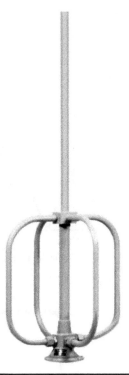

FIGURE 9.6 AS-145 DF antenna. (*South-West Research Institute.*)

However, when an unknown signal bearing is taken, the bearing is ambiguous.

An additional antenna can be used to perturb the symmetric pattern of the loop, allowing amplitude to differentiate direction. However, in modern direction finding the additional antenna is more important for measuring relative phase. A classic example of this is the AS-145 "eggbeater" antenna (Fig. 9.6) used by the U.S. Navy. It features two crossed rectangular loops and a reference omni-directional monopole.

9.1.7 Loop Array Calibration

Most loop antenna calibration measurements are made with vertically polarized radio waves, comparing amplitude and phase to a reference antenna. The amplitude and phase response of a small DF antenna are seen in Fig. 9.7 from measured data. The 180° phase reversal of the E-W loop is seen in Fig. 9.7*b*, where the phase reversal occurs when the loop has a null in its beam.

As an example of a smart antenna DF approach using complex voltages that contain both amplitude and phase, we consider simple Bartlett correlation direction finding.

Let the subscripts o represent the omni-vertical monopole response, c represent the "cosine" north-south antenna response, and

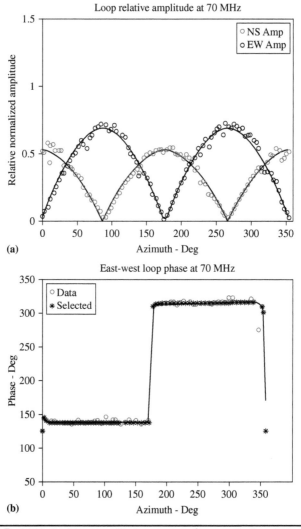

Figure 9.7 (a) Vertical loop amplitude response for a vertically polarized wave (red = north-south-oriented loop, blue = east-west-oriented loop); (b) phase for the east-west vertical loop compared to a vertical dipole antenna.

s represent the "sine" east-west antenna response. At a calibration range the antenna array of monopole and crossed loops is carefully rotated while being illuminated by an incoming plane wave. At each azimuth ϕ the complex voltages of the loop antennas are measured with respect to the monopole:

$$\hat{a}_c(\phi) = \frac{v_c}{v_o} = \frac{|v_c|e^{j(\Phi_c)}}{|v_o|e^{j(\Phi_o)}} = \frac{|v_c|}{|v_o|}e^{j(\Phi_c - \Phi_o)}, \quad \text{for } \phi = 1° \text{ to } 360° \quad (9.25)$$

$$\hat{a}_s(\phi) = \frac{v_s}{v_o} = \frac{|v_s|e^{j(\Phi_s)}}{|v_o|e^{j(\Phi_o)}} = \frac{|v_c|}{|v_o|}e^{j(\Phi_s - \Phi_o)} \quad (9.26)$$

The reference antenna response for all azimuths is

$$\hat{a}_o(\phi) = \frac{v_o}{v_o} = 1 \quad (9.27)$$

These three normalized responses form the omni, sine, cosine antenna array manifold.

$$\mathbf{a}(\phi) = \begin{bmatrix} 1 \\ \hat{a}_s(\phi) \\ \hat{a}_c(\phi) \end{bmatrix} = \begin{bmatrix} 1 & 1 & 1 & \cdots & 1 \\ \hat{a}_s(\phi_1) & \hat{a}_s(\phi_2) & \cdots & \hat{a}_s(\phi_{360}) \\ \hat{a}_c(\phi_1) & \hat{a}_c(\phi_2) & \cdots & \hat{a}_c(\phi_{360}) \end{bmatrix} \quad (9.28)$$

For antenna arrays that cover 360° in azimuth, the array calibration may occur at every 1 to 2 degrees. In Eqs. (9.25) to (9.28) we have ignored variations in polarization, elevation angle, and frequency. In the real world these are important parameters that significantly affect the antenna pattern and relative received voltages. Therefore, we expect that we may have a series of array manifolds:

$$\mathbf{a}(\phi) \rightarrow \mathbf{a}(\phi, \theta, f, E_V, E_H)$$

As a result of these additional parameters, the antenna array manifold can become quite large. For example, assume that the crossed-loop plus monopole array voltage is measured at 2° azimuth steps for 5° elevation steps from −20° to +20°. For EM surface wave signal propagation over the ocean, we need only calibrate for vertical polarization. Consider calibrating the array over 30 MHz in the high-frequency (HF) spectrum where array manifolds measurements are required every 10 kHz. The size of the omni-sine-cosine array requires

No. of measurements = elements × ϕ × θ × polarization × frequency

= 3 × 180 × 9 × 1 × 3000 = 14,580,000

Thankfully, most antenna calibration facilities are fully automated. Nevertheless, antenna calibration is a major undertaking, usually requiring many hours of calibration (see Prob. 9.4).

9.2 Adcock Dipole Antenna Array

An early but effective DF antenna that rivaled the vertical loop antenna was the Adcock antenna array, patented by F. Adcock in 1919 [10] and made popular by British radio experimentalist Smith-Rose [11]:

> It is now well known that the night errors experienced in the use of closed-coil direction-finders [vertical loops] are due to the action of the horizontally polarized component of the electric force [field]

268 Chapter Nine

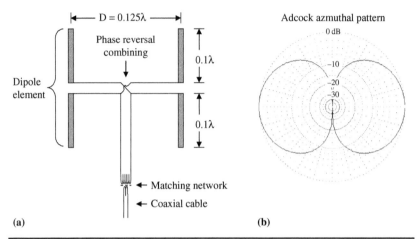

Figure 9.8 (a) Adcock dipole paired elements and (b) azimuthal response.

producing an error in observed bearing... It will be evident, therefore, that any receiving system which is unaffected by a horizontal component of electric force [field] would be free from night errors, even though the vertically polarized down coming waves [from the ionosphere] might still produce variations in received signal strength. A direction-finding receiving arrangement which fulfills this condition was patented by Adcock in 1919, but this system does not appear to have received practical consideration until Mr. Barfield and the author [Smith-Rose] experimented with it in 1926 [12].

The Adcock "U-type" DF array consists of a pair of spaced vertical dipoles (or monopoles on a ground plane) arranged to rotate about a central vertical axis (Fig. 9.8a). The Adcock elements are phase-reversed to create a beam pattern similar to the vertical loop in the horizontal plane (Fig. 9.8b). The Adcock pair was used successfully to replace vertical loop antenna, and crossed Adcock pairs (called an Adcock quadrature array) could easily be used in a Belli-Tossi system using the small radio goniometer to determine the direction of signal arrival.

From the late 1920s to the mid 1950s the Adcock array was popular for high-frequency direction finding (HFDF). During World War II the Allies created HFDF networks to triangulate and locate German ships and submarines as well as feed raw signal text to Enigma code breaking analysts. There were 17 American stations stretching from Adak, Alaska to the Caribbean, as well as 20 British and 11 Canadian stations. Their DF signal bearings were sent to plotting rooms in Washington, London, and Ottawa [13].

9.2.1 Watson-Watt DF Algorithm

Before modern digital "smart antenna" DF existed, an analog phase comparison system using the Watson-Watt algorithm was commonly used with the Adcock array. To understand the Watson-Watt algorithm, we consider a signal with modulation $m(t)$ and the received signal $s(t)$ at each of the four Adcock elements using the subscripts N, S, E, and W. The quadrature elements are set on a circle of radius $d/2$, resulting in the phased received signals of

$$S_N = m(t) e^{-jk\frac{d}{2}\sin\theta\cos(\phi+\pi/2)} = m(t) e^{+j\frac{\pi d}{\lambda}\sin\theta\sin\phi} \qquad (9.29)$$

$$S_S = m(t) e^{-jk\frac{d}{2}\sin\theta\cos(\phi-\pi/2)} = m(t) e^{-j\frac{\pi d}{\lambda}\sin\theta\sin\phi} \qquad (9.30)$$

$$S_E = m(t) e^{-jk\frac{d}{2}\sin\theta\cos(\phi-\pi)} = m(t) e^{+j\frac{\pi d}{\lambda}\sin\theta\cos\phi} \qquad (9.31)$$

$$S_W = m(t) e^{-jk\frac{d}{2}\sin\theta\cos(\phi-0)} = m(t) e^{-j\frac{\pi d}{\lambda}\sin\theta\cos\phi} \qquad (9.32)$$

The Watson-Watt DF method is to compare the signal difference of the N-S elements with the E-W difference. That is, following Adcock, the phase between antenna pairs is inverted

$$S_{NS} = m(t)\left\{ e^{+j\frac{\pi d}{\lambda}\sin\theta\sin\phi} - e^{-j\frac{\pi d}{\lambda}\sin\theta\sin\phi} \right\} \qquad (9.33)$$

and

$$S_{EW} = m(t)\left\{ e^{+j\frac{\pi d}{\lambda}\sin\theta\cos\phi} - e^{-j\frac{\pi d}{\lambda}\sin\theta\cos\phi} \right\} \qquad (9.34)$$

Recalling the exponential form of the sine function,

$$S_{NS} = 2m(t)\sin\left(\frac{\pi d}{\lambda}\sin\theta\sin\phi\right) \text{ and } S_{EW} = 2m(t)\sin\left(\frac{\pi d}{\lambda}\sin\theta\cos\phi\right) \qquad (9.35)$$

The comparison of the N-S and E-W channels gives the signal azimuth α.

$$\tan\alpha = \frac{S_N - S_S}{S_E - S_W} = \frac{S_{NS}}{S_{EW}} = \frac{\sin\left(\frac{\pi d}{\lambda}\sin\theta\sin\phi\right)}{\sin\left(\frac{\pi d}{\lambda}\sin\theta\cos\phi\right)} \qquad (9.36)$$

To simplify, we construct the Adcock quadrature array such that $d \ll \lambda$ (typically the array size uses $d \sim \lambda/8$), which allows us to approximate the sine of a small quantity as the quantity itself, resulting in

$$\tan\alpha = \frac{S_N - S_S}{S_E - S_W} \approx \frac{\sin\phi}{\cos\phi} \qquad (9.37)$$

The comparison of the N-S and E-W signals is done to extract phase, not amplitude. Equation (9.37) shows that by taking the ratio of the phase differences, the azimuth α can be estimated by a simple trigonometric function of voltage ratios.

Today Adcock DF systems using a quadrature array are popular for small, mobile DF systems in the VHF and UHF frequency bands, but we are no longer tied to Watson-Watt algorithms. The Adcock quadrature array creates three outputs corresponding to a 4-element summed omni-directional reference channel and a "cosine" and "sine" azimuthal beam channel allowing correlation DF as outlined for the loop antennas above.

9.3 Modern DF Applied to Adcock and Cross-Loop Arrays

Since the 1960s, digital approaches to DF have been most effective, all relying on the antenna array voltage covariance matrix $\mathbf{R}_{vv}$ and corresponding array manifold $\mathbf{a}(\phi)$ (or its extensions in elevation, frequency, and polarization). The algorithm can be used with loop, monopole, and dipole arrays. Here we illustrate the mathematical approach for the Adcock array and may be equally applied to the crossed-loop array, assuming that the array also has an omni-directional reference antenna. In either array, the covariance matrix of complex voltages $\mathbf{R}_{vv} = [\mathbf{v}\mathbf{v}^H]$ is based on the omni, sine, and cosine antenna relative voltages:

$$\mathbf{v} = \begin{bmatrix} 1 \\ <\hat{v}_s> \\ <\hat{v}_c> \end{bmatrix}$$

where $<\hat{v}_c> = \left\langle \dfrac{v_c}{v_o} \right\rangle = \sum\limits_{k=1}^{K} \dfrac{|v_{ck}|}{|v_{ok}|} e^{j(\Phi_{ck} - \Phi_{ok})}$, and similar for $<\hat{v}_s>$. (9.38)

The simplest correlation algorithm is Bartlett correlation. A correlation coefficient $C(\phi)$ is formed for every discrete azimuth ϕ in the antenna manifold:

$$C(\phi) = \left(\dfrac{1}{trace(\mathbf{R}_{vv})} \right) \dfrac{\mathbf{a}(\phi)^H \mathbf{R}_{vv} \mathbf{a}(\phi)}{\mathbf{a}(\phi)^H \mathbf{a}(\phi)} \quad (9.39)$$

The coefficient $C(\phi)$ is normalized to values 0-1 by the multiplication of $1/trace(\mathbf{R}_{vv})$ and the superscript H in forming the antenna manifold matrix is the Hermetian operator, which means taking the matrix complex conjugate transpose.

Once $C(\phi)$ is formed, the Bartlett DF process looks for the maximum $C(\phi)$ using standard peak finding and interpolation techniques

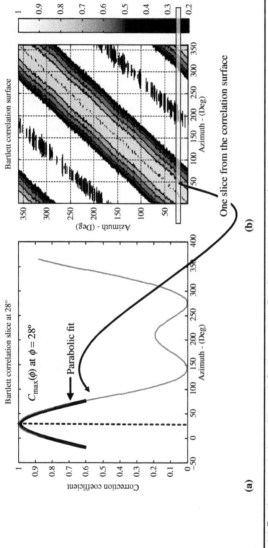

Figure 9.9 (a) Bartlett correlation with peak finding and (b) Bartlett correlation surface.

(Fig. 9.9a). In a simulation over all signal azimuth angles, we see the correlation surface (Fig. 9.9b).

Notice that the correlation peak can be approximated by a parabola of the form:

$$C(\phi) \approx a\phi^2 + b\phi + c \qquad (9.40)$$

and the maximum of the parabola that represent the DF direction is formed by

$$\frac{\partial C(\phi)}{\partial \phi} = 2a\phi + b = 0 \qquad (9.41)$$

hence

$$\phi = -\frac{b}{2a} \qquad (9.42)$$

The parabolic coefficients can be rapidly computed from sequential correlation measures near the parabolic peak. In fact, only three measures are necessary (see Prob. 9.7).

9.4 Geolocation

Until recently, the idea of geolocating radio signals was the provenance of either military intercept or navigational services. Today geolocation services are performed through many different techniques, including global positioning satellites (GPS) and their equivalents such as GLONASS. The key is measuring time of arrival from multiple satellites, and within the user handsets, solving for both position and time.

Coordinated sites, whether they be local cell towers or the World Wide Lightning Location Network (WWLLN) [14], use time difference of arrival (TDOA) hyperbolic geolocation. And low-cost geolocation estimates in cell phones can be made by listening to the received signal strength (RSS) of local transmitters.

Perhaps the oldest and most interesting is angle-of-arrival geolocation using multiple direction-finding sites or "running lines of bearing" from a moving platform such as car, ship, or airplane.

The classic method of triangulating lines of bearing goes back to Stansfield in 1947 [15]. His approach was designed for a "flat earth." In the 1970s Dennis Wangsness took a slightly different approach with a new DF algorithm for lines of bearing geolocation on a spherical earth [16]. However, examining Stansfield's original approach using modern linear algebra will prove instructive.

9.4.1 Stansfield Algorithm

We begin by examining the kth DF intercept position ($k = 1 \ldots K$) and the associated kth line of bearing toward the emitting signal at location **s** (Fig. 9.10).

Direction Finding 273

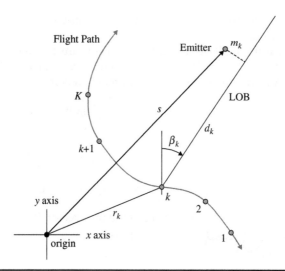

FIGURE 9.10 Schematic of multiple lines of bearing DF resulting in emitter geolocation.

The Cartesian reference system has an arbitrary origin, but within this reference system are vectors representing emitter, the DF intercept position, and the line of bearing (using distance of closest approach):

$$\mathbf{s} = \begin{bmatrix} s_x \\ s_y \end{bmatrix} \quad \text{vector position of the emitting signal (unknown)} \quad (9.43)$$

$$\mathbf{r}_k = \begin{bmatrix} r_x \\ r_y \end{bmatrix} \quad \text{vector position of the } k\text{th DF position (known)} \quad (9.44)$$

$$\mathbf{d}_k = \begin{bmatrix} d_x \\ d_y \end{bmatrix} \quad \text{vector of LOB to point of closest approach to emitter (unknown)} \quad (9.45)$$

$$\mathbf{m}_k = \begin{bmatrix} m_x \\ m_y \end{bmatrix} \quad \text{vector miss distance from LOB to emitter (unknown)} \quad (9.46)$$

For each $k = 1 .. K$ observations, we can write the emitter vector as the sum of the kth DF site position, plus the LOB vector to the point of closest approach and the miss distance vector:

$$\mathbf{s} = \mathbf{r}_k + \mathbf{d}_k + \mathbf{m}_k \quad (9.47)$$

Though the LOB vector and miss distance are unknown, their directions are known from the measured angle of arrival. Here the line of bearing (DF angle of arrival) is measured β_k clockwise from north (y axis). Therefore the kth LOB and miss vectors can be written

$$\mathbf{d}_k = d_k \begin{bmatrix} \sin \beta_k \\ \cos \beta_k \end{bmatrix} \quad \text{and} \quad \mathbf{m}_k = m_k \begin{bmatrix} -\cos \beta_k \\ \sin \beta_k \end{bmatrix} \quad (9.48)$$

where d_k and m_k are the kth LOB distance to point of closest approach and miss distance.

Because the LOB vector and miss vector are perpendicular, we expect that the dot product between them is zero. Here we express the dot product as a vector transpose times a vector:

$$\mathbf{m}_k^T \mathbf{d}_k = m_k d_k [-\cos \beta_k \quad \sin \beta_k] \begin{bmatrix} \sin \beta_k \\ \cos \beta_k \end{bmatrix} \quad (9.49)$$

giving

$$\mathbf{m}_k^T \mathbf{d}_k = m_k d_k (-\cos \beta_k \sin \beta_k + \sin \beta_k \cos \beta) = 0 \quad (9.50)$$

We could just as easily have used the unit vector of miss distance transpose, $\mathbf{u}_k^T = [-\cos \beta_k \quad \sin \beta_k]$, and using it as a multiplier on both sides of Eq. (9.47)

$$\mathbf{u}_k^T \mathbf{s} = \mathbf{u}_k^T \mathbf{r}_k + \mathbf{u}_k^T \mathbf{d}_k + \mathbf{u}_k^T \mathbf{m}_k \quad (9.51)$$

but by Eq. (9.50) we know that $\mathbf{u}_k^T \mathbf{d}_k \equiv 0$, giving

$$\mathbf{u}_k^T \mathbf{s} = \mathbf{u}_k^T \mathbf{r}_k + \mathbf{u}_k^T \mathbf{m}_k \quad (9.52)$$

Upon examination of the term $\mathbf{u}_k^T \mathbf{r}_k$, we realize that it is a known factor:

$$\mathbf{u}_k^T \mathbf{r}_k = [-\cos \beta_k \quad \sin \beta_k] \begin{bmatrix} r_{xk} \\ r_{yk} \end{bmatrix} = -r_{xk} \cos \beta_k + r_{yk} \sin \beta_k \quad (9.53)$$

The last term of Eq. (9.52) multiplies the transpose of the miss unit vector times the miss distance vector. This multiplication eliminating all directional terms leaving miss distance:

$$\mathbf{u}_k^T \mathbf{m}_k = m_k [-\cos \beta_k \quad \sin \beta_k] \begin{bmatrix} -\cos \beta_k \\ \sin \beta_k \end{bmatrix} = m_k (\cos^2 \beta_k + \sin^2 \beta_k) = m_k \quad (9.54)$$

We now write Eq. (9.52) for each kth observation ($k = 1 .. K$) as a matrix equation:

$$\begin{bmatrix} -\cos \beta_1 & \sin \beta_1 \\ \cdots & \cdots \\ -\cos \beta_K & \sin \beta_K \end{bmatrix} \mathbf{s} = \begin{bmatrix} -r_{x1} \cos \beta_1 + r_{y1} \sin \beta_1 \\ \cdots \\ -r_{xK} \cos \beta_K + r_{yK} \sin \beta \end{bmatrix} + \begin{bmatrix} m_1 \\ \cdots \\ m_K \end{bmatrix} \quad (9.55)$$

This is of the form

$$\mathbf{A}\mathbf{s} = \mathbf{X} + \mathbf{m} \quad (9.56)$$

where $\mathbf{X}$ is known, and $\mathbf{m}$ is a vector of residuals.

Direction Finding

The least-square solution requires us to assume that the sum of the square of errors m_k^2 will be minimum and the mean miss is unbiased $<m> = \sum_{k=1}^{K} m_k \Rightarrow \lim(K \to \infty) = 0$, leaving us to solve the fundamental equation:

$$\mathbf{As} = \mathbf{X} \qquad (9.57)$$

We'd like to invert the matrix $\mathbf{A}$ and immediately solve for the emitter position $\mathbf{s}$, but because $\mathbf{A}$ is not a square matrix (it is $K \times 2$), this can't be done. To make it square matrix and be invertible, multiply both sides of the equation by $\mathbf{A}^T$:

$$\mathbf{A}^T \mathbf{As} = \mathbf{A}^T \mathbf{X} \qquad (9.58)$$

The matrix product $\mathbf{A}^T \mathbf{A}$ is square and can be inverted, allowing for the solution:

$$\mathbf{s} = (\mathbf{A}^T \mathbf{A})^{-1} \mathbf{A}^T \mathbf{X} \qquad (9.59)$$

The matrix inverse of transpose product times matrix transpose has a special name called the *pseudo inverse*, sometimes written as

$$\mathbf{A}^{\#} = (\mathbf{A}^T \mathbf{A})^{-1} \mathbf{A}^T \qquad (9.60)$$

We note that unlike most other geolocation algorithm, the Stansfield algorithm summarized in Eq. (9.59) does not require iteration. This is the genius of the Stansfield algorithm.

9.4.2 Weighted Least-Square Solution

The Stansfield algorithm of Eq. (9.59) gives the unweighted least-square solution for the emitter location, but it does not answer the question of how confident we are that the true location is nearby. Nor does it handle issues such as if different DF sites have different root mean square (rms) bearing accuracies. A weighted Stansfield algorithm emerges to solve these problems, but to properly determine the weights, we must first estimate the emitter location, hence requiring improvement iterations [17]. We start by altering the fundamental Stansfield equation (9.59), multiplying each side with weight ω:

$$\omega \mathbf{As} = \omega \mathbf{X} \qquad (9.61)$$

and when multiplied by $\mathbf{A}^T \omega$ on both sides,

$$\mathbf{A}^T \omega \omega \mathbf{As} = \mathbf{A}^T \omega \omega \mathbf{X} \qquad (9.62)$$

Using $\mathbf{W} = \omega \omega$ as the weighting matrix, we again solve for emitter position:

$$\mathbf{s} = (\mathbf{A}^T \mathbf{W} \mathbf{A})^{-1} \mathbf{A}^T \mathbf{W} \mathbf{X} \qquad (9.63)$$

The weighting matrix **W** is the reciprocal of the observation (a priori) error covariance matrix:

$$\mathbf{W} = \mathbf{R}_{AOA}^{-1} \qquad (9.64)$$

where $\mathbf{R}_{AOA}$ is a K × K matrix.

Giving the weighted least-square estimate of emitter position as

$$\mathbf{s} = \left(\mathbf{A}^T \mathbf{R}_{AOA}^{-1} \mathbf{A}\right)^{-1} \mathbf{A}^T \mathbf{R}_{AOA}^{-1} \mathbf{X} \qquad (9.65)$$

If observations are independent, the covariance matrix $\mathbf{R}_{AOA}$ is a K × K diagonal matrix. The diagonal elements contain the expected squares of the statistically expected one-sigma miss distances:

$$\mathbf{R}_{AOA} = \begin{bmatrix} \gamma_1^2 \sigma_1^2 & 0 & 0 \\ 0 & \gamma_k^2 \sigma_k^2 & 0 \\ 0 & 0 & \gamma_K^2 \sigma_K^2 \end{bmatrix} \quad \text{(independent observations)} \qquad (9.66)$$

and

$$\mathbf{R}_{AOA}^{-1} = \begin{bmatrix} 1/\gamma_1^2 \sigma_1^2 & 0 & 0 \\ 0 & 1/\gamma_k^2 \sigma_k^2 & 0 \\ 0 & 0 & 1/\gamma_K^2 \sigma_K^2 \end{bmatrix} \qquad (9.67)$$

where $\gamma_k^2 = (\mathbf{s} - \mathbf{r}_k)^T (\mathbf{s} - \mathbf{r}_k)$ is the square of the estimated distance from each DF position $\mathbf{r}_k$ to the estimated emitter position at **s**

σ_k^2 = squared rms error (variance) of each DF line of bearing; this is an a priori statistical estimate of DF accuracy and is a function of such factors as DF antenna array baseline and signal-to-noise ratio

$\gamma_k \sigma_k$ = the statistical miss distance of the kth bearing that is approximated by the actual kth miss distance magnitude $|m_k|$

Note that for efficiency, the matrices **A**, **A**^{T}, and **X** have already been computed for the first Stansfield solution [Eq. (9.59)], so this improvement to **s** and the formation of $(\mathbf{A}^T \mathbf{R}_{AOA}^{-1} \mathbf{A})^{-1}$ is fairly efficient.

9.4.3 Confidence Error Ellipse

If we want to determine the quality of the Stansfield geolocation, the approach is to determine an error ellipse surrounding the geolocation that contains the true location with a specified confidence (e.g., 90% or 95% confidence …).

We need to project the error covariance matrix $\mathbf{R}_{AOA}$ into the miss distance covariance matrix $\mathbf{R}_x$, from which we derive the confidence error ellipse. The first factor of Eq. (9.65) with weighted observations is that projection transform:

$$\mathbf{R}_x = (\mathbf{A}^T \mathbf{W} \mathbf{A})^{-1} = \left(\mathbf{A}^T \mathbf{R}_{AOA}^{-1} \mathbf{A}\right)^{-1} \qquad (9.68)$$

Direction Finding

The miss distances are often modeled by a multivariate normal distribution, resulting in a confidence ellipse in which the cumulative distribution function (*cdf*) has a specified probability of containment p_o (such as $p_o = 95\%$). The relation between p_o and the cumulative exponential of the two-dimensional (bivariate) normal distribution is

$$p_o \equiv p(miss < j_o) = 1 - \exp\left(-\frac{j_o^2}{2}\right) \quad (9.69)$$

The standard deviation j_o (measured in sigma) is treated as a scaling variable. Equation (9.69) can be rewritten to express this scaling factor as a function of confidence level (cumulative probability p_o) and is calculated for common value in Table 9.1:

$$j_o = [-2\ln(1-p_o)]^{1/2} \quad (9.70)$$

The variance scaling factor j_o^2 allows scaling of the error covariance matrix $\mathbf{R}_x$ created from $\mathbf{R}_{AOA}$ that used only "one-sigma" squared AOA error. Our approach is to use the scaled miss distance covariance matrix.

$$\mathbf{R}_{jo} = j_o^2 \mathbf{R}_x \quad (9.71)$$

To find the corresponding confidence ellipse from the scaled covariance matrix, we use principal component eigenvalues and follow the same mathematical approach used in Eqs. (9.21) through (9.23). This is a standard technique applicable to many problems in which principal components are required:

$$\left|\mathbf{R}_{jo} - \lambda \mathbf{I}\right| = 0 \quad (9.72)$$

where $\mathbf{R}_{jo} = jo^2 \mathbf{R}_x = jo^2 \begin{pmatrix} a & b \\ c & d \end{pmatrix}$ derived from Eq. (9.68).

For the two-dimensional Cartesian world of Stansfield, the determinant

$$\left|\mathbf{R}_{jo} - \lambda \mathbf{I}\right| = j_o^2 \begin{vmatrix} a-\lambda & b \\ c & d-\lambda \end{vmatrix} = 0 \quad (9.73)$$

p_o (Confidence Level)	j_o (Scaling Factor)
$p_o = 50\%$	1.1774
$p_o = 68\%$	1.5096
$p_o = 90\%$	2.1460
$p_o = 95\%$	2.4477
$p_o = 99\%$	3.0349

TABLE 9.1 Confidence Probability and Sigma Scaling Factors

resulting in a second-degree polynomial that can be solved as

$$\lambda_{jo} = j_o^2 \left\{ \frac{(a+d)}{2} \pm \left[\frac{(a+d)^2}{4} - (cd-bc) \right]^{1/2} \right\} \Rightarrow \{ j_o^2 \lambda_+ \text{ and } j_o^2 \lambda_- \} \qquad (9.74)$$

The square root of the eigensolutions of λ_{jo} are proportional to the semi-major and semi-minor axes of the confidence ellipse set at the p_o level. From Eq. (9.74) the result is

$$\text{semi-major axis} = j_o \sqrt{\lambda_+} \qquad (9.75)$$

$$\text{semi-minor axis} = j_o \sqrt{\lambda_-} \qquad (9.76)$$

with orientation

$$\tan(2\theta) = \frac{(b+c)}{(a-d)} \quad \theta \text{ measured from } x\text{-axis counter clockwise} \qquad (9.77)$$

or

$$\alpha = \frac{\pi}{2} - \frac{1}{2} \tan^{-1}\left(\frac{(b+c)}{(a-d)} \right) \quad \alpha \text{ measured from } y\text{-axis clockwise} \qquad (9.78)$$

This textbook includes a MATLAB procedure called "Stansfield" that allows simulation of lines of bearing from a moving platform. A variety of parameters can be changed, including intercept geometry and rms angle-of-arrival accuracy. Typical plots from the procedure are illustrated in Fig. 9.11.

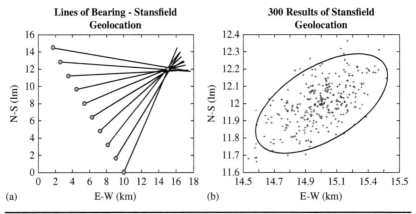

Figure 9.11 (a) Simulated lines of bearing geolocation trial; (b) 300 trial locations with 95% confidence ellipse.

9.4.4 Mahalanobis Statistics

The confidence ellipse derived from the miss distance covariance matrix was represented by the parameters of semi-major and semi-minor axes and orientation angle. In the preceding section it was treated as just one ellipse, specified by a single probability of containment p_o.

But, of course, we could create many confidence ellipses for our data. In fact, we can assume that each location with miss distance vector $\mathbf{m}_k$ resides on some confidence ellipse.

Figure 9.12 illustrates this point. At some level of confidence we have three different miss locations at positions (a), (b), and (c). They all have the same p_o confidence, but each $\mathbf{m}_k$ miss vector has a different length. Further, if we compare the miss locations (a) and (d), they lay on different confidence ellipses, yet their Euclidian miss distances are the same.

To resolve the issue of Euclidian miss distances and determine the appropriate confidence ellipse, we start by examining the Euclidian squared distance of the kth miss vector:

$$\mathbf{m}_k = \begin{bmatrix} m_x \\ m_y \end{bmatrix} \text{ has a Euclidean squared miss distance of } miss_k^2 = \mathbf{m}_k^T \mathbf{m}_k$$

(9.79)

We need a non-Euclidean measurement that stretches or compresses the Cartesian miss distance to match the longer distance along semi-major axis and shorter distance along semi-minor axis to remain on the same confidence level ellipse. This is accomplished using the error covariance matrix $\mathbf{R}_x$ (which, as we have seen, represents the "one-sigma" ellipse) to define the Mahalanobis squared distance as

$$J_k^2 = \mathbf{m}_k^T \mathbf{R}_x \mathbf{m}_k \qquad (9.80)$$

where J_k = Mahalanobis distance.

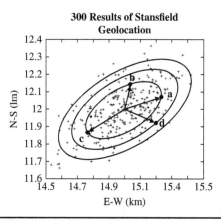

FIGURE 9.12 Confidence ellipse as continuous variable.

Chapter Nine

The Mahalanobis distance is related to the confidence ellipse probability by

$$P_k \equiv p(\text{miss} < J_k) = 1 - \exp\left(-\frac{J_k^2}{2}\right) \quad (9.81)$$

We see that each miss vector $\mathbf{m}_k$ can therefore be mapped to a confidence ellipse with a sigma scaling of

$$J_k = \left[\mathbf{m}_k^T \mathbf{R}_x \mathbf{m}_k\right]^{1/2} \quad \text{for } (k = 1 \ldots K) \quad (9.82)$$

With the set of $\{J_k\}$, we reorder the values in ascending order, expressing the ordered set as $\{J_i\}$ for $(i = 1 \ldots K)$. And now comes a further insight: The value of i can be used to express the cumulative probability of each J_k:

$$P_i = \frac{i}{K} \quad \text{for } (i = 1 \ldots K) \quad (9.83)$$

Usually the measured Mahalanobis statistic $\{J_i, P_i\}$ is plotted against the normal cumulative probability distribution $\{j_o, p_o\}$ for a range of j_o

$$p_o = 1 - \exp\left(-\frac{j_o^2}{2}\right) \quad (9.84)$$

where typically $0 \le j_o \le 4$ sigma.

Both theory and measured sigma and cumulative probability $\{j_o, p_o\}$ and $\{J_i, P_i\}$ are plotted on the same graph. If our geolocation estimates are unbiased and truly contain only the expected random normal errors modeled in the Stansfield algorithm for confidence ellipse of a specified probability (e.g., 95%), then we should get results such as Fig. 9.13.

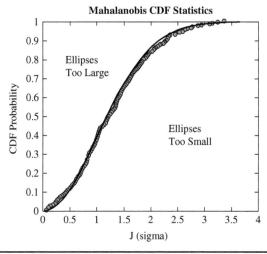

Figure 9.13 Geolocation Mahalanobis distance (grey circles) and normal cumulative distribution (black curve).

Direction Finding 281

Parameter	Value	Variance on Estimate
Location estimate	s_k	R_x
Miss vector estimate where s_o is known with an accuracy (variance) of R_o	$m_k = s_k - s_o$	$R_x + R_o$
Additional (unknown) bias and its variance	$\tilde{m}_k \equiv s_k - s_o - b$	$\tilde{R}_x \equiv R_x + R_o + R_b$
Mahalanobis miss distance	$\tilde{m}_k^T \tilde{R}_x \tilde{m}_k$	1

TABLE 9.2 Accounting for Bias and Variance

The interpretation of Fig. 9.13 showing the cumulative probability of Mahalanobis miss distances is very important during field calibration of direction-finding and geolocation systems. The process of determining and plotting the Mahalanobis miss distances is done under the assumption of zero mean error (i.e., from an observation site the mean line of bearing points toward the emitter) and the angle error is accurately represented by the rms error.

If the field system violates these assumptions, such as a corrupted antenna manifold that creates a biased line of bearing or a noisy process that changes the line of bearing rms error, then the CDF probabilities corresponding to the Mahalanobis miss distance will not lay on the expected theoretical curve, meandering off to represent miss distances too large or too small a confidence ellipse. This will become immediately obvious in the plotted graph, but the source of the error is not obvious. Bias error and unaccounted sources of noisy jitter (increasing variance) may be the root cause, but the Mahalanobis plots cannot be easily interpreted to discover the true source of the error. Table 9.2 presents an extended set of mathematics to follow bias and variance error.

9.5 References/Notes

1. Terman, F. E., *Radio Engineering*, McGraw-Hill, New York, 1947, p. 817.
2. Howeth, L. S., "History of Communications-Electronics in the United States Navy," U.S. Government Printing Office, Washington DC, 1963. Library of Congress Catalogue No. 64-62870. Chapter XXII "The Radio Direction Finder," pp. 261–265, reprinted at http://earlyradiohistory.us/1963hw22.htm.
3. Bellini, E., and A. Tosi, "A Directive System of Wireless Telegraphy," *Elec. Eng.* Vol. 2, pp. 771–775, 1907.
4. Slee, J. A., "Development of the Bellini-Tosi System of Direction Finding in the British Mercantile Marine," *J. IEE*, Vol. 62, pp. 543–550, 1924.
5. Simpson, C., *The Lusitania*, Little Brown and Company, Boston, MA, 1972.
6. Savas, T., *Hunt and Kill: U-505 and the Battle of the Atlantic*, Ed. Theodore P. Savas, Savas Beatie LLC, New York, 2005.
7. Smith-Rose, R. L., "Radio Direction-Finding by Transmission and Reception (with Particular Reference to Its Application to Marine Navigation)," *Proc. Institute of Radio Engineers*, Vol. 17, No. 3, Fig. 11, pp. 425–478, March 1929.

8. Oliver Heaviside, the electromagnetic researcher who recast Maxwell's equations and differential calculus to the form we use today, developer of transmission line theory and the coaxial cable, proposed the existence of the ionosphere. Both he and A. E. Kennelly, an American professor of electrical engineering, reasoned that Marconi's trans-Atlantic radio transmissions could only be due to free electrons in the earth's upper atmosphere. The existence of the Kennelly-Heaviside layer was confirmed two decades later by G. Briet and M. A. Tuve in the United States and by E. V. Appleton in Great Britain. Both research groups used vertically pulsed radio signals and measured the delay, but Appleton's nomenclature remains to this day: "The story of how I came to give the names D, E, and F is really a simple one. In the early work with broadcasting wavelength, I obtained reflections from the Kennelly-Heaviside layer and I used on my diagrams the letter E for the electric vector of the down coming wave. When I found in winter 1925 that I could get reflections from a higher and completely different layer, I used the letter F for the electric vector of the down coming wave. Then about the same time I got occasionally reflections from a very low height and so naturally used the letter D for the electric vector of the returning waves. Then I suddenly realized that I must name these discrete layers and being rather fearful of assuming any finality about measurements, I felt I ought not to call these layers A, B, and C, as there might be undiscovered layers, both below and above them. I therefore felt that the original designations for the electric field vector **D**, **E**, and **F** might be used for the layers themselves."
See:
Dellinger, J. H., "The Role of the Ionosphere in Radio Wave Propagation," *AIEE Transactions*, Vol. 58, pp. 803–822, Nov. 1939.
Heaviside, O., "Telegraphy," *Encyclopedia Britannica*, 10th ed., Vol. 33, Dec. 19, 1902, p. 215.
Kennelly, A. E., "On the Elevation of the Electrically Conducting Strata of the Earth's Atmosphere," *Electrical World and Engineer*, p. 473, March 15, 1902.
Silberstein, R., "The Origin of the Current Nomenclature for the Ionospheric Layers," *J. Atmos. Terr. Physics.*, JATP 0259, p. 382.
9. Canck, M. H., "Radio Waves and Sounding the Ionosphere," *Antenne X*, No. 123, July 2007.
10. Adcock, F., "Improvements in Means for Determining the Direction of a Distant Source of Electro-Magnetic Radiation," British Patent No. 130490, 1919.
11. Smith-Rose, R. L., "Radio Direction-Finding by Transmission and Reception (with Particular Reference to Its Application to Marine Navigation)," *Proc. Institute of Radio Engineers*, Vol. 17, No. 3, pp. 425–478, 466, March 1929.
12. Smith-Rose, R. L., and R. H. Barfield, "The Cause and Elimination of Night Errors in Radio Direction Finding," *J. IEE*, Vol. 64, pp. 831–838, 1926.
13. Erskine, R., "Shore High-Frequency Direction-Finding in the Battle of the Atlantic: An Undervalued Intelligence Asset," *Journal of Intelligence History*, ISSN 1616–1262, Vol. 4, No. 2, Winter 2004.
14. World Wide Lightning Location Network (WWLLN), Univ. of Washington: http://webflash.ess.washington.edu/ [2014]. Rodger, C. J., S. W. Werner, J. B. Brundell, N. R. Thomson, E. H. Lay, R. H. Holzworth, and R. L. Dowden, "Detection Efficiency of the VLF World-Wide Lightning Location Network (WWLLN): Initial Case Study," *Annales Geophys.*, 24, pp. 3197–3214, 2006.
Corbosiero, K. L., S. F. Abarca, F. O. Rosales, and G. B. Raga, "The World Wide Lightning Location Network: Network Overview, Evaluation, and Its Application to Tropical Cyclone Research," available at www.atmos.ucla.edu/~kristen/presentations/Corbosiero_TexasAM_11.pdf [2014].
15. R. G. Stansfield, "Statistical Theory of DF Fixing," *Journal of IEE*, vol. 94, no. 15, pp. 762–770, 1947.
16. D. Wangsness, "A New Method of Position Estimation Using Bearing Measurements," *IEEE Trans. on Aerospace and Electronic Systems*, pp. 959–960, Nov. 1973.
17. Price, M. G. "Linear Least Squares Estimation," private publication, Jan. 1, 2006.

9.6 Problems

9.1 Using the area of a circle, write Eqs. (9.6), (9.7), and (9.8) in terms of the loop area rather than circular loop radius.

9.2 During World War I the vertical loop antenna was manually rotated to determine signal direction. Once a signal was found in the frequency band, the loop was rotated quickly until the signal was *null* and direction determined. (The null is sharper and easier to determine.) The loop antenna (its x-z plane) points 45° northeast.
   (a) What is the azimuth of the incoming signal?
   (b) Sketch a top-down view of the antenna loop, the 0° elevation radiation pattern, and the relative direction of the incoming daytime signal when the loop gives a null output signal.

9.3 A crossed-loop system performs skywave DF on a signal that comes from +30° azimuth from north ($\phi = 60°$). If the skywave has a 45° elevation angle ($\theta = 45°$) and has a horizontal electric field component 40 percent as large as the vertical, what is the error in measured DF?

9.4 Imagine that you are calibrating HF an AS-145 (vertical monopole, crossed vertical loop) antenna on a ship. You are able to make an omni-sine-cosine vector calibration DF measurement on one frequency in 333 ms. Surrounding water attenuates horizontal polarized radio waves so only $E_{vertical}$ calibration is required, and only one elevation near 0° is available since the transmitter is located on land 10 nautical miles away at a shore facility. It takes approximately 20 minutes for the ship to make a full 360° turn. How long does it take to calibrate the ship's AS-145 DF antenna over 3 to 30 MHz of the HF spectrum with antenna manifolds created every 50 kHz measured every 2° in azimuth?

9.5 In 1923 the Poldu transmitter operated at 3092 kHz. During the 1920s the Adcock quadrature array receiving antennas was practical. With $d = \lambda/8$, determine the antenna pattern for 0° elevation angle. What is the antenna spacing in meters? What are the normalized Watson-Watt antenna pattern equations for an elevation of 0° ($\theta = 90°$)?

9.6 For the Adcock quadrature array with $d = \lambda/8$, plot the DF error between the full solution and the Watson-Watt approximation for 30° elevation angle (polar zenith angle $\theta = 60°$). Use the MATLAB Procedure Watson_Watt.m.

9.7 Assume that an array has been calibrated at 2° intervals ($\phi = 0, 2, ..., 358$). Using the Bartlett correlation response, a rough peak finding algorithm finds the highest correlation response at ϕ_0. The correlation responses of adjacent azimuths at ϕ_{-2} and ϕ_{+2} are also measured giving the set $C(\phi_{-2})$, $C(\phi_0)$, and $C(\phi_{+2})$. Using ϕ_0 as the reference azimuth, determine the maximum correlation C_{max} and the corresponding best estimate of DF azimuth ϕ_{max} using the parabolic approximation.

9.8 Using the equations developed in Prob. 9.7 (or the MATLAB procedure Parabolic_Fit.m), plot the parabolic fit and determine the peak DF direction if the highest correlation occurs at $\phi_0 = 40°$ with a calibration step of 2° and measured correlation coefficients are

$$C(\phi_{-2}) = 0.5, C(\phi_0) = 0.9, \text{ and } C(\phi_{+2}) = 0.2$$

9.9 Derive the cumulative distribution function (*cdf*)

$$P_o \equiv p(miss < j_o) = 1 - \exp\left(-\frac{j_o^2}{2}\right)$$

from the bivariate normal probability distribution function (*pdf*) where the probability of miss at distance r is given by

$$p(r) = \frac{1}{2\pi} \exp\left(-\frac{r^2}{2}\right)$$

9.10 What happens when AOA bias is introduced into the Stansfield LOB geolocation estimates? Are the confidence ellipses of true emitter containment too large or too small?

CHAPTER 10
Vector Sensors

10.1 Introduction

Much of the well-known work on direction finding and beamforming presented in this text has involved *scalar* sensor arrays composed of uniform antennas that are capable of extracting only one component from the incident electric or magnetic field for processing. Improved arrival angle estimation accuracy and interfering source separation is possible by exploiting the *complete* vector field using a polarization sensitive antenna called a *vector sensor*. Vector sensors have found application across a number of disciplines such as acoustics, seismology, and remote sensing. For example, in acoustic applications, the acoustic pressure field as well as the three components of the particle velocity are measured with a vector sensor composed of orthogonally oriented scalar hydrophones. In electromagnetic applications, vector sensors are used to acquire all six Cartesian field components—three electric (E) and three magnetic (H)—of an electromagnetic plane wave with a combination of orthogonally oriented electric dipoles and magnetic loops as shown in Fig. 10.1.

The ideal vertical electric dipole can only detect the vertical electric field whereas the ideal horizontal loop can only respond to the vertical magnetic field. The electromagnetic vector sensor nests three or six orthogonal antennas in order to sense all E, or all H, or all E and H fields simultaneously. With this sensor composition, the Poynting vector ($\bar{S}$) can be computed via the cross product of the electric and magnetic field vectors ($\bar{S} = \bar{E} \times \bar{H}$). The Poynting vector indicates the propagation direction of the incident plane wave, which can also be converted into a convenient angle-of-arrival representation within the coordinate system. By replacing scalar antennas with vector sensor antennas in spatially displaced arrays, the arrival angle can be estimated via two independent approaches: the cross-product direction-finding technique applied at each vector sensor and the traditional direction-finding techniques studied in Chap. 7 that exploit the spatial phase delay between vector sensor antennas in the array. Ambiguous scalar

---

This chapter was written by Jeffrey D. Conner, Senior Research Engineer, Georgia Tech Research Institute.

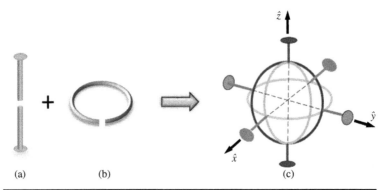

Figure 10.1 Electromagnetic vector sensor composed of orthogonally oriented electric and magnetic dipoles: (a) electric dipole, (b) magnetic dipole (loop), (c) six-axis electromagnetic vector sensor.

array arrival angle estimates derived by traditional techniques can be disambiguated with the help of the vector sensor antennas and the cross-product direction-finding technique.

Many variations exist on the vector sensor antenna configuration. The early direction-finding system developed by Watson-Watt [1] could be considered a vector sensor as some configurations were composed of orthogonal crossed loop antennas oriented along the x and y planes with a z-directed electric dipole as a "sense" antenna. The loop antennas were used to compute two ambiguous estimates of a source's arrival angle in the x-y plane (180° apart) that were disambiguated by the sign of the E_z field in order to satisfy the right-hand rule for plane wave propagation. The processing performed by Watson-Watt is a simplification of the vector sensor processing we study in this chapter. In the years to follow, research by Compton [2] into the performance of three crossed electric dipoles called the *tripole* laid the groundwork for the fundamental research into the use of the full six-axis vector sensor for direction of arrival estimation [3–5] and beamforming [6, 7].

Electromagnetic vector sensor design and associated signal processing continues to be a rich and active research area. Many flavors of vector sensors have been designed and built by industry, government, and academia. Commercial vector sensors have been developed by companies such as Orbit/FR (formerly Flam and Russell), Quasar Federal Systems (QFS), and Invertix. Orbit/FR was the original developer of the Compact Array Radiolocation Technology (CART) vector sensor that is a similar style to that shown in Fig. 10.1. This is a simple design in which all antennas are aligned with the axes of a Cartesian coordinate system. Invertix designed a similar vector sensor [8], as shown in Fig. 10.2a, except that the antenna elements are rotated in the Cartesian coordinate system. Figure 10.2b illustrates a vector sensor designed by the Australian Department of Defence that is composed of three mutually orthogonal octagonal loops that cover

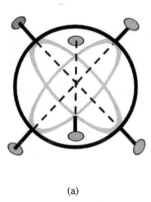 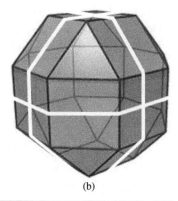

Figure 10.2 Example vector sensor configurations: (a) Invertix rotated CART, (b) Giselle antenna.

the surface of a rhombicuboctahedron called the *Giselle antenna* [9]. There are two orthogonal vertical loops and one horizontal loop in order to sense all three-magnetic field components to estimate the two-dimensional arrival signal angles.

One practical application of electromagnetic vector sensors is ground-based remote sensing of radio atmospheric signals, including manmade and the naturally caused sferics produced by lightning strikes. Vector sensors are one component of the Worldwide Lightning Network (WWLN), whereby the arrival angle of the sferic estimated with a vector sensor is converted into a line of bearing (LOB) and the intersection of several LOBs from multiple vector sensor sites gives an indication of the geolocation of the lightning source. The sferic signal occurs in the very-low-frequency (VLF) range up to about 20 kHz. The long wavelength at VLF makes application of traditional spatially displaced arrays challenging because the array baseline would be very large (10 to 100 km). The vector sensor, on the other hand, has the advantage that all elements are collocated at a single site. However, the individual dipole and loop elements of the vector sensor antenna shown in Fig. 10.1 can only realistically be constructed to be electrically small at VLF frequencies. Electrically small dipoles and loops typically have omni-directional patterns in either the azimuth or elevation planes. The orthogonal configuration of antennas keep mutual coupling at a minimum. Another modern application of vector sensors has been in the geolocation of emitters in the congested high-frequency (HF) spectrum [8, 10]. Again, the main advantage is the small footprint of the vector sensor, compared to traditional spatially displaced arrays that require very large apertures (tens of meters). Advanced signal processing techniques can extract two-dimensional arrival angle estimates for line-of-sight (LOS), over-the-horizon (OTH), and skywave propagating signals. The polarization sensitivity of vector sensors further permits discrimination of multiple sources from the same arrival angle by their polarization separation. Vector sensors have been integrated onto unmanned aerial vehicles (UAVs) [11, 12], where they can provide a two-dimensional arrival angle estimate with a very small physical antenna array footprint.

The growth in vector sensor applications is due to the many advantages of vector sensor antennas compared to traditional spatially displaced arrays, which include the following:

- Six vector measurements in a single antenna
- Polarization sensitivity and diversity
- 2D angle-of-arrival (azimuth and elevation) and polarization state estimation with a single vector sensor and a single time snapshot
- More accurate than spatially displaced arrays of the same footprint
- No spatial undersampling ambiguities (i.e., grating lobes)
- No synchronization required between separate sensors due to collocation of elements
- Superior co-channel/co-direction interference mitigation performance
- Wide field of view
- Outstanding low-frequency (< 2 GHz) performance

Implementing the ideal vector sensor and fully adhering to the ideal assumptions can be challenging. Some of the challenges associated with the design and implementation of vector sensor antennas include the following:

- Difficulty in construction such that all elements are collocated with a common phase center
- Costs associated with an increased number of receiver channels for a single antenna
- Nonnegligible mutual coupling between supposed orthogonal antenna elements
- Antenna pattern variations within and between individual elements

The remainder of this chapter focuses on the presentation of the vector sensor fundamentals for direction-finding and beamforming applications to include (1) the vector sensor antenna array response and beampattern, (2) cross-product and super-resolution direction-finding techniques, (3) vector sensor beamforming properties, and (4) Cramer-Rao bounds on direction-finding performance.

10.2 Vector Sensor Antenna Array Response

In this section, we derive the steering vector (discussed at length in Chap. 4) for an array of vector sensors. The steering vector is the parametric representation of the vector sensor response to a uniform plane-wave signal originating from any angle of arrival for each constituent antenna element in the array. This definition will serve as the basis for all future analysis in this chapter and provides a framework for analyzing more advanced vector sensor geometries whose construction is arbitrarily composed and oriented.

10.2.1 Single Vector Sensor Steering Vector Derivation

Figure 10.3 depicts a full six-axis CART electromagnetic vector sensor antenna and corresponding coordinate system. CART is the traditional physical configuration for vector sensor processing and is composed of three-electric dipole antennas and three-loop antennas (magnetic dipoles). Dipole antennas are used to measure the electric-field vector components (E_x, E_y, and E_z) of a transverse electromagnetic wave (TEM) incident upon the vector sensor, whereas the loop antennas—the dual antenna of an electric dipole—are used to measure the corresponding magnetic field components H_x, H_y, and H_z. The antenna elements of each dipole triplet are orthogonal to each other and aligned with the principal axes $(\hat{x}, \hat{y}, \hat{z})$ of a Cartesian coordinate system. The corresponding spherical coordinate system $(\hat{r}, \hat{\theta}, \hat{\phi})$ is shown as well, where $0 \leq \theta \leq \pi$ denotes the elevation angle measured from the +z axis and $0 \leq \phi < 2\pi$ refers to the azimuthal angle measured from the +x axis toward the +y axis. The CART vector sensor configuration is sensitive to the two-dimensional angle of arrival as well as the polarization of the incident plane wave. The goal from here will be to derive the vector sensor steering vector by expressing the individual Cartesian electric and magnetic field components for all six

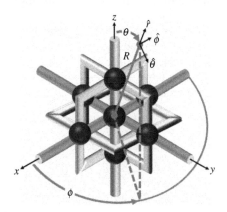

FIGURE 10.3 CART vector sensor antenna and coordinate system definition.

axes, $E_{x'}$, $E_{y'}$, $E_{z'}$, $H_{x'}$, $H_{y'}$ and H_z in terms of the angle of arrival (θ, ϕ) and polarization state (γ, η) of an incident transverse electromagnetic plane wave.

First, consider the electric field phasor representation $\tilde{E}(r)$ for a transverse electromagnetic wave propagating in the $-\hat{r}$ direction (corresponding to an incoming signal) within a right-handed spherical coordinate system with unit vectors $\hat{\phi}, \hat{\theta}, -\hat{r}$ in that order given as

$$\tilde{E}(r) = [E_{\phi 0}\hat{\phi} + E_{\theta 0}\hat{\theta}]e^{jkr} \tag{10.1}$$

where $k = 2\pi/\lambda$ is the wavenumber of the transverse electromagnetic wave, $E_{\phi 0}$ and $E_{\theta 0}$ are complex amplitudes of the incident electric field defined as

$$E_{\phi 0} = |E_{\phi 0}|, \quad E_{\theta 0} = |E_{\theta 0}|e^{j\eta} \tag{10.2}$$

where $-\pi \leq \eta \leq \pi$ is the polarization phase difference of $E_{\phi 0}$ relative to the phase of $E_{\theta 0}$. The total electric field phasor is then given by

$$\tilde{E}(r) = [\,|E_{\phi 0}|\hat{\phi} + |E_{\theta 0}|e^{j\eta}\hat{\theta}]e^{jkr} \tag{10.3}$$

and the corresponding instantaneous electric field is

$$E(r,t) = \mathcal{R}e\,[\tilde{E}(r)e^{j\omega t}] = |E_{\phi 0}|\cos(\omega t + kr)\hat{\phi} + |E_{\theta 0}|\cos(\omega t + kr + \eta)\hat{\theta} \tag{10.4}$$

If η is equal to 0 or $\pm\pi$, then the wave is said to be linearly polarized. If the amplitudes $|E_{\phi 0}|, |E_{\theta 0}|$ are equal and η equals $\pm\pi/2$, then the resulting wave is said to be circularly polarized. In the most general case, $|E_{\theta 0}|, |E_{\phi 0}|$, and η are all nonzero, and the tip of the electric field vector traces an ellipse in the θ-ϕ plane as shown in Fig. 10.4a.

The shape of the ellipse and its handedness are determined by the value of the polarization ratio $\tan\gamma e^{j\eta}$ where $\tan\gamma = |E_{\theta 0}|/|E_{\phi 0}|$ where $0 \leq \gamma \leq \pi/2$ is defined as the auxiliary angle. Alternatively, the polarization ellipse can be described by the ellipticity angle $-\pi/4 \leq \alpha \leq \pi/4$ and the orientation (or rotation) angle $0 \leq \beta \leq \pi$. As seen in Fig. 10.4b, the Poincaré sphere representation illustrates the relationships between these two polarization state descriptions where the pairs (γ, η) and (α, β) separately and uniquely specify the state of polarization of a point on the sphere, M. These polarization state pairs are related mathematically as [3, 13]

$$\cos 2\gamma = \cos 2\alpha \cos 2\beta \tag{10.5}$$

$$\tan\eta = \tan 2\alpha \csc 2\beta \tag{10.6}$$

or inversely as

$$\tan 2\beta = \tan 2\gamma \cos\eta \tag{10.7}$$

$$\sin 2\alpha = \sin 2\gamma \sin\eta \tag{10.8}$$

Vector Sensors   291

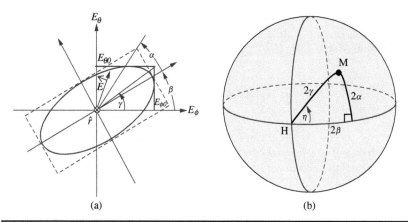

FIGURE 10.4 (a) Polarization ellipse for transverse electromagnetic wave in spherical coordinates and (b) Poincaré sphere representation of the parameters α, β, γ, and η.

The angle γ is limited to the range $0 \leq \gamma \leq \pi/2$; thus the sign of β in Eq. (10.7) is governed by the sign of the $\cos\eta$ term where

$$\gamma > 0 \text{ if } \cos\eta > 0$$
$$\gamma < 0 \text{ if } \cos\eta < 0 \tag{10.9}$$

For our purposes, the electric field components will be expressed in terms of γ and η as (to be consistent with the literature of Compton and Li [3])

$$E_\phi = E_o \cos\gamma, \; E_\theta = E_o \sin\gamma e^{j\eta} \tag{10.10}$$

The total electric field phasor is now expressed as

$$\tilde{E}(r) = E_o[\cos(\gamma)\hat{\phi} + \sin(\gamma)e^{j\eta}\hat{\theta}]e^{jkr} \tag{10.11}$$

or in vector format as

$$\bar{e} = \begin{bmatrix} E_\phi \\ E_\theta \end{bmatrix} = E_o \begin{bmatrix} \cos(\gamma) \\ \sin(\gamma)e^{j\eta} \end{bmatrix} e^{jkr} \tag{10.12}$$

The electric field vector components expressed in spherical coordinates can be converted to Cartesian using the standard identity

$$\begin{bmatrix} \hat{x} \\ \hat{y} \\ \hat{z} \end{bmatrix} = \begin{bmatrix} \sin\theta\cos\phi & \cos\theta\cos\phi & -\sin\phi \\ \sin\theta\sin\phi & \cos\theta\sin\phi & \cos\phi \\ \cos\theta & -\sin\theta & 0 \end{bmatrix} \begin{bmatrix} \hat{r} \\ \hat{\theta} \\ \hat{\phi} \end{bmatrix} \tag{10.13}$$

to yield the electric field vector response in Cartesian coordinates for the four parameters (θ, ϕ, γ, η) and an amplitude E_o

$$\bar{e} = \begin{bmatrix} E_x \\ E_y \\ E_z \end{bmatrix} = E_o \begin{bmatrix} \cos\theta\cos\phi & -\sin\phi \\ \cos\theta\sin\phi & \cos\phi \\ -\sin\theta & 0 \end{bmatrix} \begin{bmatrix} \sin(\gamma)e^{j\eta} \\ \cos(\gamma) \end{bmatrix} e^{jkr} \tag{10.14}$$

Chapter Ten

The corresponding magnetic field components, H_x, H_y, and H_z are computed from the direction of propagation $-\hat{r}$ and the electric field phasor as

$$\bar{h} = \frac{1}{Z_o}(-\hat{r}) \times \bar{e} = \frac{1}{Z_o}\begin{bmatrix} E_\phi \hat{\theta} \\ (-E_\theta)\hat{\phi} \end{bmatrix} e^{jkr} \quad (10.15)$$

where Z_o is the intrinsic impedance of the transmission medium. Converting the spherical components for the magnetic field vector into Cartesian coordinates results in

$$\bar{h} = \begin{bmatrix} H_x \\ H_y \\ H_z \end{bmatrix} = \frac{E_o}{Z_o} \begin{bmatrix} -\sin\phi & -\cos\theta\cos\phi \\ \cos\phi & -\cos\theta\sin\phi \\ 0 & \sin\theta \end{bmatrix} \begin{bmatrix} \sin(\gamma)e^{j\eta} \\ \cos(\gamma) \end{bmatrix} e^{jkr} \quad (10.16)$$

Concatenating the electric and magnetic field vector components, normalizing the amplitude scaling, and assuming the vector sensor is located at the origin (i.e., $r = 0$) results in the general form of the vector sensor steering vector given as

$$\bar{a}(\theta,\phi,\gamma,\eta) = \begin{bmatrix} E_x \\ E_y \\ E_z \\ H_x \\ H_y \\ H_z \end{bmatrix} = \begin{bmatrix} \cos\theta\cos\phi & -\sin\phi \\ \cos\theta\sin\phi & \cos\phi \\ -\sin\theta & 0 \\ -\sin\phi & -\cos\theta\cos\phi \\ \cos\phi & -\cos\theta\sin\phi \\ 0 & \sin\theta \end{bmatrix} \begin{bmatrix} \sin(\gamma)e^{j\eta} \\ \cos(\gamma) \end{bmatrix} = \bar{\Theta}(\theta,\phi)\bar{p}(\gamma,\eta)$$

(10.17)

There are a few observations worth mentioning about the vector sensor steering vector and how it compares to the spatially displaced array geometries we have studied before. First, the size of the steering vector is 6×1 meaning that a single six-axis vector sensor contains six scalar antenna elements that are co-located and co-centered at a single location. As a result, there are no time delay induced phase shifts between elements unlike in spatially displaced arrays making the vector sensor steering vector independent of changes with frequency. In addition, the steering vector is sensitive to polarization state (γ, η); thus it is theoretically possible to separate signals arriving from the same direction (θ, ϕ) based on the diversity of their polarization states. As we see later on, it is possible to separate the estimation of the angle of arrival from the polarization state estimation, which helps simplify the search to only two dimensions and allows us to use many of the direction-finding algorithms studied in Chap. 7.

Finally, this model assumes that there is no mutual coupling between the axes; however, in practice some mutual coupling always exists and vector sensor designers strive to reduce this coupling to a negligible effect.

10.2.2 Vector Sensor Array Signal Model and Steering Vector

The response of a six-axis vector sensor to an incoming arbitrary plane wave is characterized by the angle of arrival (θ, ϕ), the polarization state (γ, η) and an amplitude E_o as given by Eq. (10.17). Assume that M signals arrive with unique arrival angles and polarization states $\bar{\psi}_m = (\theta_m, \phi_m, \gamma_m, \eta_m)$ such that the instantaneous response of the single vector sensor is given by

$$\bar{x}_m(t) = \bar{g}(\theta_m, \phi_m) \underbrace{\bar{a}(\theta_m, \phi_m, \gamma_m, \eta_m)}_{\bar{a}(\bar{\psi}_m)} \underbrace{E_{o,m} e^{j(\omega_m t + \varphi_m)}}_{s_m(t)} + \bar{n}(t) \qquad (10.18)$$

where, ω and φ are the carrier frequency and phase of the incoming signal $s(t) = E_o e^{j(\omega t + \varphi)}$, and $\bar{n}(t)$ is a 6×1 additive complex valued noise vector drawn from a zero mean Gaussian random variable with standard deviation σ_n and covariance $\sigma_n \bar{I}$, where $\bar{I}$ is the identity matrix. Equation (10.18) also contains the term $\bar{g}(\theta, \phi)$ that is a 6×1 vector containing the E- and H-field responses of the short dipoles and loops. Typically, this gain term $\bar{g}(\theta, \phi)$ is not included because of the duality of the short electric dipole and loop responses and the presumed normalization of the intrinsic impedance Z_o between the E- and H-fields. Therefore, Eq. (10.18) reduces to

$$\bar{x}_m(t) = \bar{a}(\bar{\psi}_m) s_m(t) + \bar{n}(t) \qquad (10.19)$$

Next, consider an array of L such vector sensors positioned at (x_l, y_l, z_l) with respect to the coordinate origin forming a spatially distributed array of vector sensors. The spatial phase factor for the lth vector sensor is given by

$$q_l(\theta, \phi) = e^{j 2\pi (\bar{r}_l^T \bar{u}(\theta, \phi))/\lambda} = e^{j 2\pi (x_l \sin\theta \cos\phi + y_l \sin\theta \sin\phi + z_l \cos\theta)/\lambda} \qquad (10.20)$$

where $\bar{r}_l = [x_l, y_l, z_l]^T$ is the 3×1 sensor location vector and $\bar{u}(\theta, \phi) = [\cos\theta \sin\phi, \sin\theta \sin\phi, \cos\theta]^T$ is the unit vector in the source direction.

The steering vector for the lth vector sensor in the array due to the mth signal is then

$$\bar{v}_{vs,l}(\bar{\psi}_m) = q_l(\theta_m, \phi_m) \bar{a}(\bar{\psi}_m) = q_l(\theta_m, \phi_m) \bar{\Theta}(\theta_m, \phi_m) \bar{p}(\gamma_m, \eta_m) \qquad (10.21)$$

and the response of the vector sensor array is generated by concatenating the L steering vectors

$$\bar{v}_{vs}(\bar{\psi}_m) = \begin{bmatrix} \bar{v}_{vs,1}(\bar{\psi}_m) \\ \vdots \\ \bar{v}_{vs,L}(\bar{\psi}_m) \end{bmatrix} \qquad (10.22)$$

The complete normalized instantaneous response for an array of vector sensors is then given as

$$\bar{x}_m(t) = \bar{v}_{vs}(\bar{\psi}_m)s_m(t) + \bar{N}(t) \quad (10.23)$$

where $\bar{N}(t)$, of size $6L \times 1$, is the concatenation of the 6×1 noise vector $\bar{n}(t)$ for the L vector sensors. One should recognize how the form of Eq. (10.23) is similar to the spatially distributed arrays with non-polarization-sensitive (i.e., scalar) antennas we have studied before.

10.3 Vector Sensor Direction Finding

10.3.1 Cross-Product Direction Finding

Recall from fundamental electromagnetic theory that the definition of a uniform plane wave is such that it propagates in a direction orthogonal to the electric and magnetic fields subject to the right-hand rule as shown in Fig. 10.5. This fact motivated the design of the vector sensor array and the derivation of the array steering vector. By measuring all six electric and magnetic field components, the direction of propagation can be computed completely through a simple vector cross-product calculation as shown in Eq. (10.24).

$$\bar{k} = \begin{bmatrix} k_x \\ k_y \\ k_z \end{bmatrix} = \frac{\bar{e} \times \bar{h}^*}{\|\bar{e}\| \cdot \|\bar{h}\|} = \begin{bmatrix} \sin\theta\cos\phi \\ \sin\theta\sin\phi \\ \cos\theta \end{bmatrix} \stackrel{\text{def}}{=} \begin{bmatrix} u \\ v \\ w \end{bmatrix} \quad (10.24)$$

The direction of propagation computed in Eq. (10.24) has been normalized by the Frobenius norm |⋅|, which results in the incident wave's direction of propagation expressed in terms of its direction cosines (u, v, w), which can be related to the elevation and azimuth angles.

Solving Eq. (10.24) for the arrival angle gives

$$\theta = \cos^{-1}(k_z) = \sin^{-1}\left(\sqrt{k_x^2 + k_y^2}\right)$$

$$\phi = \tan^{-1}(k_y/k_x) \quad (10.25)$$

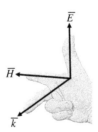

FIGURE 10.5 Right-hand coordinate system representation illustrating relationship between the direction of propagation to the electric and magnetic fields.

Once the arrival angles are estimated, the polarization parameters can be found by first taking the pseudo-inverse of Eq. (10.17), resulting in

$$\bar{p} = \begin{bmatrix} p_1 \\ p_2 \end{bmatrix} = [\bar{\Theta}^H(\theta,\phi)\bar{\Theta}(\theta,\phi)]^{-1}\bar{\Theta}^H(\theta,\phi)\,\bar{a} \qquad (10.26)$$

where $\bar{a} = [E_x, E_y, E_z, H_x, H_y, H_z]^T$ from Eq. (10.17). The corresponding polarization parameters are then given as

$$\gamma = \tan^{-1}(|p_1/p_2|)$$
$$\eta = \angle p_1 - \angle p_2 \qquad (10.27)$$

Next, let us consider a simple example to highlight some features of the cross-product direction-finding technique. A signal $s(t)$ of the form

$$s(t) = A\sin(2\pi f_c t)(e^{-at} - e^{-bt}) \qquad (10.28)$$

is incident upon a single six-axis vector sensor positioned at the origin of the coordinate system defined in Fig. 10.3. The signal $s(t)$, shown in Fig. 10.6, is a damped sinusoid with amplitude $A = 2$, frequency $f_c = 1$ kHz, and damping factors $a = 225$, $b = 1000$.

The signal is arriving at an angle of $(\theta, \phi) = (90°, 60°)$ corresponding to $s(t)$ propagating in the x-y plane with polarization state $(\gamma, \eta) = (45°, 0°)$. Complex white Gaussian noise is added to each of the six vector sensor channels with a signal-to-noise ratio (SNR) of 30 and 10 dB corresponding to noise standard deviations of 0.01 and 0.1, respectively.

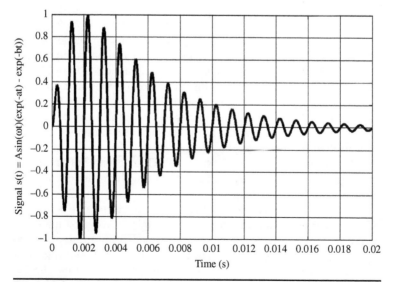

FIGURE 10.6 Damped sinusoidal signal.

Chapter Ten

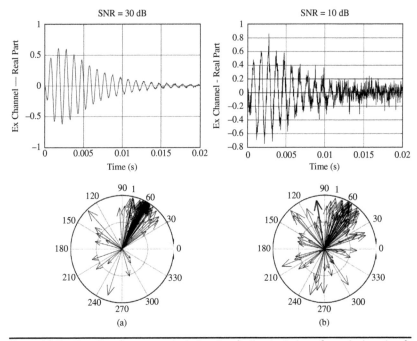

FIGURE 10.7 Results of cross-product direction-finding technique for arrival angle of $(\theta, \phi) = (90°, 60°)$ and for SNRs of (a) 30 dB, (b) 10 dB.

Equation (10.24) is then evaluated to determine the pointing direction at each time sample and the results are shown in Fig. 10.7.

We see from the results in Fig. 10.7 that a 3D angle of arrival can be estimated with only a single data snapshot and that the time samples with the strongest amplitude values provide the most reliable estimate of the angle of arrival. Multiple "reliable" angle estimates can be averaged together over some time window to yield a more robust estimate of the arrival angle. This SNR requirement is one limitation in using the cross-product technique for angle estimation. This motivates the development of more sophisticated vector sensor processing techniques capable of averaging, smoothing, and/or mitigating interference in the measurements. Many of the techniques studied in the previous chapters such as MUSIC, MVDR, or ESPRIT are capable of improving the angle estimation process and are studied with greater detail in the next section.

One additional benefit of the cross-product technique is that not all six vector sensor axes are required to obtain an angle estimate. For example, a pair of loop antennas in a crossed configuration such as the one shown in Fig. 10.8 is often used by the remote sensing, amateur radio, and search and rescue communities for estimating the azimuthal arrival angle only.

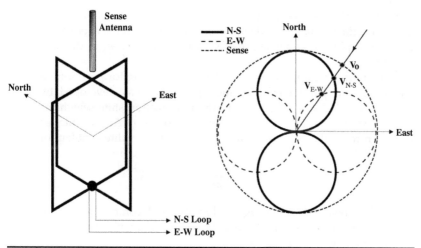

FIGURE 10.8 Crossed loop direction-finding configuration and pattern.

The crossed loop is constructed with one loop oriented along the north-south direction and the other loop oriented orthogonally along the east-west loop. The loops have orthogonal "figure 8" patterns as shown in Fig. 10.8 where the peak of the north-south loop occurs at the null of the east-west loop and vice versa. The arrival angle can be estimated by taking the arctangent of the voltages between the two loops as

$$\phi = \arctan\left(\frac{V_{N-S}}{V_{E-W}}\right) \tag{10.29}$$

A sense antenna is included to help resolve the ambiguous arrival quadrant for the angle estimate. Here the sense antenna is represented as a dipole that measures the z-directed E-field component E_z. Evaluating the cross product of Eq. (10.24) yields the direction of propagation which is then used to determine the correct quadrant.

10.3.2 Super Resolution Direction Finding

Many of the well-known super resolution techniques such as MVDR and MUSIC can also be applied to the vector sensor. The difference between the traditional super resolution implementation for spatially distributed arrays and that of the vector sensor is that for the vector sensor one must also search for the two polarization state parameters (γ, η) in addition to the two angle parameters (θ, ϕ). Performing an efficient search over a 4D space can be challenging. There exist several techniques for partitioning the search of the angle estimates from the polarization state. One such technique is detailed in [7] whereby spatio-polarization beams computed using linearly constrained minimum variance (LCMV) beamforming decouples the angle estimation from the polarization estimation by passing only a single signal of interest along certain angular/polarization dimensions. Vector sensor beamforming implementations are discussed in greater detail in the next section.

An alternative approach to LCMV beamforming for partitioning the angle and parameter estimation problems was developed by Ferrara and Parks in [14]. Therein, the polarization state search for MVDR, MUSIC, and the adapted angular response (AAR) techniques is shown to be the solution to a generalized eigenvalue problem of the form

$$\bar{A}\bar{v} = \lambda \bar{B}\bar{v} \tag{10.30}$$

where λ corresponds to the eigenvalues of the decomposition and $\bar{v}$ the associated eigenvectors. This can be solved easily with MATLAB's eig function as

$$[\text{v},\text{lamda}] = \text{eig}(\text{A},\text{B}) \tag{10.31}$$

Recall that in traditional cases, in which the polarization of an incident signal is known or the antennas are not polarization sensitive, we can generate the traditional MVDR and MUSIC pseudo-spectrums and perform a 2D search over angle with

$$P_{MVDR}(\theta, \phi) = \frac{\bar{v}_{vs}^H(\theta, \phi)\bar{v}_{vs}(\theta, \phi)}{\bar{v}_{vs}^H(\theta, \phi)\bar{R}_{xx}^{-1}\bar{v}_{vs}(\theta, \phi)} \tag{10.32}$$

$$P_{MUSIC}(\theta, \phi) = \frac{\bar{v}_{vs}^H(\theta, \phi)\bar{v}_{vs}(\theta, \phi)}{\bar{v}_{vs}^H(\theta, \phi)\bar{E}_N \bar{E}_N^H \bar{v}_{vs}(\theta, \phi)} \tag{10.33}$$

where $\bar{v}_{vs}(\theta, \phi) = \bar{\Theta}(\theta, \phi)$ is the steering vector for a single vector sensor given in Eq. (10.21). When the polarization state is unknown, for any fixed arrival angle, a maximization can be performed first over the polarization state $\bar{p}$. For MVDR the pseudo-spectrum P_{MVDR} with polarization diversity now becomes

$$P_{MVDR}(\bar{\psi}) = \max_{\bar{p}} \frac{(\bar{p}^H(\gamma, \eta)\bar{\Theta}^H(\theta, \phi))\cdot(\bar{\Theta}(\theta, \phi)\bar{p}(\gamma, \eta))}{(\bar{p}^H(\gamma, \eta)\bar{\Theta}^H(\theta, \phi))\cdot \bar{R}_{xx}^{-1}\cdot(\bar{\Theta}(\theta, \phi)\bar{p}(\gamma, \eta))} \tag{10.34}$$

The maximizing vector $\bar{p}_{max}$ corresponding to the incident polarization state is then the associated eigenvector that satisfies the generalized eigenvalue problem given by

$$\underbrace{\bar{\Theta}^H(\theta, \phi)\bar{\Theta}(\theta, \phi)}_{\bar{A}}\bar{p}_{max} = \lambda_{max} \underbrace{\bar{\Theta}^H(\theta, \phi)\bar{R}_{xx}^{-1}\bar{\Theta}(\theta, \phi)}_{\bar{B}}\bar{p}_{max} \tag{10.35}$$

where the terms $\bar{A}$ and $\bar{B}$ equired for Eq. (10.31) are shown. The polarization state $\bar{p}_{max}$ can be found by searching for the maximum eigenvalue λ_{max}; then the individual polarization state parameters (γ, η) can be solved using a form similar to Eq. (10.27) as

$$\gamma = \tan^{-1}\left|\frac{p_{max}(1)}{p_{max}(2)}\right|, \eta = \angle\left(\frac{p_{max}(1)}{p_{max}(2)}\right) = \angle p_{max}(1) - \angle p_{max}(2) \quad (10.36)$$

where $\tan^{-1}(\cdot)$ is the four quadrant inverse tangent and the $\angle$ extracts the phase angle from the complex valued eigenvector ratio in $\bar{p}_{norm} = [\tan\gamma e^{j\eta} \quad 1]^T = [(p_{max}(1)/p_{max}(2)) \quad 1]^T$.

Next, let us revisit the earlier example in Sec. 10.3.1 except this time we use the vector sensor version of MVDR described above to estimate an angle of arrival of $(\theta, \phi) = (30°, -50°)$ and polarization state $(\gamma, \eta) = (75°, 45°)$ and a noise standard deviation of 0.01. First, a MATLAB meshgrid of angles is created and Eq. (10.35) is evaluated at each angle pair (θ_k, ϕ_k) to produce the 3D surface shown in Fig. 10.9.

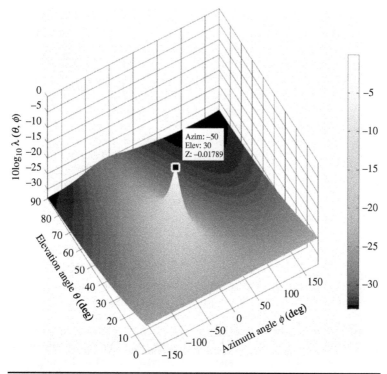

Figure 10.9 VS-MVDR pseudo-spectrum for a single signal arriving at $(\theta, \phi) = (30°, -50°)$.

A 2D peak search is performed over all angle pairs to locate the arrival angle estimate. For this peak, the polarization state is then estimated using the corresponding eigenvector and Eq. (10.36),

$$\bar{p}_{max} = \begin{bmatrix} -1 \\ -0.19 + j0.19 \end{bmatrix}, \bar{p}_{norm} = \begin{bmatrix} \tan\gamma \, e^{j\eta} \\ 1 \end{bmatrix}$$

$$= \begin{bmatrix} 2.63 + j2.63 \\ 1 \end{bmatrix} \Rightarrow (\gamma, \eta) = (74.97°, 44.94°) \quad (10.37)$$

10.4 Vector Sensor Beamforming

Vector sensor beamforming exploits the spatial and polarization diversity of incident signals by weighting the complex electromagnetic field components as opposed to the traditional spatial filtering we have studied before. Recall that in spatially displaced arrays, the arrival angle information is embedded within the time delays between the array elements, which is characterized by spatial phase factors of the form $e^{jk\bar{r}^T\bar{u}(\theta,\phi)}$, where $k = 2\pi/\lambda$ is the wavenumber and λ is the wavelength. As such, spatially displaced arrays are often referred to as *phased arrays* and by nature its performance is an explicit function of frequency. Therefore, the elements must be positioned such that they do not violate Nyquist sampling spatially in order to prevent spurious responses at unintended angles, such as grating lobes. An ideal single vector sensor antenna, on the other hand, has the phase center of all its elements collocated. Thus there are no spatial phase factors present, and the vector sensor behavior is independent of frequency resulting in no spurious lobes.

Consider a single vector sensor with steering vector $\bar{v}_{vs}(\bar{\psi}_d)$ from Eq. (10.21), which has been steered toward a desired parameter vector $\bar{\psi}_d = [\theta_d, \phi_d, \gamma_d, \eta_d]$ with arrival angle (θ_d, ϕ_d) and polarization state (γ_d, η_d). The beampattern $g(\bar{\psi}_d, \bar{\psi}_i)$ is then the response of the vector sensor $\bar{v}_{vs}(\bar{\psi}_d)$, when steered toward $\bar{\psi}_d$, to a signal incident upon the vector sensor with a parameter vector $\bar{\psi}_i$ given as [6]

$$g(\bar{\psi}_d, \bar{\psi}_i) = |\bar{v}_{vs}(\bar{\psi}_i)^H \bar{v}_{vs}(\bar{\psi}_d)|^2 / 4 \quad (10.38)$$

where the beampattern reaches its maximum value when the parameter vectors $\bar{\psi}_d$ and $\bar{\psi}_i$ are equal. The right-hand side of Eq. (10.38) has been normalized by a factor of 4 so that when $\bar{\psi}_d = \bar{\psi}_i$, the maximum value is unity. The vector sensor response for cases when $\bar{\psi}_d \neq \bar{\psi}_i$ is a complicated function dependent upon not only the difference in arrival angle but also the polarization state. To help simplify the beampattern analysis, consider the case where $\phi_d = \phi_i = 0°$ and $\theta_d = 90°$ corresponding to the desired signal's angle of arrival being aligned with the x-z plane. Using Eq. (10.21), the term $\bar{v}_{vs}(\bar{\psi}_i)^H \bar{v}_{vs}(\bar{\psi}_d)$ in Eq. (10.38) simplifies to

$$\bar{v}_{vs}(\bar{\psi}_i)^H \bar{v}_{vs}(\bar{\psi}_d) = (1+\sin\theta_i)\left[\cos\gamma_d \cos\gamma_i + \sin\gamma_d \sin\gamma_i e^{j(\eta_d - \eta_i)}\right] \quad (10.39)$$

Next, evaluating $|\bar{v}_{vs}(\bar{\psi}_i)^H \bar{v}_{vs}(\bar{\psi}_d)|^2$ and simplifying with trigonometric identities yields

$$|\bar{v}_{vs}(\bar{\psi}_i)^H \bar{v}_{vs}(\bar{\psi}_d)|^2 = \frac{(1+\sin\theta_i)^2}{2}[1+\cos 2\gamma_d \cos 2\gamma_i$$
$$+\sin 2\gamma_d \sin 2\gamma_i \cos(\eta_d - \eta_i)] \quad (10.40)$$

This expression can be further simplified with the help of the work by Compton on the *tripole* [2]—an antenna composed of three orthogonal electric dipoles and no magnetic loops—and the Poincaré sphere in Fig. 10.10. The polarizations of the desired and incident signals are represented as points M_d and M_i on the Poincaré sphere where the angles $2\gamma_d$ and $2\gamma_i$ form the sides of a spherical triangle with arc $M_d M_i$. The angle $\eta_d - \eta_i$ is the angle opposite the arc $M_d M_i$. The expression within the brackets in Eq. (10.40) is simplified by applying the spherical cosine law, resulting in

$$\cos 2\gamma_d \cos 2\gamma_i + \sin 2\gamma_d \sin 2\gamma_i \cos(\eta_d - \eta_i) = \cos(M_d M_i) \quad (10.41)$$

Equation (10.35) is therefore equivalent to

$$|\bar{v}_{vs}(\bar{\psi}_i)^H \bar{v}_{vs}(\bar{\psi}_d)|^2 = \frac{(1+\sin\theta_i)^2}{2}\left[1+\cos(M_d M_i)\right]$$
$$= (1+\sin\theta_i)^2 \cos^2\left(\frac{M_d M_i}{2}\right) \quad (10.42)$$

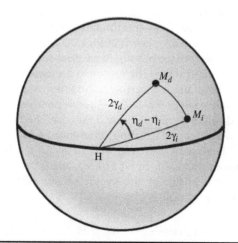

FIGURE 10.10 Poincaré sphere illustrating relationship between spherical angles and distances between two distinct polarizations.

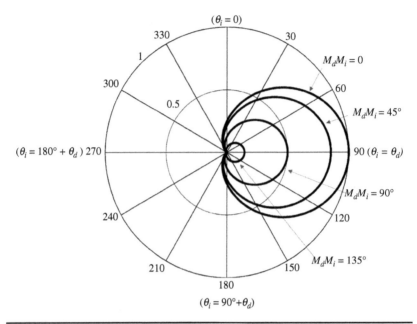

Figure 10.11 Vector sensor beampattern cross-section in x-z plane ($\phi_d = \phi_i = 0$) as a function of incident elevation angle θ_i for varying polarization separation $M_d M_i$.

and Eq. (10.38) is then given by

$$g(\bar{\psi}_d, \bar{\psi}_i) = \frac{(1+\sin\theta_i)^2}{4} \cos^2\left(\frac{M_d M_i}{2}\right) \quad (10.43)$$

The result in Eq. (10.43) shows that when the signals arrive at the same azimuth angle, the beampattern is dependent upon the difference in arrival elevation angle and the polarization separation $M_d M_i$ on the Poincaré sphere. Two- and three-dimensional plots of the vector sensor beampattern are shown in Figs. 10.11 and 10.12. As seen in Fig. 10.11, the vector sensor beampattern cross-section of Eq. (10.42) is shaped like a cardioid where the response decreases with an increase in either the arrival angle θ_i or the polarization separation $M_d M_i$. A beampattern null exists at $\theta_i = 180° + \theta_d$ corresponding to the direction opposite the desired elevation angle. Likewise, the beampattern has no response when the polarization difference $M_d M_i = 180°$. We also see in the 3D beampattern in Fig. 10.12 that no grating lobes exist because the maximum value in Eq. (10.38) only occurs when $\bar{\psi}_d = \bar{\psi}_i$.

The 3-dB beamwidth of the vector sensor when both signals arrive from the same angle is given by [6]

$$2\cos^{-1}\left(\sqrt{2}/\cos\frac{M_d M_i}{2} - 1\right), \text{ if } M_d M_i \in [0, 90°] \quad (10.44)$$

$$0, \text{ if } M_d M_i \in [90°, 180°]$$

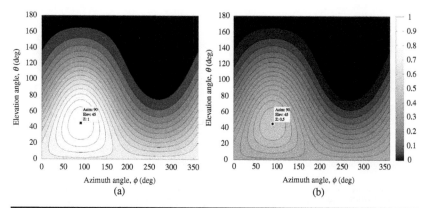

FIGURE 10.12 Three-dimensional vector sensor beampattern for desired angle $(\theta_d, \phi_d) = (45°, 90°)$ and polarization state $(\gamma_d, \eta_d) = (45°, 90°)$ corresponding to right-hand circular for (a) matched ($\bar{\psi}_i = \bar{\psi}_d$) and (b) unmatched right-hand circular and horizontal polarization where $\theta_d = \theta_i$, $\phi_d = \phi_i$, $\gamma_i = 0°$, $\eta_i = 0°$ corresponding to a polarization separation $M_d M_i$ of 90°.

The corresponding signal-to-interference-plus-noise ratio (SINR) is given by [6]

$$SINR = \sigma_d^2 \left[\frac{2}{\sigma_n^2} - \frac{(1+\sin\theta_i)^2}{(\sigma_n^2 + \sigma_{i,u}^2)} \left(\frac{\sigma_{i,u}^2}{2\sigma_n^2} + \frac{\sigma_{i,c}^2 \cos^2 \frac{M_d M_i}{2}}{2\sigma_{i,c}^2 + \sigma_n^2 + \sigma_{i,u}^2} \right) \right]$$ (10.45)

where σ_d^2 is the power of the desired signal, σ_n^2 is the uncorrelated noise power, σ_i^2 is the power of the incident signal that may contain both unpolarized and completely polarized components with powers $\sigma_{i,u}^2$ and $\sigma_{i,c}^2$. If we assume that the incident signal arrives is in the same direction as the desired signal and is completely polarized, then $\sigma_{i,u}^2 = 0$ and Eq. (10.45) reduces to

$$SINR_{vs} = 2SNR_d \left[\frac{1 + 2SNR_i \sin^2\left(\frac{M_d M_i}{2}\right)}{1 + 2SNR_i} \right]$$ (10.46)

which can be compared to the SINR of the tripole antenna [2] given as

$$SINR_{tripole} = SNR_d \left[\frac{1 + SNR_i \sin^2\left(\frac{M_d M_i}{2}\right)}{1 + SNR_i} \right]$$ (10.47)

where $SNR_d = \sigma_d^2 / \sigma_n^2$ and $SNR_i = \sigma_{i,c}^2 / \sigma_n^2$ are the signal-to-noise ratios of the desired and incident signals.

304 Chapter Ten

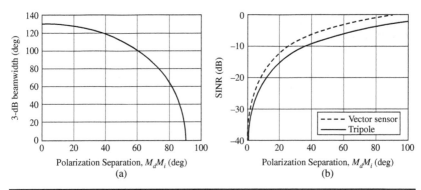

FIGURE 10.13 Effect of polarization separation on the vector sensor (a) 3-dB beamwidth and (b) SINR for $SNR_d = 0$ dB and $SNR_i = 40$ dB.

Plots of the 3-dB beamwidth and SINR versus polarization separation $M_d M_i$ are shown in Fig. 10.13. In Fig. 10.13a, the beamwidth is shown to be approximately 130° when the polarizations of the desired and incident signals are the same; however, as the polarization separation increases the beamwidth decreases until it is exactly zero for polarization separations greater than 90°. Figure 10.13b is a plot of the SINR for the vector sensor and tripole of Eqs. (10.46) and (10.47), where $SNR_d = 0$ dB and $SNR_i = 40$ dB. As can be seen, generally an increase in polarization separation improves SINR for both antennas. This demonstrates the ability of the vector sensor and tripole to offer interference protection by differentiating among signals with the same arrival angle, but with sufficient polarization separation. In the case of the vector sensor, the lowest SINR is equal to $2SNR_d / (1 + 2SNR_i)$, which is higher than the tripole's lowest SINR, but not by much for the example in Fig. 10.13; however, we see that the improvement in SINR occurs more quickly for the vector sensor than the tripole and reaches zero decibels when the separation $M_d M_i = 90°$.

10.5 Vector Sensor Cramer-Rao Lower Bound

The Cramer-Rao lower bound for the vector sensor follows a similar derivation as for the spatially displaced array due to their similar forms as seen in Fig. 10.13. However, in the vector sensor case the parameter vector $\bar{\psi} \triangleq [\theta, \phi, \gamma, \eta]$ has the additional polarization state parameters that are often treated as nuisance parameters when solving for the bounds on the angle of arrival.

In general, the multi-parameter Cramer-Rao lower bound is a bound on the covariance matrix of any unbiased estimate of $\bar{\psi}$ defined as

$$\bar{C}(\bar{\psi}) \triangleq \bar{J}^{-1} \tag{10.48}$$

where $\bar{J}$ is the Fisher Information Matrix (FIM) whose elements are (Eq. 8.34 of [15])

$$J_{ij} = B \cdot tr\left[\bar{R}_{xx}^{-1} \frac{\partial \bar{R}_{xx}}{\partial \bar{\psi}_i} \bar{R}_{xx}^{-1} \frac{\partial \bar{R}_{xx}}{\partial \bar{\psi}_j}\right] + 2Re\left[\frac{\partial \bar{\mu}^H}{\partial \bar{\psi}_i} \bar{R}_{xx}^{-1} \frac{\partial \bar{\mu}^H}{\partial \bar{\psi}_j}\right] \quad (10.49)$$

where tr is the trace operator of a matrix.

Here, $\bar{R}_{xx}$ is the autocorrelation of the vector sensor signal model in Eq. (10.23),

$$\bar{R}_{xx} = E\{\bar{x} \cdot \bar{x}^H\} = \sigma_s^2 \bar{v}_{vs}(\bar{\psi}) \bar{v}_{vs}^H(\bar{\psi}) + \sigma_n^2 \bar{I} \quad (10.50)$$

and where the signal $s(t)$ has power σ_s^2 and B temporal snapshots along with zero-mean Gaussian noise of power σ_n^2 uncorrelated with the signal $s(t)$. Under this assumption, the second term in Eq. (10.49) is equal to zero, reducing the elements of the Fisher Information Matrix to

$$J_{ij} = B \cdot tr\left[\bar{R}_{xx}^{-1} \frac{\partial \bar{R}_{xx}}{\partial \bar{\psi}_i} \bar{R}_{xx}^{-1} \frac{\partial \bar{R}_{xx}}{\partial \bar{\psi}_j}\right] \quad (10.51)$$

The parameter vector $\bar{\psi}$ can contain any number of unknown parameters that we wish to estimate. Some of these parameters may be of interest whereas others may be unwanted or *nuisance parameters*. For example, we may wish to estimate the angle of arrival and determine the Cramer-Rao bound on estimation performance; however, the polarization state (γ, η) as well as signal and noise powers, carrier frequency, etc. may also be unknown, but are considered nuisance parameters because they influence the Cramer-Rao bound calculation for the angle of arrival. In order to mitigate the influence of nuisance parameters on the Cramer-Rao bound, the parameter vector $\bar{\psi}$ is partitioned into wanted and unwanted parameters as in [15]

$$\bar{\psi} = \begin{bmatrix} \bar{\psi}_w \\ \bar{\psi}_u \end{bmatrix} \quad (10.52)$$

where $\bar{\psi}_w = [\theta, \phi]^T$ are the wanted parameters and $\bar{\psi}_u = [\gamma, \eta, \ldots]^T$ are all the unwanted parameters. For simplicity, it is assumed that only the polarization state parameters are unknown, whereas all other parameters such as the signal power, noise power, frequency, etc. are known.

The partitioned $\bar{J}$ matrix is therefore of size 4×4 and is written as

$$\bar{J} = \begin{bmatrix} \bar{J}_{\bar{\psi}_w \bar{\psi}_w} & \bar{J}_{\bar{\psi}_w \bar{\psi}_u} \\ \bar{J}_{\bar{\psi}_u \bar{\psi}_w} & \bar{J}_{\bar{\psi}_u \bar{\psi}_u} \end{bmatrix} \quad (10.53)$$

and the Cramer-Rao bound for the wanted parameters is then computed as

$$\bar{C}(\bar{\psi}_w) = \left[\bar{J}_{\bar{\psi}_w \bar{\psi}_w} - \bar{J}_{\bar{\psi}_w \bar{\psi}_u} \bar{J}_{\bar{\psi}_u \bar{\psi}_u}^{-1} \bar{J}_{\bar{\psi}_u \bar{\psi}_w}\right]^{-1} \quad (10.54)$$

The interested reader is referred to [16] for a rigorous closed-form expression of the vector sensor Cramer-Rao lower bound.

In summary, the Cramer-Rao bound can be computed with the following steps,

1. Derive the FIM entries for Eq. (10.51).
2. Partition $\bar{J}$ in blocks corresponding to the wanted and unwanted nuisance parameters as in Eq. (10.53).
3. Compute the Cramer-Rao bound using Eq. (10.54) for each wanted parameter.

For example, if the wanted parameters were the arrival angles (θ, ϕ) and the unwanted parameters were the polarization state, then the FIM of Eq. (10.53) is a 4×4 matrix given as

$$\bar{J} = \begin{bmatrix} J_{\theta\theta} & J_{\theta\phi} & J_{\theta\gamma} & J_{\theta\eta} \\ J_{\phi\theta} & J_{\phi\phi} & J_{\phi\gamma} & J_{\phi\eta} \\ J_{\gamma\theta} & J_{\gamma\phi} & J_{\gamma\gamma} & J_{\gamma\eta} \\ J_{\eta\theta} & J_{\eta\phi} & J_{\eta\gamma} & J_{\eta\eta} \end{bmatrix} = \begin{bmatrix} \bar{J}_{(\theta,\phi)(\theta,\phi)} & \bar{J}_{(\theta,\phi)(\gamma,\eta)} \\ \bar{J}_{(\gamma,\eta)(\theta,\phi)} & \bar{J}_{(\gamma,\eta)(\gamma,\eta)} \end{bmatrix} \quad (10.55)$$

and Eq. (10.54) would then result in the matrix $\bar{C}(\theta, \phi)$ of size 2×2 given by

$$\bar{C}(\theta, \phi) = \begin{bmatrix} C_{\theta\theta} & C_{\theta\phi} \\ C_{\phi\theta} & C_{\phi\phi} \end{bmatrix} = \left[\bar{J}_{(\theta,\phi)(\theta,\phi)} - \bar{J}_{(\theta,\phi)(\gamma,\eta)} \bar{J}_{(\gamma,\eta)(\gamma,\eta)}^{-1} \bar{J}_{(\gamma,\eta)(\theta,\phi)} \right]^{-1} \quad (10.56)$$

The relevant derivatives for computing the FIM entries in Eq. (10.51) required to evaluate Eq. (10.56) are [11]

$$\frac{\partial \bar{R}_{xx}}{\partial \psi_i} = \sigma_s^2 \frac{\partial \bar{v}_{vs}(\bar{\psi})}{\partial \psi_i} \bar{v}_{vs}^H(\bar{\psi}) + \sigma_s^2 \bar{v}_{vs}(\bar{\psi}) \frac{\partial \bar{v}_{vs}^H(\bar{\psi})}{\partial \psi_i} \quad (10.57)$$

$$\frac{\partial \bar{v}_{vs,l}(\bar{\psi})}{\partial \theta} = \left[\frac{\partial \bar{q}_l(\theta,\phi)}{\partial \theta} \bar{\Theta}_l(\theta,\phi) + \bar{q}_l(\theta,\phi) \frac{\partial \bar{\Theta}_l(\theta,\phi)}{\partial \theta} \right] \bar{p}(\gamma,\eta)$$

$$\frac{\partial \bar{v}_{vs,l}(\bar{\psi})}{\partial \phi} = \left[\frac{\partial \bar{q}_l(\theta,\phi)}{\partial \phi} \bar{\Theta}_l(\theta,\phi) + \bar{q}_l(\theta,\phi) \frac{\partial \bar{\Theta}_l(\theta,\phi)}{\partial (\phi)} \right] \bar{p}(\gamma,\eta) \quad (10.58)$$

$$\frac{\partial \bar{q}_l(\theta,\phi)}{\partial \theta} = j \frac{2\pi}{\lambda} \text{diag} \left\{ \bar{r}_i^T \frac{\partial \bar{u}(\theta,\phi)}{\partial \theta} \right\} \bar{q}_l(\theta,\phi)$$

$$\frac{\partial \bar{q}_l(\theta,\phi)}{\partial \phi} = j \frac{2\pi}{\lambda} \text{diag} \left\{ \bar{r}_i^T \frac{\partial \bar{u}(\theta,\phi)}{\partial \phi} \right\} \bar{q}_l(\theta,\phi) \quad (10.59)$$

$$\frac{\partial \bar{v}_{vs,l}}{\partial \gamma} = \left[\bar{q}_l(\theta,\phi) \bar{\Theta}_l(\theta,\phi) \right] \frac{\partial \bar{p}(\gamma,\eta)}{\partial \gamma}$$

$$\frac{\partial \bar{v}_{vs,l}}{\partial \eta} = \left[\bar{q}_l(\theta,\phi) \bar{\Theta}_l(\theta,\phi) \right] \frac{\partial \bar{p}(\gamma,\eta)}{\partial \eta} \quad (10.60)$$

$$\frac{\partial \bar{p}(\gamma,\eta)}{\partial \gamma} = \begin{bmatrix} \cos\gamma \; e^{j\eta} \\ -\sin\gamma \end{bmatrix}$$

$$\frac{\partial \bar{p}(\gamma,\eta)}{\partial \eta} = \begin{bmatrix} j\sin\gamma \; e^{j\eta} \\ 0 \end{bmatrix} \tag{10.61}$$

As an example, let us evaluate the vector sensor Cramer-Rao lower bound on the arrival angle (θ, ϕ) with the polarization state (γ, η) as the unwanted parameters. The bound is evaluated as a function of SNR for the case where a single CW signal of the form $s(t) = e^{j2\pi f_c t}$ with $f_c = 10$ kHz, $f_s = 10 f_c$ and $B = 100$ snapshots is incident upon a single vector sensor at the coordinate system origin with arrival angle $(\theta, \phi) = (45°, -150°)$ and polarization state $(\gamma, \eta) = (75°, 45°)$. The SNR is varied from -20 to 20 dB where the signal power $\sigma_s^2 = 1$ and the corresponding noise power is calculated as

$$\sigma_n^2 = \frac{\sigma_s^2}{10^{\frac{SNR}{10}}} \tag{10.62}$$

The resulting square root of the bounds on the azimuth $\left(C_{\phi\phi}^{1/2}\right)$ and elevation $\left(C_{\theta\theta}^{1/2}\right)$ arrival angles from Eq. (10.56) versus SNR are shown in Fig. 10.14. Also included for comparison are the corresponding angle estimation error standard deviations for the MVDR estimator discussed in Sec. 10.3.2. Twenty separate trials were performed for each SNR and the standard deviation of the difference between the true and estimated angle was computed. From the results we can see that as SNR increases that, the standard deviation on the angle estimation error asymptotically approaches the Cramer-Rao lower bound. For low SNR we see that, the MVDR values settle to a

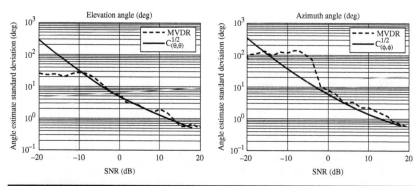

FIGURE 10.14 Example vector sensor Cramer-Rao lower bound versus SNR and MVDR angle-of-arrival estimator. From the results, it is clear that as SNR increases, MVDR converges toward the Cramer-Rao lower bound.

constant; however, the bound does not flatten out even though the variance on angle error exceeds what one would get for uniform angle error over $\pm\pi$, or in the case of elevation angle between 0 and $\pi/2$. As a result, the bound in this region can be considered "valid" once it is smaller than the uniform error assumption.

10.6 Acknowledgment

The author would like to acknowledge Cathy Crews for her help in reviewing the chapter content and for her mentoring over the years.

10.7 References

1. Volakis, J., "Direction Finding Antennas and Systems," in *Antenna Engineering Handbook*, New York, McGraw-Hill, 2007, pp. 11–12.
2. Compton, R. T., "The Tripole Antenna: An Adaptive Array with Full Polarization Flexibility," *IEEE Transactions on Antennas and Propagation*, Vol. AP-29, No. 11, pp. 944–952, 1981.
3. Li, J., and R. Compton, "Angle and Polarization Estimation Using ESPRIT with a Polarization Sensitive Array," *IEEE Transactions on Antennas and Propagation*, Vol. 39, No. 9, pp. 1376–1383, 1991.
4. Li, J., "Direction and Polarization Estimation Using Arrays with Small Loops and Short Dipoles," *IEEE Transactions on Antennas and Propagation*, Vol. 41, No. 3, pp. 379–387, 1993.
5. Nehorai, A., and E. Paldi, "Vector-Sensor Array Processing for Electromagnetic Source Localization," *IEEE Transactions on Signal Processing*, Vol. 42, No. 2, pp. 376–398, 1994.
6. Nehorai, A., K. C. Ho, and B. Tan, "Minimum-Noise-Variance Beamformer with an Electromagnetic Vector Sensor," *IEEE Transactions on Signal Processing*, Vol. 47, No. 3, pp. 601–618, 1999.
7. Wong, K., and M. D. Zoltowski, "Self-Initiating MUSIC-Based Direction Finding and Polarization Estimation in Spatio-Polarizational Beamspace," *IEEE Transactions on Antennas and Propagation*, Vol. 48, No. 8, pp. 1235–1245, 2000.
8. Robey, F., "High Frequency Geolocation and System Characterization (HF Geo) Proposer's Day Brief," *IARPA*, 22 June 2011. [Online]. Available: http://www.iarpa.gov/images/files/programs/hfgeo/HFGeo_Proposers_Day_Brief.pdf. [Accessed Feb. 22, 2015].
9. Martinsen, W., "Giselle: A Mutually Orthogonal Triple Twin-Loop Ground Symmetrical Broadband Receiving Antenna for the HF Band," *Defence Science and Technology Organization*, July 2009. [Online]. Available: http://www.dtic.mil/cgi-bin/GetTRDoc?Location=U2&doc=GetTRDoc.pdf&AD=ADA508699. [Accessed Feb. 22, 2015].
10. San Antonio, G., W. H. Lee, and M. Parent, "High Frequency Vector Sensor Design and Testing," in 2013 US National Committee of URSI National, Boulder, CO, 9–12 Jan. 2013.
11. Mir, H. S., and J. D. Sahr, "Passive Direction Finding Using Airborne Vector Sensors in the Presence of Manifold Perturbations," *IEEE Transactions on Signal Processing*, Vol. 55, No. 1, pp. 156–164, 2007.
12. Appadwedula, S., and C. Keller, "Direction-Finding Results for a Vector Sensor Antenna on a Small UAV," *Fourth IEEE Workshop on Sensor Array and Multichannel Processing*, pp. 74–78, 2006.
13. Deschamps, G., "Techniques for Handling Elliptically Polarized Waves with Special Reference to Antennas: Part II—Geometrical Representation of the Polarization of a Plane Electromagnetic Wave," *Proceedings of the IRE*, Vol. 39, No. 5, pp. 540–544, 1951.

14. Ferrara, E., and T. M. Parks, "Direction Finding with an Array of Antennas Having Diverse Polarizations," *IEEE Transactions on Antennas and Propagation*, Vol. 31, No. 2, pp. 231–236, 1983.
15. Van Trees, H. L., *Optimum Array Processing (Detection, Estimation and Modulation Theory, Part IV)*, New York: Wiley-Interscience, 2002.
16. Wong, K. T., and Y. Xin, "Vector Cross-Product Direction-Finding with an Electromagnetic Vector-Sensor of Six Orthogonally but Spatially Noncollocating Dipoles/Loops," *IEEE Transactions on Signal Processing*, Vol. 59, No. 1, pp. 160–171, 2011.

10.8 Problems

Polarization

10.1 For the electric field of a uniform plane wave given by

$$E(r,t) = 3\cos(\omega t - kr + 30°)\hat{\theta} + 4\cos(\omega t - kr + 45°)\hat{\phi}$$

(a) Determine the polarization state in terms of (γ, η) and (α, β) expressed in degrees.
(b) Plot the locus of the electromagnetic field $E(0, t)$ (i.e., E_ϕ vs. E_θ).
(c) From the plot, is the plane wave linearly, circularly, or elliptically polarized?

Direction Finding

10.2 A plane wave signal is incident upon an electromagnetic vector sensor at the origin of a spherical coordinate system. The following electric and magnetic fields were measured:

$$\bar{a} = \begin{bmatrix} E_x \\ E_y \\ E_z \\ H_x \\ H_y \\ H_z \end{bmatrix} = \begin{bmatrix} 0.4096 \\ 0.7094 \\ 0.1485 - j0.5540 \\ -0.0742 + j0.2770 \\ -0.1286 + j0.4798 \\ 0.8192 \end{bmatrix}$$

Compute the angle of arrival (θ, ϕ) and polarization state (γ, η) in degrees using the cross-product direction finding technique for a vector sensor.

10.3 Modify the MATLAB code from Fig. 10.9 to include a second co-channel signal $s_2(t) = \cos(\omega t)$ with arrival angle $(\theta, \phi) = (65°, 100°)$ and polarization state $(\gamma, \eta) = (10°, -125°)$ and noise standard deviation 0.01. Plot the MVDR pseudo-spectrum for the two signals in the form of Fig. 10.9.

10.4 Modify the MATLAB code from Fig. 10.9 to produce
  (a) A plot of the 3D MUSIC spectrum in a format similar to the MVDR result in Fig. 10.9.
  (b) A plot of the cross sections of the MVDR and MUSIC pseudo-spectrums for $\theta = 30°$ versus azimuth from $-180°$ to $180°$. Normalize the peak values of each pseudo-spectrum to 0 dB.

Beamforming

10.5 Prove the result of Eq. (10.40) starting with Eq. (10.38) for the case where $\theta_d = 90°$ and $\phi_d = \phi_i = 0°$. Hint: Use the following trigonometric identities:

$$e^{jx} = \cos(x) + j\sin(x)$$
$$\sin^2(x) + \cos^2(x) = 1$$
$$\sin(2x) = 2\cos(x)\sin(x)$$
$$\cos^2 x \cos^2 y + \sin^2 x \sin^2 y = \frac{1}{2}(1 + \cos 2x \cos 2y)$$

10.6 Compute the polarization separation $M_d M_i$ in degrees for polarization states of the desired and incident signals equal to $\gamma_d = 45°$, $\gamma_i = 75°$, $\eta_d = 32°$, $\eta_i = 25°$.

10.7 Create a 2D polar plot of the vector sensor beampattern using Eq. (10.43) for $\theta_i = 0$ to $360°$ and polarization separation $M_d M_i = \pi/3$.

10.8 Derive the closed-form result for the SINR of the full six-axis vector sensor in Eq. (10.46) starting from Eq. (10.45), where $\sigma_{i,u}^2 = 0$, $\theta_i = 90°$, $SNR_d = \sigma_d^2/\sigma_n^2$, and $SNR_i = \sigma_{i,c}^2/\sigma_n^2$.

10.9 Equation (8.8) presents a method for computing beamforming weights subject to a constraint vector $\bar{u} = [u_1 \; u_2 \; \cdots \; u_N]^T$, whose entries u_n are equal to 1 for desired signals and 0 for interfering signals. The weight vector $\bar{w}$ that satisfies these constraints is given by

$$\bar{w}^H = \bar{u}^T \bar{A}^H \left[\bar{A} \cdot \bar{A}^H + \sigma_n^2 \bar{I}\right]^{-1}$$

where $\sigma_n^2 = 10^{-9}$ is a small constant added to the diagonal of $\bar{A} \cdot \bar{A}^H$ to better condition the matrix prior to its inversion.

Consider a single vector sensor where a desired signal arrives at $(\theta_d, \phi_d) = (90°, 60°)$ and one interferer arrives at $(\theta_i, \phi_i) = (90°, 210°)$. Both signals have the same polarization state of $(\gamma, \eta) = (45°, 75°)$. Therefore, the vector $\bar{u} = [1 \; 0]^T$ and the matrix $\bar{A}$ are

$$\bar{A} = \left[\bar{\Theta}(\theta_d, \phi_d)\bar{p}(\gamma, \eta) \quad \bar{\Theta}(\theta_i, \phi_i)\bar{p}(\gamma, \eta)\right]$$

For this scenario, perform the following:
(a) Compute the normalized weight vector $\bar{w}_{norm}$ where $\bar{w}$ has been normalized by its 2-norm value with the help of MATLAB's "norm" function.
(b) Create a 2D plot of the normalized vector sensor beampattern for the cross-section $\theta = 90°$ and $\phi = 0°$ to $360°$ with the following overlaid within the figure:
  - The vector sensor beampattern in dB from Eq. (10.38) steered toward (θ_d, ϕ_d).
  - The beamsteered pattern in dB using the weight vector $\bar{w}$ computed in (a) to add a null at (θ_i, ϕ_i).
  - Mark both the arrival angle of the desired and interfering signals on the two curves.

Cramer-Rao Bound

10.10 For example in Fig. 10.14, and using the same parameters, for example 100 snapshots, etc., use the MUSIC direction finding technique code created in question 10.4 to plot the standard deviation of azimuth and elevation arrival angle estimation errors for signal-to-noise ratios from -20 to 20 dB. Compute errors using a minimum of 20 Monte-Carlo trials for each signal-to-noise ratio. Overlay the Cramer-Rao lower bound and MVDR direction finding technique results from Fig. 10.14 for comparison.

10.11 For the example in Fig. 10.14, modify the Cramer-Rao lower bound MATLAB calculation such that the bound is computed for the estimation of the polarization state parameters (γ, η) versus signal-to-noise ratios from -20 to 20 dB using the same parameters, for example 100 snapshots, etc. [Hint: The polarization state parameters become the wanted parameters and arrival angles are the nuisance parameters in Eq. (10.52).] Using MVDR, compute the standard deviation of the error estimates for the two polarization state parameters γ and η and overlay the computed Cramer-Rao lower bound for each.

CHAPTER 11
Smart Antenna Design

11.1 Introduction

Up to this point in the book, the primary focus has been on the signal-processing "brains" and algorithms behind smart antennas. Not much attention has been paid to the impact of antenna and array geometry designs on smart antenna performance. Often in the literature, smart antenna performance is evaluated with simplifying assumptions about the antennas and arrays such as isotropic antennas, uniform array spacing, and/or linear geometries. This is done to help make the problem more tractable much like the assumption of white Gaussian-distributed noise statistics in smart antenna algorithm development. Modern smart antennas operating in real-world environments include sophisticated antennas and arrays designed to either conform or reconfigure to their immediate surroundings to optimize performance. These complex geometries are challenging to design because either analytic solutions may not exist or performing an exhaustive search for the solution is intractable. Instead, a marriage is formed among global optimization algorithms, computational electromagnetics, and computer-processing power to numerically analyze these complex geometries in situ.

Global optimization algorithms seek to find the universally best solution(s) in the presence of multiple suboptimal, local solutions subject to the constraints on the optimization. The smart antenna techniques studied in Chap. 8 assume the optimization problem is convex, which means there is only a single unique solution in the search space. Therefore, local optimization algorithms based on gradient descent can be used to find the optimal solution quickly and accurately. For more complex problems, like the ones considered in this chapter, the assumption that the optimization is convex might not be true and as a result we turn to global optimization algorithms to help solve the problem.

This chapter was written by Jeffrey D. Conner, Senior Research Engineer, Georgia Tech Research Institute.

We concentrate on two specific global optimization algorithms: (1) the genetic algorithm and (2) the cross-entropy (CE) method. The genetic algorithm—based on Charles Darwin's theory of evolution and natural selection—is part of a class within the broader global optimization discipline called *evolutionary optimization algorithms*, which are inspired by the evolutionary, biological, and cognitive processes of nature. On the other hand, the CE method is part of a class of probabilistic and stochastic algorithms, which search for a representation of the random process from which the optimal solution is sampled. Algorithms like the CE method are considered by some to be in the same class of algorithm as genetic algorithms; however, probabilistic algorithms such as the CE method tend to emerge as the result of mathematic rigor including well-understood convergence properties and performance bounds under set conditions, which contrast with the meta-heuristic nature of evolutionary algorithms. The similarities and differences between these two algorithms become apparent in the next section.

Computational electromagnetics involves modeling the interaction of electromagnetic fields and waves within physical structures and the surrounding environment through computationally efficient approximations to Maxwell's equations. Typical computational electromagnetic techniques include the method of moments (MoM), finite difference time domain (FDTD), and the finite element method (FEM). Details of the theory behind these techniques are not covered in this chapter; however, we make use of an open-source MoM solver called *Numerical Electromagnetics Code* (NEC), which is a popular code for modeling metallic wire antennas and complex wire structures. Its use allows us to incorporate complex electromagnetic effects such as mutual coupling between wires and the presence of ground planes into the design process, which are often neglected in the literature. We will use MATLAB as a wrapper to integrate the genetic algorithm and CE method with the NEC code to optimize smart antenna designs. This creates a powerful tool for designing real-world smart antennas, one in which the designer does not need to have expertise in the physics of the problem in order to generate solutions.

The remainder of this chapter focuses on introducing the reader to the areas of global optimization and computational electromagnetics techniques in the design of antennas and arrays for smart antenna applications. In Sec. 11.2, detail is provided on the genetic algorithm and the CE method for global optimization. This includes some comments on convergence performance and accuracy as well as analysis of some simple problems to help illustrate the fundamental characteristics of the algorithms. Section 11.3 covers the design of nonuniform array geometries using global optimization algorithms to shape the array radiation pattern through either antenna element thinning or nonuniform spacing between antenna elements. Section 11.4 discusses the use of global optimization algorithms as adaptive nulling algorithms. Keys to consider include the convergence speed, computational costs, and solution accuracy compared to the traditional adaptive algorithms presented in Chap. 8. Section 11.5 introduces the Numerical Electromagnetic Code (NEC), its integration with MATLAB, and its use in smart antenna design. Two simple examples are presented, first for designing a dipole antenna and then for generating the radiation pattern of a monopole array. These examples, along with examples in Secs. 11.3 and 11.4, provide the foundation for evaluating real-world nonuniform array geometries and adaptive nulling for modern smart antennas. Section 11.6 presents a unique application of the integration between global optimization algorithms and computational electromagnetics techniques to design a space-filling wire antenna called the *crooked wire antenna* [1]. This antenna is designed by a genetic algorithm to have uniform hemispherical coverage with maximal gain over a broad range of frequencies while conforming to a fixed volume. The result is an odd antenna design that could not have been achieved with traditional design methodologies. The chapter concludes with some thoughts on current trends and future research in the area of smart antenna design.

11.2 Global Optimization Algorithms

There are many degrees of freedom from which to choose when addressing the antenna and array design problem such as element location, complex weighting, antenna type, size, etc. The feasible region of solutions is typically extremely large, and as a result, it would be practically impossible to exhaustively search through all possible solutions to arrive at an optimal result. Therefore, optimization techniques are used to reduce the search time required to find an acceptable solution. Traditionally, the synthesis of antennas and arrays requires expert knowledge and significant insight into the physics of the problem in order to exploit it. However, with improvements in the personal computer and computational electromagnetics along with new optimization techniques that imitate the evolutionary, biological, and cognitive processes of nature, it has become much easier to generate practical solutions without requiring significant expertise.

As seen in Chap. 8, traditional techniques used for solving the antenna and array design problem were deterministic in nature, using a measurable and quantifiable knowledge of the array physics to determine an optimal result. The user develops cost function criteria based on parameters of interest to exploit knowledge of the deterministic qualities of the antenna or array. Typical deterministic optimization algorithms applied to the design problem include Newton's method, simplex method, least squares, least mean squares (LMS), conjugate gradient method (CGM), as well as many other constrained and unconstrained linear/nonlinear programs. The deterministic methods usually travel along directions of descent about a computational cost surface toward an optimal solution. There is a distinct disadvantage to this technique in that one still needs expert knowledge of the array physics in order to quantify the right cost. The solutions stemming from deterministic algorithms require good initial guesses, can become unstable, get trapped in local minima, and potentially require long convergence times.

In contrast to conventional deterministic optimization techniques, global optimization methods governed by stochastic principles have become a popular tool for solving antenna and array design problems. Stochastic optimization algorithms use populations of random solutions that are scored based on their fitness to the desired solution and are then improved in the future by exploiting the previous solutions which performed the best. Unlike deterministic algorithms, one does not need expert knowledge of the antenna's physics to achieve an optimal result. At each moment in time a random population of potential optimal solutions is created and its fitness to the desired solution is evaluated to direct the algorithm toward the optimal solution. Stochastic algorithms do not follow the computational cost surfaces like gradient-based methods; rather they possess the ability to jump around the solution space and escape from the trappings of local minima to converge to a globally optimal solution.

The advantages of population-based stochastic algorithms include the following:

- Optimization with a mixture of discrete and continuous variables. Discrete optimization can result from encoding the continuous variables. Mixtures can contain binary values, integers, etc.
- A wide area of the feasible region is searched by generating a population of candidate solutions, and then the algorithm converges toward a local neighborhood of the global optimum solution.
- By using a population of candidate solutions, parallel processing on multiple platforms is possible.
- Multiextremal, multiobjective, multivariable optimization problems can be solved without the use of derivative information.
- Avoid trappings of local extrema.
- Nonlinear fitness functions can be used, which are not applicable to conventional algorithms.
- One does not need to be an expert in the physics of the antenna array or on its interactions with the surrounding environment in order to exploit it.

The disadvantages are related to the speed of convergence, accuracy, and computational costs:

- May only converge to a small neighborhood around the global optimum.

- May require a large number of cost function calculations, which can be time consuming when coupled with computational electromagnetics code.
- It may not be clear when to terminate the optimization resulting in excess computation and little improvement in the solution.

In this section, we focus on two algorithms: (1) the genetic algorithm and (2) the CE method. These algorithms represent two popular techniques within global optimization research in the areas of *evolutionary* and *probabilistic* techniques and have been applied to a wide range of problems in electromagnetics and smart antennas.

Evolutionary algorithms are population-based metaheuristic optimization algorithms inspired by the evolutionary, biological, and cognitive processes of nature. The genetic algorithm is based on Charles Darwin's theory of natural selection and biological evolution. For a given population, the best-fit individuals, called *parents*, are selected to reproduce offspring through crossover, mutation, and recombination operations. The fitness of these offspring is then evaluated and the best performers replace the least-fit members of the population (i.e., survival of the fittest). This process is continued over several iterations, called *generations*, and the best performing individual converges toward the globally optimal solution. Some additional examples of evolutionary algorithms applied to electromagnetics and smart antenna design include the following:

- *Particle swarm optimization:* Developed by James Kennedy and Russell Eberhart in 1995 [2], it based on the cognitive behavior between individuals in a swarm. For example, such swarms would be schools of fish or flocks of birds. The application of PSO in electromagnetics research has been extensive [3–5].
- *Ant colony optimization:* Developed by Marco Dorigo in 1992, it is inspired by how ants forage for food through their interaction with pheromones and scent chemicals to help promote the optimum path to the food source for others to follow [6].
- *Simulated annealing:* Based on the annealing process in metallurgy involving the heating and cooling of materials to increase its ductility which is translated as a slow decrease in the probability of accepting least-fit solutions as the algorithm explores the search space [7].
- *Harmony search:* Based on the improvisational process of musicians to search for a harmonious state.
- *Bees algorithm:* Based on the foraging behavior of honey bees and pollination [8].

- *Firefly algorithm:* Based on the communication role of bioluminescent interaction between fireflies while mating [9].
- *Cuckoo search:* Inspired by the obligate brood parasitism of some cuckoo species to lay eggs in the nests of other birds [10].
- *Bat algorithm:* Based on the echolocation of microbats with varying pulse rates of emission and loudness [11].
- *Bacterial foraging algorithms:* Inspired by the foraging behavior of Escherichia (E.) coli bacteria in the human intestine [12].

11.2.1 Description of Algorithms

The Genetic Algorithm

The genetic algorithm (GA)—developed by John Holland in the 1960s and 1970s—is based on Charles Darwin's theory of natural selection, biological evolution, and survival of the fittest. The fundamental relationship between Darwin's theory and its use as an optimization algorithm starts with the encoding of the optimization variables into strings called *chromosomes*. In the natural world, chromosomes refer to the thread-like structures inside cells that carry our DNA—the building blocks of the genetic makeup for the human body composed of *genes* that make each living creature unique. Chromosomes usually come in pairs, with half passed to offspring from the mother and the other half coming from the father. A collection of chromosomes forms a *population*. In Darwin's theory, natural selection acts to preserve and accumulate advantageous features of previous generations into the current population, whereas offspring with disadvantaged features gradually die off removing inferior genes from the gene *pool*. However, uncontrollable and random changes in the evolved genetic sequence called *mutations* may occur and introduce diversity into the population. These concepts together form the typical framework for the genetic algorithm:

1. *Encoding:* Create a "genetic" representation of the solution space. This typically involves encoding individual chromosomes with binary or continuous variables.
2. *Mating:* Create a mating pool of parents, perform selection, and produce offspring.
3. *Mutation:* Introduce random mutations into the parents in the population.
4. *Fitness:* Evaluate the performance of the population relative to some fitness (or cost) criteria.
5. *Natural selection:* Perform natural selection by keeping the best-fit (i.e., *elite*) chromosomes while discarding all others and replacing them with the children in the next generation.

Encoding The choice of encoding for a particular problem can have a big impact on the algorithm performance and the implementation of the remaining steps in the algorithm. Traditional genetic algorithms used binary encoding. Binary encoding was chosen as there is a finite number of solutions in the search space equal to $2^{N_{bits} \cdot N_{var}}$, where N_{bits} is the number of binary bits used to encode a single variable and N_{var} is the total number of variables to encode. As a result, there is a tradeoff between the fidelity of the final solution and how close this final solution is to the globally optimal solution. Advancements to the genetic algorithm have included the use of continuous and mixed integer encoding for variables. The appeal of the genetic algorithm is the way in which you can define disparate representations of features with a common numerical encoding in order to perform optimization. For example, if you were describing a human by his or her features, you might label them as male or female with blond/brunette/red hair who is XX feet tall weighing YY pounds. This is a mixture of contextual and numerical attributes about a particular person. In the genetic algorithm, these unique features could be encoded with a 3-bit binary number and grouped into a single chromosome as [0 1 0 1 1 0 1 0 1 0 0 1]. This is a distinct advantage over traditional convex optimization techniques.

Mating The implementation of the mating step is the element of the genetic algorithm with the greatest variability. The first component is the *selection* of the parents from the population for mating. The two most popular techniques are called *roulette wheel* and *tournament selection*. In the case of roulette wheel selection—like roulette wheels in a casino—different spots on the wheel have different odds of winning. In the genetic algorithm, the mating pool is composed of the N_{sel} most-fit chromosomes, which are assigned a slice of the wheel, where the odds of winning for a given chromosome is based on its overall fitness. Typically, the roulette wheel odds are assigned in one of two ways: (1) cost-based or (2) rank-based. In a cost-based approach, the parents are selected based on their fitness—the higher their fitness, the greater chance they have of being selected. The cost-based roulette wheel is generated as

1. Sort all chromosomes in the mating pool based on their respective costs from lowest to highest.
2. Calculate the sum of all chromosome costs and normalize the individual costs in step 1. This results in normalized costs in the range of 0 to 1.
3. Draw a uniform random number r in the range of 0 to 1.
4. Starting from the first chromosome in the sorted population, compute the cumulative sum for each chromosome location. When the cumulative sum is greater than the number r, stop and return the chromosome at this location.

320 Chapter Eleven

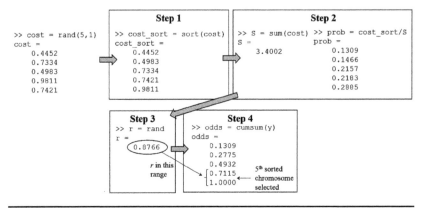

FIGURE 11.1 Example cost-based roulette wheel selection approach.

An example illustrating the step-by-step approach to cost-based roulette wheel selection is shown in Fig. 11.1.

The rank-based approach is very similar to the cost-based approach, except that the odds are fixed for all generations, whereas the cost-based approach changes from generation to generation based on the changes in population fitness. The assignment of odds in the rank-based roulette wheel is somewhat arbitrary. As an example, one possible assignment of the probabilities for a mating pool of size 4 could be

```
>> Npool = 4; prob = (1:Npool)/sum(1:Npool)
prob =
   0.1000  0.2000  0.3000  0.4000
>> odds = cumsum(prob)
odds =
   0.1000  0.3000  0.6000  1.0000
```

Typically, the rank-based approach is favored over the cost-based approach for a number of reasons. First, in the cost-based approach, the roulette wheel needs to be recomputed in each new generation. It is also possible in the cost-based approach that the best chromosome may dominate a large percentage of the roulette wheel reducing greatly the selection odds for other chromosomes. This may cause uniformity in the population, resulting in the algorithm converging too fast to a less accurate solution. How the odds map to the fitness of the chromosomes in the mating pool has a great impact on convergence speed, solution accuracy, and consistency. Intuition, as in theory would suggest that the chromosomes with the highest fitness should receive the highest odds in the selection process in order to encourage the best solutions; however, as shown in Fig. 11.2, this has unintended consequences. The balance to strike is between the separation of the overall best chromosome and the mating pool average. These values should not converge together too quickly, if at all. With separation of the population average

Smart Antenna Design 321

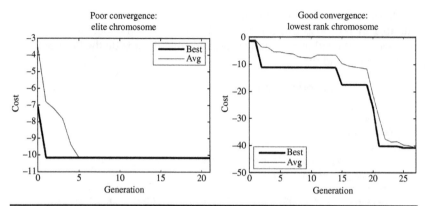

FIGURE 11.2 Examples of convergence based on the assignment of odds in the roulette wheel. (Left) Poor convergence: caused by assigning largest odds to elite chromosome. (Right) Good convergence: caused by assigning largest odds to lowest-ranked chromosome in mating pool.

and best chromosome, the algorithm continues to search for solutions in other areas of the solution space. This can be achieved by assigning the largest odds of selection in the roulette wheel to the lowest-rank (i.e., worst-fit) chromosome in the mating pool.

After the parent selection process is completed and a mother and a father chromosome have been chosen, the next step is to produce offspring. This is accomplished through a process called *crossover*. In crossover, N_{mom} elements are contributed by the mother chromosome and N_{dad} by the father, where $N_{dad} = 1 - N_{mom}$. There are many ways in which N_{mom} and N_{dad} elements can be contributed. A simple way is to randomly draw a *single point* crossover within the $N_{var} = N_{mom} + N_{dad}$ locations of a chromosome and take the first N_{mom} elements from the mother and the remaining N_{dad} from the father. One could produce additional offspring by taking the first N_{dad} from the father and the remaining N_{mom} from the mother.

```
>> crossover_pt = randi (10,1)
crossover_pt =
    6
>> mom = rand(1,10)
mom =
    0.8168 0.5303 0.9310 0.2392 0.4178 0.0537 0.6302 0.0212
    0.1350 0.4894
>> dad = rand(1,10)
dad =
    0.6426 0.9023 0.5793 0.5582 0.9366 0.8762 0.6342 0.2043
    0.4596 0.3015
>> offspring = [mom(1:crossover_pt) dad(crossover_pt+1:end)]
offspring =
    0.8168 0.5303 0.9310 0.2392 0.4178 0.0537 0.6342 0.2043
    0.4569 0.3015
```

Last, in the case of continuous variables, the single-point crossover variable in the offspring can be computed as a blend between the mother and the father to produce solutions that lie outside the bounds of the mother and the father:

$$\text{offspring} = \text{mom} + \alpha\,(\text{mom} - \text{dad}) \tag{11.1}$$

```
>> alpha = rand
   alpha =
       0.9153
>> offspring(crossover_pt) = mom(crossover_pt)+alpha*(mom
(crossover_pt)-dad(crossover_pt))
offspring =
   0.8168 0.5303 0.9310 0.2392 0.4178 -0.6992 0.6342 0.2043
   0.4569 0.3015
```

This encourages further exploration of the solution space, but the offspring must be checked to ensure they are still feasible and within the specified bounds. Many variations exist on the crossover mechanism to generate offspring, including two-point, N-point, uniform, arithmetic, quadratic, etc. The interested reader is referred to [13, 14] for further information.

The last mechanism for producing offspring is mutation. After the crossover mechanism has generated new offspring from mothers and fathers in the population, a percentage of all the variables in the population are randomly changed (i.e., mutated). In the case of binary variables, this is simply flipping the bit polarity from 0 to 1 or vice versa. For continuous variables, this would be some random value within the defined limits of the solution space. An example MATLAB implementation is given in Fig. 11.3.

Convergence Check The final component of the genetic algorithm is the stopping criteria. There are many different ways of performing this. Some typical limits would be set on the total number of generations or a threshold on a minimum acceptable cost for the best performer. One could also terminate once the best performer and average cost in the population have converged to the same value as no significant variability will likely occur in future generations. This condition could happen early on in the process, so one must take care in ensuring that more than one successive generations have this condition.

Example Next, let's consider a simple demonstration of the genetic algorithm for continuous variables for the minimization of the N-dimensional trigonometric test function given by [15]

$$S(x) = 1 + \sum_{n=1}^{N} 8\sin^2(\eta(x_n - x_n^*)^2) + 6\sin^2(2\eta(x_n - x_n^*)^2) + \mu(x_n - x_n^*)^2 \tag{11.2}$$

Smart Antenna Design

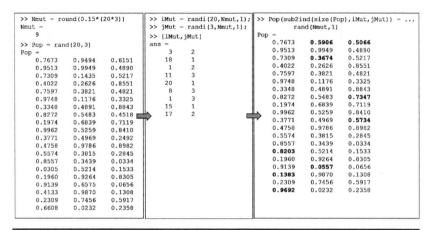

FIGURE 11.3 Example of mutation operation in continuous genetic algorithm for mutation rate of 15 percent, which is equal to nine mutations (in **bold**) in a population size of 60.

where $\eta = 7$, $\mu = 1$, $x_n^* = 0.9$. The optimal solution for this problem is defined as $\overline{x}^* = [0.9, 0.9, \ldots, 0.9]$, resulting in a global minimum $S(x^*) = 1$. Figure 11.4 illustrates this function in two dimensions (i.e., $N = 2$). As is seen, the cost surface contains many local minima, but only one global minimum. This is a challenging test function, especially for the traditional optimization techniques discussed in Chap. 8.

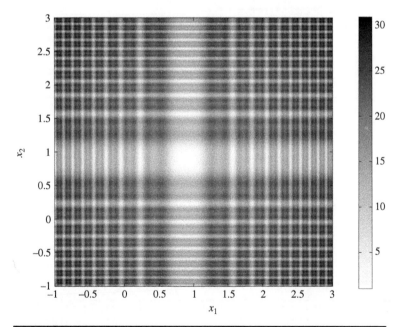

FIGURE 11.4 Two-dimensional trigonometric test function cost surface where the minimum value equals 1 at $(x_1, x_2) = (0.9, 0.9)$.

For this example, the problem dimension N was set to 10. The genetic algorithm settings included a population size of 20, a mutation rate of 12 percent, mating pool of size 4, with roulette wheel selection and single point crossover. The odds for the roulette wheel were set to [0, 0.1, 0.3, 0.6, 1.0], where the range [0, 0.1] = 10 percent were the odds of the best fit chromosome and [0.6, 1.0] = 40 percent were the odds of the last chromosome in the mating pool. This was done to minimize the chance of early convergence due to favoring the elite chromosome.

The results of the genetic algorithm optimization procedure are shown in Fig. 11.5. The top subplot includes the progression of the best chromosome as it converges toward the optimal solution as $\bar{x}^* =$ [0.9, 0.9, ... , 0.9]. The final chromosome for this single instance is given in Table 11.1. As can be seen, this solution is within a small neighborhood of the optimal solution. The bottom subplot contains the progression of the score for the best chromosome and the population average. Early on, when the score values in the population are high, the impact of the mutation rate is not apparent; however, as the best chromosome starts to converge toward the global minimum,

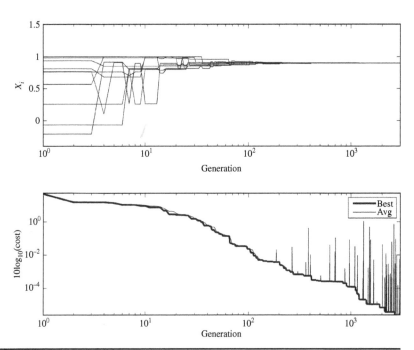

Figure 11.5 Single optimization result of the trigonometric test function of Eq. (11.2) by the genetic algorithm. (Top) Progression of the best chromosome for each generation toward the optimal solution; (bottom) progression of the best and average score for each generation.

x_1	x_2	x_3	x_4	x_5
0.9000	0.9003	0.8999	0.9004	0.9001
x_6	x_7	x_8	x_9	x_{10}
0.8999	0.9002	0.9000	0.9001	0.8995

TABLE 11.1 Result of a Single Instance of the Genetic Algorithm for the Trigonometric Test Function Example

mutations change the population average, so a greater deviation of the population average and best chromosome is observed.

Cross-Entropy Method

The cross-entropy (CE) method is a general stochastic optimization technique based on a fundamental principle of information theory called *cross entropy* (or Kullback-Leibler divergence) [16, 17]. CE was first introduced in 1997 by Reuven Y. Rubinstein of the Technion, Israel Institute of Technology, as an adaptive importance sampling for estimating probabilities of rare events [18] and was extended soon thereafter to include both combinatorial and continuous optimization [19]. The CE method has successfully optimized a wide variety of traditionally hard test problems, including the max-cut problem, traveling salesman, quadratic assignment problem, and n-queen. Additionally, CE has been applied to antenna pattern synthesis [20, 21], queuing models for telecommunication systems [22], DNA sequence alignment [23], scheduling, vehicle routing [24], learning Tetris [25], direction of arrival estimation [26], speeding up the backpropagation algorithm [27], updating pheromones in ant colony optimization [28], blind multiuser detection [29], optimizing MIMO capacity [30], and many others.

To illustrate how the CE procedure is implemented, assume for the time being that the score $S(\bar{x})$ is to be maximized over all states $\bar{x} \in \chi$, where $\bar{x} = [x_1, ..., x_M]^T$ is a vector of candidate solutions defined on the feasible set χ. The global maximum of $S(\bar{x})$ is represented by

$$S(\bar{x}^*) = \gamma^* = \max_{\bar{x} \in \chi} S(\bar{x}) \qquad (11.3)$$

The probability that the score function $S(\bar{x})$ evaluated at a particular state x is close to γ^* is classified as a rare event. This probability can be determined from an *associated stochastic problem* (ASP)

$$P_v(S(\bar{x}) \geq \gamma) = E_v I_{\{s(\bar{x}) \geq \gamma\}} \qquad (11.4)$$

where P_v is the probability measure that the score is greater than some value γ close to γ^*, $\bar{x}$ is the random variable produced by a probability distribution function $f(\cdot, v)$, E_v is the expectation operator, and $I_{\{\cdot\}}$ is a set of indices, where $S(\bar{x})$ is greater than or equal to γ.

Calculating the right-hand side of Eq. (11.4) is a nontrivial problem and can be estimated using a log-likelihood estimator with parameter v

$$\hat{v}^* = \arg\max_{v} \frac{1}{M_s} \sum_{i=1}^{M_s} I_{\{S(x_i)\geq \gamma\}} \ln f(x_i, v) \qquad (11.5)$$

where x_i is generated from $f(\cdot, v)$ and M_s is the number of samples where $S(x_i) > \gamma$ and $M_s \leq M$.

As γ becomes close to γ^*, most of the probability mass is close to $\bar{x}^*$ and is an approximate solution to Eq. (11.3). One important requirement is that as γ becomes close to γ^* that $P_v(S(\bar{x}) \geq \gamma)$ is not too small otherwise; the algorithm will result in a suboptimal solution. Therefore, there is a tradeoff between γ being arbitrarily close to γ^* while maintaining accuracy in the estimate of v.

The CE method efficiently solves this estimation problem by adaptively updating the estimate of the optimal density $f(\cdot, v^*)$, thus creating a sequence of pairs $\{\hat{\gamma}^{(t)}, \hat{v}^{(t)}\}$ at each iteration t in an iterative procedure that converges quickly to an arbitrarily small neighborhood of the optimal pair $\{\gamma^*, v^*\}$.

The iterative CE procedure for estimating $\{\gamma^*, v^*\}$ is given by

1. *Initialize parameters:* Set initial parameter $\hat{v}^{(0)}$, choose a small value ρ, set population size M, smoothing constant α, and set iteration counter $t = 1$.

2. *Update $\hat{\gamma}^{(t)}$:*

    Given $\hat{v}^{(t-1)}$, let $\hat{\gamma}^{(t)}$ be the $(1-\rho)$-quantile of $S(\bar{x})$ satisfying

    $$P_{v(t-1)}(S(\bar{x}) \geq \gamma^{(t)}) \geq \rho \qquad (11.6)$$

    $$P_{v(t-1)}(S(\bar{x}) \leq \gamma^{(t)}) \geq 1 - \rho \qquad (11.7)$$

    with $\bar{x}$ sampled from $f(\cdot, \hat{v}^{(t-1)})$. Then, the estimate of $\gamma^{(t)}$ is computed as

    $$\hat{\gamma}^{(t)} = S_{(\lceil (1-\rho)M \rceil)} \qquad (11.8)$$

    where $\lceil \cdot \rceil$ rounds $(1-\rho)M$ toward infinity.

3. *Update $\hat{v}^{(t)}$:*

    Given, $\hat{v}^{(t-1)}$, determine $\hat{v}^{(t)}$ by solving the CE program

    $$\hat{v}^{(t)} = \max_{v} \frac{1}{M_s} \sum_{i=1}^{M_s} I_{\{S(x_i)\geq \hat{\gamma}^{(t)}\}} \ln f(x_i, v) \qquad (11.9)$$

4. *Optional step* (smooth update of $\hat{v}^{(t)}$):

To decrease the probability of the CE procedure converging too quickly to a suboptimal solution, a smoothed update of $\hat{v}^{(t)}$ can be computed.

$$\hat{v}^{(t)} = \alpha \hat{v}^{(t)} + (1-\alpha)\hat{v}^{(t-1)} \qquad (11.10)$$

where $\hat{v}^{(t)}$ is the estimate of the parameter vector computed with Eq. (11.9), $\hat{v}^{(t-1)}$ is the parameter estimate from the previous iteration, and α (for $0 < \alpha \leq 1$) is a constant smoothing coefficient. By setting $\alpha = 1$, the update will not be smoothed.

5. Set $t = t + 1$ and repeat steps 2 to 4 until some stopping criterion is satisfied.

What ultimately is produced is a family of pdfs $f(\cdot, \hat{v}^{(0)})$, $f(\cdot, \hat{v}^{(1)})$, $f(\cdot, \hat{v}^{(2)}), \ldots, f(\cdot, \hat{v}^*)$ that are directed by $\hat{\gamma}^{(1)}, \hat{\gamma}^{(2)}, \hat{\gamma}^{(3)}, \ldots, \hat{\gamma}^*$ toward the neighborhood of the optimal density function $f(\cdot, v^*)$. The pdf $f(\cdot, \hat{v}^{(t)})$ acts to carry information about the best samples from one iteration to the next. The CE parameter update of Eq. (11.9) ensures that there is an increase in the probability that these best samples will appear in subsequent iterations. At the end of a run as $\hat{\gamma}$ is closer to γ^*, the majority of the samples in $\bar{x}$ will be identical and trivially so too are the values in $S(\bar{x})$.

The initial choice of $\hat{v}^{(0)}$ is arbitrary given that the choice of ρ is sufficiently small and K is sufficiently large so that $P_v(S(\bar{x}) \geq \gamma)$ does not vanish in the neighborhood of the optimal solution. This vanishing means the pdf degenerates too quickly to one with unit mass, thus freezing the algorithm in a suboptimal solution.

The procedure above was characteristic of a one-dimensional problem. It can be easily extended to multiple dimensions by considering a population of candidate solutions $\bar{X} = [\bar{x}_1, \ldots, \bar{x}_N]$ with $\bar{x}_n = [x_{1,n}, \ldots, x_{M,n}]^T$. The pdf parameter is extended to a row vector $\bar{v} = [v_1, \ldots, v_N]$ which is then used to independently sample the columns of matrix $\bar{X}$, and consequently Eq. (11.9) is calculated along the columns of $\bar{X}$.

Although the CE method was presented as a maximization problem, it is easily adapted to minimization problems by setting $\hat{\gamma} = S_{(\lceil \rho M \rceil)}$ and updating the parameter vector with those samples x_i, where $S(x_i) \leq \hat{\gamma}$.

The difference between discrete and continuous optimization with CE is simply the choice of pdf used to fill the candidate population. The most typical choice of pdf for continuous optimization is the Gaussian (or normal) distribution where v in $f(\cdot, v)$ is represented by the mean μ and variance σ^2 of the distribution. Samples from this distribution can be extracted using MATLAB's `randn` function. Other popular choices include the shifted exponential distribution, double-sided exponential, and beta distribution. Many other continuous distributions are reasonable, although distributions from the *natural exponential family* (NEF) are typically chosen as convergence to a unit mass can be guaranteed and the CE program of Eq. (11.9) can be solved analytically. The update equations that satisfy Eq. (11.5) for continuous optimization are

$$\hat{\mu} = \frac{\sum_{i=1}^{M_s} I_i x_i}{\sum_{i=1}^{M_s} I_i}, \hat{\sigma}^2 = \frac{\sum_{i=1}^{M_s} I_i (x_i - \hat{\mu})^2}{\sum_{i=1}^{M_s} I_i} \qquad (11.11)$$

which are simply just the sample mean and sample variance of those *elite* samples, i where the objective function $S(x_i) \geq \gamma$. The worst performing elite sample is then used as the threshold parameter $\gamma^{(t+1)}$ for Eq. (11.8) in the next iteration. The result presented in Eq. (11.11) is one of the simplest, most intuitive, and versatile CE parameter estimates available. In this case, as $\hat{\gamma}^{(t)}$ becomes close to γ^* the samples in the population will become identical; thus the variance of the sample population will begin to decrease toward zero, resulting in a Gaussian pdf having unit mass about the sample mean of the population. The final location of $\hat{\gamma}^*$ will be represented by this final mean, that is, $\hat{x}^* = \hat{\mu}^*$ as $\hat{\sigma}^2 \to 0$.

The CE method is modified for combinatorial optimization problems by choosing a density function that is binary in nature. The most popular choice is the Bernoulli distribution, **Ber**(p), with success probability p represented by the pdf,

$$f(x;p) = p^x(1-p)^{1-x}, x \in \{0,1\} \qquad (11.12)$$

where $f(x; p)$ equals p when $x = 1$ and $1 - p$ when $x = 0$. Samples from the Bernoulli distribution with Bernoulli success probability vector $\bar{p}$ can be drawn in MATLAB using

```
Ber_p = (rand(Npop,NVar) <= repmat(p,Npop,1))
```

The update equations that satisfy Eq. (11.5) for combinatorial optimization are

$$\hat{p} = \frac{\sum_{i=1}^{M_s} I_{\{S(x_i) \geq \gamma\}} x_i}{\sum_{i=1}^{M_s} I_{\{S(x_i) \geq \gamma\}}} \qquad (11.13)$$

An exact mathematical derivation of the convergence properties of the CE method for continuous optimization is still an open problem; however, from experience the convergence properties of the continuous form CE method appear similar to its combinatorial counterpart, but better behaved. The convergence of the combinatorial form of CE has been studied extensively and a good treatment of the theory begins with a simplified form based on parameter update by the single best performer in the population [15, 31]. More generally, the convergence properties of combinatorial optimization problems based on parameter update by the best $\lceil \rho K \rceil$ performers in the population were derived in [32]. The results presented were specific to problems with unique optimal solutions where the candidate population is evaluated by a deterministic scoring function. The main conclusion is that when using a constant smoothing parameter (as is most commonly implemented), convergence of the CE method to the optimal solution is represented by

1. The sampling distribution converging with probability 1 to a unit mass located at some random candidate $x^{(t)} \in X$.
2. And, the probability of locating the optimal solution being made arbitrarily close to one (at the cost of a slow convergence speed).

This is accomplished by selecting a sufficiently small value of smoothing constant, α. It is suggested that guaranteeing the location of the unique optimal solution with probability 1 can only be achieved by using smoothing coefficients that decrease with increasing time. Examples of such smoothing sequences are

$$a^{(t)} = \frac{1}{Mt}, \frac{1}{(t+1)^\beta}, \frac{1}{(t+1)\log(t+1)^\beta}, \beta > 1 \qquad (11.14)$$

When choosing a smoothing constant, there is a tradeoff between speed of convergence and achieving the optimal solution with high probability. Regardless, the sampling distribution will always converge to a unit mass at some candidate $x^{(t)} \in X$ when using a smoothing constant. Generally, the speed of convergence experienced using constant smoothing parameters is generally faster than decreasing smoothing schemes. Also, for the last two smoothing techniques of Eq. (11.14) location of the optimal solution may be guaranteed with probability one, but convergence to a unit mass with probability one is not. It is not known whether a smoothing technique exists where both the sampling population converges to a unit mass with probability 1 and the optimal solution is located with probability 1. The authors of [32] suggest that from their experience convergence of the sampling population to a unit mass with probability 1 and locating the optimal solution with probability 1 are mutually exclusive events.

Next, we investigate some simple optimization problems to help illustrate the properties and performance of the CE method. First, consider the maximization of the score function $S(x)$ as presented in [19, 15], given by

$$S(x) = e^{-(x-2)^2} + 0.8e^{-(x+2)^2}, \; x \in \mathbb{R}^1 \qquad (11.15)$$

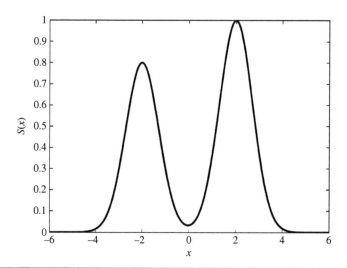

FIGURE 11.6 Plot of the score function S(x) for continuous CE optimization example. The goal is to locate the global maximum of $S(x^*) = 1$ at $x = 2$.

A plot of the score function is shown in Fig. 11.6. Observe from this figure that the function $S(x)$ has two maxima located at $x = -2$ and 2, respectively. The maximum located at $x = -2$ is a local maximum, whereas the maximum located at $x = 2$ is the global maximum whose value is equal to 1. The location of the maximum at $x = 2$ is the goal for this example. The optimization will be performed using the continuous form CE procedure. The initial parameters chosen are $\mu^{(0)} = -6$, $\sigma^{2(0)} = 100$, $\alpha = 0.7$, $\rho = 0.1$, and $K = 100$. The stopping criterion is defined as $\max(\sigma^{2(t)}) < \varepsilon$, where ε is set to eps = 2.2204×10^{-16} corresponding to the floating-point relative accuracy in the MATLAB programming environment. Typically, very little improvement is observed once the variance decreases below 10^{-6}. The value $\rho K = 10$ uses the top 10 percent best performers in the population when updating the distribution parameters. The initial position of the mean is arbitrary given that ρ is small, K is large, and the variance is sufficiently large such that the choice of smoothing coefficient does not freeze the procedure in a suboptimal solution. One could bias early solutions by initializing the mean close to the optimal solution.

The results of this optimization example are presented in Figs. 11.7 through 11.9. Examining Fig. 11.9b (corresponding to the value of x that produced the worst score evaluation at each iteration) reveals that the CE procedure located the global maximum in no worst than 14 iterations. With a population size of 100, the optimal position can be located at worst with 1400 evaluations of the

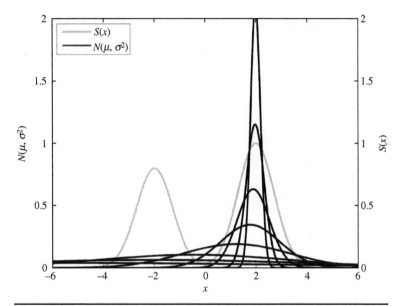

FIGURE 11.7 Progression of Gaussian pdf as the distribution converges to the optimum. As $\hat{\gamma}$ becomes close to γ^*, the variance of the Gaussian pdf decreases and the majority of its probability mass is deposited about the mean, thereby locating the optimal solution, which in this case is $x = 2$.

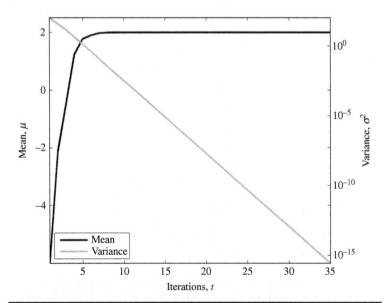

FIGURE 11.8 Progression of the mean and variance of the Gaussian distribution versus the number of iterations. Note that as the variance decreases toward 0, the mean settles to a constant value, $x = 2$.

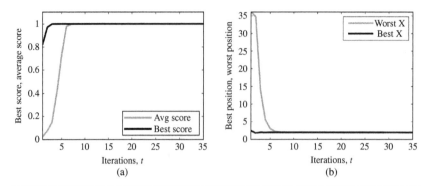

FIGURE 11.9 (a) Progression of the best and average population score versus number of iterations. (b) Progression of the best and worst population position versus number of iterations. As the Gaussian pdf degenerates to a unit mass, the best and average score become equal as all samples in the population are identical.

score function. Of note is the progression of the Gaussian distribution toward a unit mass as seen in Figs. 11.7 and 11.9a. As the variance of the Gaussian distribution begins to decrease toward zero (or as $\hat{\gamma}^{(t)}$ becomes close to γ^*), the values of both the best and average score in the population become identical. Also, this decrease in variance positions most of the probability mass close to the optimal solution $x^* = 2$; thus as the mass nears unity, the final position and mean are themselves equal to each other. Furthermore, the estimate of the optimal score $\hat{\gamma}^{(t)}$ begins to equal the best score in the population and eventually all score values in the population become equal to the best score. This is suggested by the convergence of the best and average scores to the same value. Overall, the smoothing coefficient value of 0.7 was low enough to ensure (on average) that the simulation converged to the optimal solution, but could have been chosen larger in order to improve the convergence speed. A value of 0.9 was about the highest one could choose and still locate the optimal solution. However, accuracy in locating the optimal solution is sacrificed for speed of convergence. One thing of note is the exponential convergence behavior of the alpha smoothing procedure shown in Fig. 11.8. The values of the variance are plotted on a log scale and notice that the variances values are linearly decreasing with increasing t, confirming this assertion.

Next, consider the example first presented in [15] where the elements of an unknown binary sequence are estimated using the combinatorial form of the CE method. Although the elements of the binary sequence are unknown, it is assumed that a deterministic scoring function, $S(\bar{x})$ can be measured in response to user defined inputs, $\bar{x}$. Using information

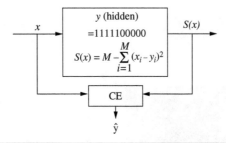

FIGURE 11.10 Blackbox experiment to determine unknown binary sequence.

about these user defined inputs and the corresponding scoring response, CE can be utilized to form an estimate of the unknown sequence $\hat{y}$ as shown in Fig. 11.10.

The unknown binary sequence y is the 10-element sequence

$$\bar{y} = [1\ 1\ 1\ 1\ 1\ 0\ 0\ 0\ 0\ 0] \quad (11.16)$$

The known scoring function is given by

$$S(\bar{x}) = M - \sum_{i=1}^{M}(x_i - y_i)^2 \quad (11.17)$$

where $M = 10$ is the total number of elements in the sequence.

The scoring function attempts to minimize the sum square error between $\bar{x}$ and $\bar{y}$. The result is then subtracted from the total number of elements, M, in order to create a maximization problem. Therefore, the global maximum of $S(\bar{x}^*) = \gamma^*$ is equal to 10 corresponding to all elements of $\bar{x}$ and $\bar{y}$ being equal.

For this example, the Bernoulli probability density function is used with the optimal success probabilities estimated using Eq. (11.12). The initial success probabilities of the CE procedure are set to $p^{(0)} = 0.5$ for each element of $\bar{y}$. An initial value of $p = 0.5$ provides all elements of the initial population with an equally likely chance of being 0 or 1. Values of p less than 0.5 bias solutions toward 0, whereas values of p greater than 0.5 bias solutions toward 1. The population size is set at $K = 100$ with worst elite sample selection parameter ρ set equal to 0.1. The alpha smoothing coefficient was set to 0.9 in this example to show the fast convergence of the CE procedure. The procedure was stopped when all elements of $\hat{p}$ were either 0 or 1 and the average score of the population was equal to γ^*, representing complete degeneration of the pdf to a unit mass. The results of the optimization are shown in Fig. 11.11.

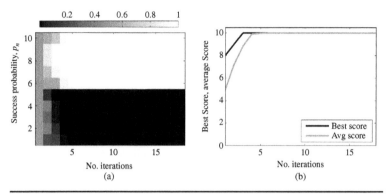

FIGURE 11.11 Results of CE method optimization for the blackbox experiment in Fig. 11.10. (*a*) Progression of Bernoulli success probabilities for each iteration. (*b*) Progression of best and average population scores at each iteration.

11.3 Optimizing Smart Antenna Arrays

Periodic antenna arrays are antenna element configurations with uniform spacing between elements. The periodic array represents the discretization of a continuous line source aperture at periodic locations in space. This discretized array samples the incident wavefront at specific locations producing radiation patterns governed by the Nyquist sampling criterion. The periodicity determines how the sampling is specified and as a result introduces limitations on the practical implementation of periodic arrays. For example, array patterns of periodic arrays with element spacings greater than one-half wavelength sample the wave front at a rate greater than Nyquist and are considered overspecified. This overspecification reduces beam efficiency and beam scanning in the visible pattern region due to the introduction of grating lobes (mainlobes at angular locations other than desired). Additionally, this periodicity further restricts the maximum attainable beam efficiency for a given spacing by bounding the minimum peak sidelobe power and mainlobe beamwidth. Practical implementation of periodic arrays can be difficult due to increased cost and mechanical complexity of arrays because a larger number of elements are required to improve upon these limitations. Introducing *aperiodic* (i.e., nonuniform) element spacings helps relax these limitations and offers greater flexibility. An aperiodic array can be produced from a periodic array of N elements in one of two ways:

1. *Thinning*: Eliminating a subset of active elements in the periodic array
2. *Nonuniform spacing*: Shifting the geometric locations of periodic elements to produce aperiodic spacings

Shifting the geometric locations of the elements implies nonuniform element spacing. The total number of elements in the array is kept the same, but the total aperture length may increase or decrease depending on these nonuniform element spacings. Changing the length of the array will affect the radiation pattern beamwidth and sidelobe power providing more flexibility in overcoming the limitations of periodically spaced arrays. Previously, the synthesis of array geometries has been investigated using genetic algorithms [33, 34], simulated annealing [35, 36], particle swarm optimization [5, 37], and the CE method [20].

Thinning antenna arrays is the strategic elimination of a subset of active elements in the array in order to maintain similar radiation properties as the full array, but using a smaller number of elements in doing so. For applications such as satellite communications, radar, and interferometers for astronomy that require a highly directive array, but one with moderate gain, the active elements of the array can be thinned without significantly affecting the array's radiation properties. In particular, the beamwidth of the array is proportional to the largest dimension of the array aperture. Therefore, for a constant array length, removing elements will proportionally increase the gain of the array while leaving the beamwidth relatively unchanged. The number of elements in the array must be large in order to justify using numerical techniques in designing thinned arrays. The distinction that the "number of elements be large" dictates that the number is large enough such that it would be practically impossible to exhaustively search through all possible combinations and test which combination proved best. The design of aperiodic arrays by element thinning is typically accomplished through either statistics or optimization routines. Bounds on achievable designs by statistical array thinning of large arrays are presented by Lo [38] and Steinberg [39] with the design method by Skolnik [40] one of the more popular statistical techniques. In modern times the genetic algorithm [41, 42], simulated annealing [43], particle swarm optimization [37], ant colony optimization [44], and the CE method [20] have all been used to thin large arrays with great success and can achieve designs not predicted by statistical theory.

11.3.1 Thinning Array Elements

The optimal thinning of array elements for a given objective is achieved by performing array weight control. Consider the uniformly spaced N element linear array of Fig. 11.12.

The array is weighted with a series of coefficients $\bar{w} = [w_1, w_2, \ldots, w_{N/2}] \in [0, 1]$. These weights are binary coefficients applied linearly to the array and represent amplitude excitation to each element of the array. By setting weight w_n of element n to a value of 1, element n is active (i.e., turned 'on'), whereas if element n is weighted with a value

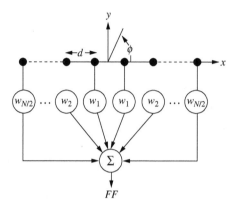

Figure 11.12 x axis, weighted linear antenna array of N uniformly spaced elements.

of 0, element n is inactive (i.e., 'off') and does not contribute to the array radiation pattern. Eliminating elements in this manner introduces an aperiodic spacing between elements. The problem then becomes how to choose a specific combination of 0's and 1's in order to satisfy a given objective.

The example considered here was first presented in [41] for thinning linear arrays with a genetic algorithm to minimize the peak sidelobe level. A more detailed example exists in [4], which we analyze here for comparison of the genetic algorithm and CE method solutions. The discrete nature of the problem requires the combinatorial form of the CE method and the binary form of the genetic algorithm.

Consider the symmetric, uniform linear array in Fig. 11.12, where $N = 52$ elements. Because the array is symmetric, the number of weights for optimization is equal to 26; however, we fix $w_1 = 1$ as low-sidelobe tapers always have a maximum at their center and $w_{N/2} = 1$ to preserve the array length so that the mainlobe beamwidth does not change during the optimization. As a result, the weights for 4 elements in the array are fixed; thus the dimensionality of the optimization problem is reduced to 24. In this case, the total number of possible combinations to exhaustively search is equal to $2^{24} = 16,777,216$.

The goal of the optimization is to thin the total number of active elements while minimizing the peak sidelobe level of the array in Fig. 11.12 whose far-field pattern is given by

$$FF_n(u) = \frac{EP(u)}{FF_{max}} 2 \sum_{n=1}^{N/2} w_n \cos((n-0.5)kdu) \qquad (11.18)$$

where N = number of elements in the array = 52
w_n = amplitude weight of element n, $w_n \in [0, 1]$
d = spacing between elements of original uniform array = 0.5λ
k = wavenumber = $2\pi/\lambda$
$u = \cos\phi$, $0 \le \phi \le 180°$ with 1000 equally spaced sample points between $[0, 1]$
$EP(u)$ = element pattern = 1 for isotropic sources
FF_{max} = peak value of far-field pattern = $2\sum_n w_n$

The score function is defined as the maximum sidelobe power of the far-field magnitude in the sidelobe region $\lambda/Nd \le |u| \le 1$ of the original uniformly spaced array, where λ/Nd is the location of the first null in u-space,

$$\text{Score} = 20\log_{10} \max(|FF_n(u)|), \text{ for } \frac{\lambda}{Nd} \le |u| \le 1 \quad (11.19)$$

All 2^{24} combinations were exhaustively evaluated using MATLAB's combn function from the MATLAB Central File Exchange. Each combination was evaluated using Eq. (11.19) and only one combination corresponded to the global minimum with a sidelobe level of –18.632 dB. This value came from combination 16,776,823 corresponding to the one-sided weight vector $\bar{w}$ given in Table 11.2.

The resulting array has 44 active elements, which is a 15 percent reduction in the total number of elements in the array while producing a pattern with a peak sidelobe level 30 percent smaller than the original 13.2 dB sidelobe level for the full array.

Next, we test the ability of the genetic algorithm and CE method to locate this solution to Eq. (11.19). The parameters of the two algorithms are given in Table 11.3 and were chosen using experience with the algorithms.

Two sample results showing the optimally thinned array and their respective performance are shown in Fig. 11.13 for the genetic algorithm and Fig. 11.14 for the CE method. Figure 11.15 illustrates the distribution of optimized results for 500 independent trials.

Element n	1	2	3	4	5	6	7	8	9	10	11	12	13
w_n	1	1	1	1	1	1	1	1	1	1	1	1	1
Element n	14	15	16	17	18	19	20	21	22	23	24	25	26
w_n	1	1	1	0	0	1	1	1	0	1	1	0	1

TABLE 11.2 Weight Vector w_n of a Thinned Uniform Linear Array to Minimize the Peak Sidelobe Level

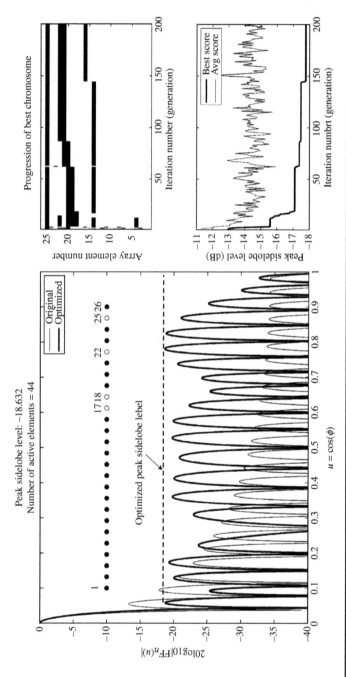

Figure 11.13 Sample result for array thinning with the genetic algorithm. In the result shown, the algorithm located the globally optimum solution.

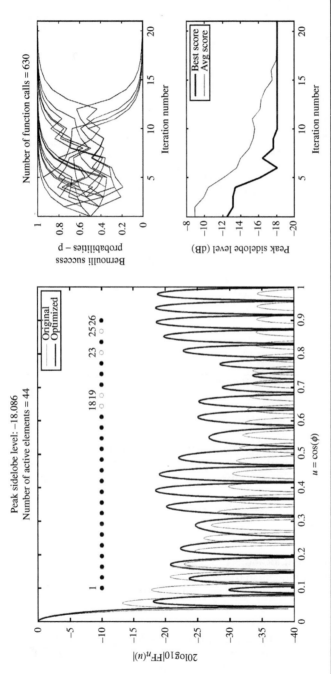

FIGURE 11.14 Example of CE method optimization for thinned linear array. For this result, the CE method converged to a small neighborhood around the global minimum.

340 Chapter Eleven

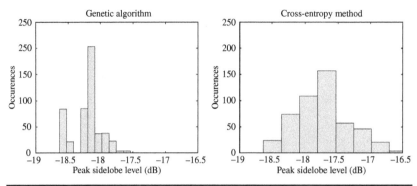

FIGURE 11.15 Distribution of optimized peak sidelobe levels for thinned array example. (Left) Genetic algorithm, (right) CE method.

Genetic Algorithm		Cross-Entropy	
Population Size	20	Population Size	20
Selection	Roulette	Smoothing Parameter α for μ, σ^2	1, 0.7
Crossover	Single-point	Sample Selection Parameter, ρ	0.1
Mutation Rate	0.15	Initial Success Probabilities, $\bar{p}$	0.5
Mating Pool	4		

TABLE 11.3 Algorithm Settings for Thinning Scenario

From this figure, it is clear that for the settings in Table 11.3 that the genetic algorithm settles on the optimal solution more often than the CE method and also has a smaller variance on the final solution.

11.3.2 Optimizing Array Element Positions

In the array thinning process the introduction of aperiodic spacing between elements to achieve a low sidelobe level is quantized by the 2^N possible element location combinations, where N is the total number of antenna elements in the array. By altering the spacing between elements, one can achieve greater control over the shape of the radiation pattern.

The spacing between elements can be defined in terms of the interelement spacings d_n, or the absolute element locations x_n as shown in Fig. 11.16. Setting up the optimization procedure to optimize either interelement spacing or absolute element locations requires some additional constraints to ensure that features such as the mainlobe shape and beamwidth are preserved. Given that the variables are

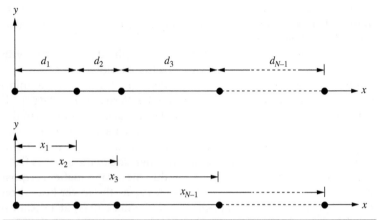

FIGURE 11.16 Nonuniform array spacing definitions based on (top) interelement spacings d_n and (bottom) absolute locations x_n.

bounded on some interval $[l, u]$ optimizing interelement spacing will provide greater control in the optimization procedure, but may require an additional constraint on the total array length in order to preserve the mainlobe shape and beamwidth. Choosing to optimize the absolute element locations provides better control over the total array length, but may cause some element positions to overlap one another within the bounds of the array, length, and as a result the total number of elements in the array may change from one iteration to the next. This could be an added side benefit of thinning elements while optimizing the element spacing.

The array factor for a symmetric array for nonuniform element spacing is given by

$$FF_n(u) = \frac{EP(u)}{FF_{max}} 2 \sum_{n=1}^{N/2} w_n \cos((n-0.5)kd_n u) \tag{11.20}$$

where $d_n = \beta(d/\lambda)$, $0 \le \beta \le 1$ bounds the maximum interelement spacing.

The array factor in terms of absolute element locations is given by

$$FF_n(u) = \frac{EP(u)}{FF_{max}} 2 \sum_{n=1}^{N/2} w_n \cos(kx_n u) \tag{11.21}$$

where $0 \le x_n \le L$ for an array with length L. All remaining terms are as defined in Eq. (11.18).

The score function to minimize the peak sidelobe level with nonuniform element locations is the same as for element thinning given by Eq. (11.19). As seen from Figs. 11.13 and 11.14, this score function leads to sidelobes of relatively constant height. We investigate here the ability of nonuniformly spaced array elements to achieve a similar sidelobe level reduction.

As an example, consider a uniformly weighted linear array with $N = 32$ isotropic antenna elements positioned along the x axis and symmetric about the y axis as shown in Fig. 11.12. The goal is to minimize the peak sidelobe level of the far-field pattern $FF(u)$, $\lambda/Nd \leq u \leq 1$, where $d = 0.5\lambda$ corresponding to the default spacing of a uniform linear array. The absolute element locations x_n as a function of wavelength will be the variables in this optimization. As such, the separation between the first two elements on either side of the y axis is fixed at 0.5λ to ensure that one array pair within the array satisfies Nyquist spatial sampling. This value will also serve as the lower bound on element positions. Element position $\pm N/2$ will be fixed at a distance of $\pm(N-1)/2 \cdot d = 7.75\lambda$ from the origin corresponding to the one-sided array length of a uniform linear array to minimize mainlobe distortion. This value will serve as the upper bound on element positions. Here, the CE method is used for the optimization. The design parameters for the technique are summarized in Table 11.4. Note that the initial values for the mean of the Gaussian distributions for each element location were set to $(n - 0.5) d/\lambda$, for $n = 1, 2, \ldots, N/2$, which initially biases the solution to be about the original position of element n in the uniformly spaced array.

The results for one optimization instance of the CE method are shown in Fig. 11.17. The final optimized element positions are given in Table 11.5 and visualized in Fig. 11.18.

The peak sidelobe level of this array is -17.53 dB, which is about a 4-dB improvement over the uniform linear array peak sidelobe level of -13.2 dB. As stated previously, there is a chance when optimizing the absolute element locations that the optimized positions may overlap one another. This is the case, as is seen in Table 11.5, for elements 7 and 8. What this result actually represents is a single element, but with twice the amplitude. The two elements at this location can

Symbol	Quantity	Value
α	Smoothing parameter	0.7
ρ	Sample selection parameter	0.1
K	Population size	100
x_n	Limits on element position n	$[0, (N-1) d/2]$
$\mu^{(0)}$	Initial mean of element positions	$(n - 0.5) d/\lambda$
$\sigma^{2(0)}$	Initial variance of element positions	100

TABLE 11.4 CE Design Parameters for Symmetric, Linear Array Scenario

Smart Antenna Design

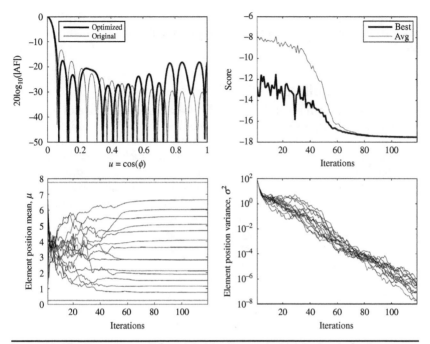

FIGURE 11.17 Example result for minimizing the peak sidelobe level of a linear array using nonuniform element locations and the cross-entropy method.

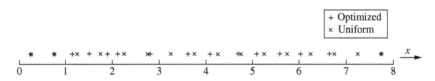

FIGURE 11.18 Plot of the optimized element locations (+) versus the uniform linear array locations (x). From this image one can see the degree of deviation of the optimized array from the uniform case.

x_1	x_2	x_3	x_4	x_5	x_6	x_7	x_8
0.250	0.749	1.148	1.507	1.901	2.117	**2.801**	**2.816**
x_9	x_{10}	x_{11}	x_{12}	x_{13}	x_{14}	x_{15}	x_{16}
3.613	4.088	4.701	5.093	5.590	6.034	6.641	7.750

Shown in **bold** are two elements that are closely spaced.

TABLE 11.5 Optimized Element Locations x_n Normalized by the Wavelength λ for the Nonuniform Peak Sidelobe Level Optimization Performed by the CE Method

be replaced by a weight $w_7 = 2$, and $w_8 = 0$ in Eq. (11.21) with all other weights equal to 1. The resulting array pattern will be nearly identical to that of Fig. 11.17.

11.4 Adaptive Nulling

Adaptive beamforming and nulling of different sources is one of the primary functions of a smart antenna in order to improve the signal-to-interference-noise (SINR) ratio of the array output. This is studied extensively in Chap. 8 for traditional techniques such as least mean squares (LMS), sample matrix inversion (SMI), recursive least squares (RLS), and the conjugate gradient method (CGM). Here, we examine the ability of global optimization algorithms such as the genetic algorithm and CE method to perform the same function.

Generally, the array is blind to the sources and directions of the interference, so the adaptation is performed by minimizing the observed output power of the array in response to an assumed desired signal present in the mainlobe of the pattern and interferers arriving in the sidelobes of the pattern as shown in Fig. 11.19. The goal is to determine the optimum array weights to improve the SINR of the array output by placing nulls in the directions of the interferers.

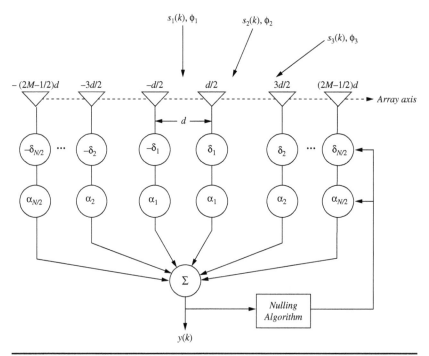

FIGURE 11.19 Uniform linear array for adaptive nulling.

The total array output voltage due to N_{sig} signals arriving from angles ϕ_i with voltage $s_i(k)$ at time sample k is given by

$$AF = \frac{1}{\sum_m w_m} \sum_{i=1}^{N_{sig}} s_i(k) EP(\phi_i)$$

$$\times \left[w_M^* e^{-j\frac{(2M-1)}{2}kdu_i} + \ldots + w_1^* e^{-j\frac{1}{2}kdu_i} + w_1 e^{j\frac{1}{2}kdu_i} + \ldots + w_M e^{j\frac{(2M-1)}{2}kdu_i} \right]$$

(11.22)

Here, the array has an even number of elements such that $2M = N =$ total number of array elements, which are symmetric about the center of the array. The weights $w_m = a_m e^{j\delta_m}$ are also symmetric about the array center, except that the phase value δ_m of the weights w_m^* on the left-hand side have been negated in order to produce nulls in the pattern [45]. Furthermore, the phase value is constrained to be less than $\pi/8$ radians to limit the distortion of the mainlobe in the pattern. Likewise, a low sidelobe amplitude taper computed from a Chebyshev or Taylor window function can be applied.

The total output power of the array is then calculated as

$$\text{Total output power} = 20\log_{10}|AF| \quad (11.23)$$

Consider the example in Table 11.6 from [46] for comparison of the performance between the genetic algorithm and the CE method. Three signals corresponding to a desired user at u-space location $u = 0$ and two 30-dB interferers located at $u = 0.62, 0.72$ are incident upon a symmetric, 40-element uniform linear array shown in Fig. 11.19. A 30-dB Chebyshev amplitude weights α_n are added to the array to introduce a low sidelobe taper.

The settings of each algorithm are summarized in Table 11.7. These parameters we chosen to balance convergence speed and solution accuracy.

Number of sources	1
Number of interferers	2
Source power	0 dB
Interferer power	30 dB
Source DOA	$u = 0$
Interferer DOA	$u = 0.62, 0.72$
Number of elements	40
Element spacing	0.5λ
Amplitude weights	Chebyshev, 30 dB

TABLE 11.6 Parameters for Adaptive Nulling Scenario

Genetic Algorithm		Cross-Entropy	
Population Size	20	Population Size	100
Selection	Roulette	Smoothing Parameter, μ, σ^2	0.7
Crossover	Single-point	Sample Selection Parameter, ρ	0.1
Mutation Rate	0.15		
Mating Pool	4		

TABLE 11.7 Algorithm Settings for Adaptive Nulling Scenario

Figure 11.20 contains an example result showing the adapted array pattern with two nulls in the direction of the interferers while preserving the mainlobe direction toward the desired user. The mainlobe has become slightly distorted in terms of beamwidth, attenuation, and point direction, but the total degradation in output power for the desired signal is very close to 0. Also, we see that the array pattern is no longer symmetric. In order to add nulls in the directions of the interferer while minimizing mainlobe distortion, the sidelobe levels are free to increase and null locations are allowed to move.

Figure 11.21 illustrates an example of the convergence of the genetic algorithm and CE method for the phase-only adaptive nulling scenario. We see that both algorithms are capable of converging to a close neighborhood of the global optimum solution of 0 dB corresponding to the power of the desired user. Overall, we see that the CE method was able to locate a score 20 dB lower than the genetic algorithm within the same number of iterations, but at the cost of an increase in the number of score function evaluations due to the increased population size in the CE method. As seen in Fig. 11.22, this reduced cost is due to a deeper null depth in the directions of each interferer. We see that in the case of the genetic algorithm, the final

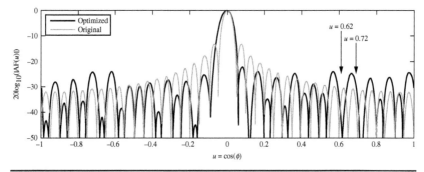

FIGURE 11.20 Example phase-only adaptive nulling result for the problem in Table 11.8. Two deep nulls are produced in the direction of the interferers ($u = 0.62$, 0.72) while minimizing the distortion of the mainlobe in the direction of the desired user ($u = 0$).

TABLE 11.8 Phase Weights for Adaptive Nulling Example in Fig. 11.20

Element n	21	22	23	24	25	26	27	28	29	30
δ_n (deg)	18.2	6.3	5.8	11.7	9.2	3.8	12.9	11.6	8.7	4.8
Element n	31	32	33	34	35	36	37	38	39	40
δ_n (deg)	9.5	10.3	12.5	8.1	16.5	7.1	13.5	15.4	12.4	7.0

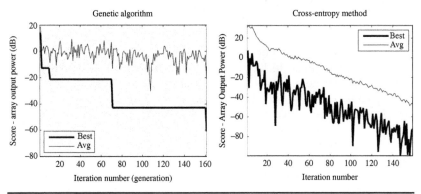

FIGURE 11.21 Example of convergence for genetic algorithm and cross-entropy method for phase-only adaptive nulling scenario.

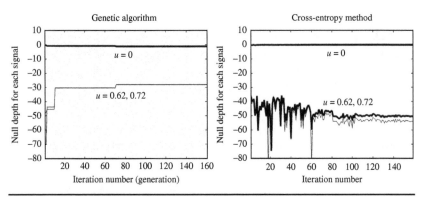

FIGURE 11.22 Progression of null depth for each signal for single instance of phase-only adaptive nulling scenario.

null depths are essentially equal, which one would expect as the power of the two interferers is equal. The CE method had smaller mainlobe distortion in the direction of the desired user compared to the genetic algorithm.

Of note is the difference in the convergence of the two algorithms. The exponential convergence of the CE method is apparent in Fig. 11.21 and we see that the population average follows the slope of the best performer in the population. The optimization was terminated prior to the best and average population scores converging together like in Fig. 11.14 as the maximum variance of the phase weights had decreased below a value of 10^{-6}. For the genetic algorithm, we see the expected trend where the best chromosome in each generation converges toward the global minimum, while the population average covers a wide range of potential solutions.

11.5 NEC in Smart Antenna Design

Numerical Electromagnetics Code (NEC) is a computer program that computes the electromagnetic response and interaction of antennas and other structures composed of metallic wires using the MoM computational electromagnetic technique. It was originally developed by Lawrence Livermore National Laboratory and the University of California back in 1981 [47]. The current version of the code is called *NEC4* and its use is licensed through Lawrence Livermore and is under U.S. export control. The latest version available in the public domain without requiring a license is *NEC2*. The use of NEC2 in the public domain is extensive, its availability is widespread, with many independent and complementary compilations for all operating systems as well as commercial and free-ware GUI-based wrapper programs around the NEC2 engine. Here, we use MATLAB as the wrapper for combining NEC2 with the optimization techniques in Sec. 11.2 to evaluate realistic smart antennas design.

11.5.1 NEC2 Resources

The main resource for NEC2 containing useful information on the history, background descriptions of the numerical methods applied in NEC2, and user guides on program operation is

```
http://www.nec2.org/
```

Once there, the user can find links to other useful sources of NEC2 information, including a link to the "Unofficial NEC Archives" available at

```
http://nec-archives.pa3kj.com
```

wherein one will find additional links to several versions of NEC2 code compiled for different operating systems and programming languages. Here, we use the Windows executable C++ version of NEC2 called *nec2++* available at

```
http://www.pa3kj.com/PA3NEC_Archive/nec2++.exe
```

Another useful piece of software is called *4nec2*, which is a graphical tool for creating, analyzing, and viewing geometries and data products from NEC2/NEC4 files. A screenshot of the 4nec2 environment is shown in Fig. 11.23. 4nec2 is available for download at

```
http://www.qsl.net/4nec2/
```

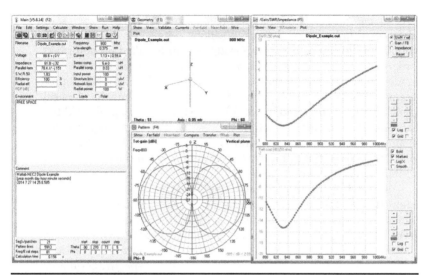

FIGURE 11.23 4nec2 software graphical user interface and postprocessing products. Clockwise from left: Main 4nec2 window, 3D geometry display, SWR/return loss graph, 2D polar radiation pattern.

In this section, we focus on the integration of MATLAB with the core NEC2 engine; however, 4nec2 gives the user the ability to create and verify the .nec input files before and after the design process. This can be useful for troubleshooting MATLAB scripts used to create the input files prior to optimization.

11.5.2 Setting Up the NEC2 Simulation

The general structure of the NEC2 input file is based on the old punched card format used by early computers. Figure 11.24 includes an example of a punched card along with a card reader machine. The punched card is a write-once medium that encoded data onto a single

 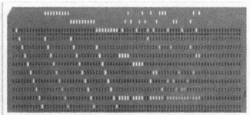

FIGURE 11.24 (Left) Punched card reader[1]. (Right) 80-column punched card.

---

[1]Image courtesy of Richard Smith, https://www.flickr.com/photos/smith/5786129343, licensed under Creative Commons 2.0.

card by *punching* holes along lines with 80 columns. The presence or absence of holes in predefined positions represented data and/or commands to the card reader. Groups of cards called *decks* formed programs and collections of data, which were then fed into the card reader to compute the result of the deck program. Early NEC2 programs were written using punched cards; and over time the physical punch card was replaced by an ASCII text file, which preserved the same convention and format for expressing data and programs. An example ASCII text file program is shown in Fig. 11.25. The input ASCII file is indicated by the *.nec* extension and the output file extension containing the solution is *.out*.

Within the input file, each line represents a separate card identified by the first two columns. For example, in the program of Fig. 11.25 there are comment cards (CM, CE), geometry cards (GW, GE), frequency cards (FR), excitation cards (EX), radiation pattern cards (RP), and program execution control cards (XQ, EN). Many other cards are available to generate specific geometry structures (helix, patches, mesh surface) and to include effects of ground planes, loading, networks, and transmission lines, etc. The order in which the cards are specified is important to how the NEC2 program is executed. A typical program has the following order:

1. Specify comments with CM card and indicate termination of comment section with CE card
2. Specify geometry [i.e., wires (GW card)] and indicate termination of geometry with GE card. GE = 1 indicates a ground plane is present.
3. Specify program control cards such as frequency (FR card), excitation type (EX card), radiation pattern (RP card), and (if any) a ground plane (GN card).
4. Execute the program with XQ card.
5. Indicate completion of program with EN card.

```
CM Matlab NEC2 Dipole Example
CM [year month day hour minute seconds]
CM 2014 7 27 14 25 8.585
CE
GW 1 21 0 0 -0.083278 0 0 0.083278 0.001
GE 0
FR 0 81 0 0 800 2.5
EX 0 1 11 0 1 0
RP 0 73 1 1001 -90 0 5 5
XQ
EN
```

Figure 11.25 Modern ASCII text file for defining the NEC2 program. Shown is a program that computes the radiation pattern for a dipole antenna.

Smart Antenna Design

A detailed description of available cards, their attributes, and their definitions is available in part III of the nec2.org user's guide:

http://www.nec2.org/part_3/toc.html

For each card, there are 80 columns consistent with the traditional punched card format. Several columns are grouped together to form a specific value/entry into the NEC2 program. The entry format for a particular card must be adhered to closely for accurate reading by the "card reader" (i.e., NEC2 executable). As an example, the frequency card has the following input specification (shown in Fig. 11.26).

The first two columns are always reserved for the card type, which for this example is "FR." Columns 3 to 5 are reserved for an integer value (I) defining the type of frequency stepping to be used where a value of 0 indicates *linear stepping* between adjacent frequencies like MATLAB's linspace function and a value of 1 indicates multiplicative stepping. Next, columns 6 to 10 contain an integer value for the number of frequency steps. If the field is blank, a value of one is assumed. Columns 21 to 30 specify a floating point value for the starting frequency in megahertz and columns 31 to 40 specify a floating point value for the frequency stepping increment also in megahertz. For example, in the dipole antenna program of Fig. 11.25, line 7 corresponds to the FR card with 81 linear frequency steps starting at 800 MHz with 2.5-MHz steps. Next, we discuss some often used cards such as the GW card to create wires, the EX card to create excitation sources, and the RP card to compute the radiation pattern.

The format of the GW card for specifying wires is shown in Fig. 11.27. The wire is defined by its two Cartesian end points (XW1, YW1, ZW1) and (XW2, YW2, ZW2), the wire radius, and a unique numeric label called a *tag* assigned to each wire in the program. Each wire is further divided into small *segments* for input to the MOM computational electromagnetic technique used by NEC2. The segment length is typically much smaller than a wavelength (e.g., $\leq \lambda/12$) and the wire radius is much smaller than the segment length (e.g., $\leq \lambda/100$) in order to

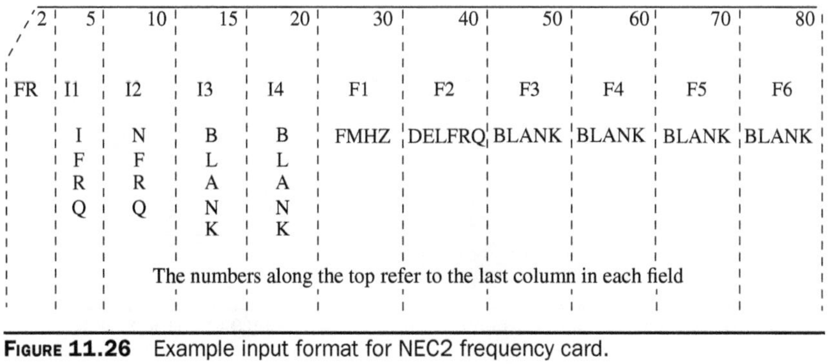

2	5	10	15	20	30	40	50	60	70	80
FR	I1	I2	I3	I4	F1	F2	F3	F4	F5	F6
	INFRQ	NFRQ	BLANK	BLANK	FMHZ	DELFRQ	BLANK	BLANK	BLANK	BLANK

The numbers along the top refer to the last column in each field

FIGURE 11.26 Example input format for NEC2 frequency card.

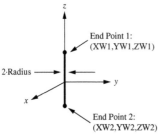

GW - Wire Specification Card		
Columns	Parameter	Description
1–2	GW	Card Label
3–5	I1 – ITG	Wire Tag Number
6–10	I2 – NS	Number of Segments for Wire
11–20	F1 – XW1	X-coordinate of Wire endpoint 1 in meters
21–30	F2 – YW1	Y-coordinate of Wire endpoint 1 in meters
31–40	F3 – ZW1	Z-coordinate of Wire endpoint 1 in meters
41–50	F4 – XW2	X-coordinate of Wire endpoint 2 in meters
51–60	F5 – YW2	Y-coordinate of Wire endpoint 2 in meters
61–70	F6 – ZW2	Z-coordinate of Wire endpoint 2 in meters
71–80	F7 – RAD	Wire radius in meters

FIGURE 11.27 NEC2 wire specification card (GW) and wire parameter definitions.

satisfy the standard thin wire approximation. NEC2 also supports an extended thin-wire kernel approximation (EK card) for thicker wire radii. For the example .nec input file in Fig. 11.25, there is a single GW card specifying a wire in the z-direction of length 0.1665 meter with tag 1, 21 segments, and a wire radius of 1 mm.

Figure 11.28 depicts the inputs for the excitation card (EX). There are five distinct excitation sources available, with variable input to the EX card for each: (1) voltage source (applied E-field source), (2) incident plane-wave, linear polarization, (3) incident plane-wave, right-hand elliptic polarization, (4) incident plane-wave, left-hand elliptic polarization, (5) elementary current source, and (6) voltage source (current-slope-discontinuity). The applied E-field voltage source is most commonly used and is further defined in Fig. 11.28. The voltage source resides in the middle of a wire segment. In the case of a dipole antenna, the user would define an odd number of segments such that the voltage source will be as close to the center of the dipole wire as possible. For a monopole antenna the user would locate the source on the segment closest to the ground plane. Multiple excitation cards can exist within a single program to specify multiple sources of varying types. For example, an antenna array of N-dipoles

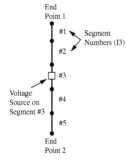

EX - Excitation Card			
Columns	Parameter	Description	Typical Value
1–2	EX	Card Label	EX
3–5	I1–Type	Type of excitation to use	0: Voltage source (Applied E–field source)
6–10	I2–TagNr	Value dependent on type in I1	IF I1 = 0: Wire tag number of source
11–15	I3–SegNr	Value dependent on type in I1	IF I1 = 0: Wire segment number of source for wire tag number
16–20	I4	Value dependent on type in I1	IF I1 = 0: '00' Column 19 = 0 (no action) Column 20 = 0 (no action)
21–30	F1–Vreal	Real part of the voltage in volts	= 1 V
31–40	F2–Vimag	Imaginary part of the voltage in volts	= 0

FIGURE 11.28 NEC2 excitation card (EX) and example voltage source implementation on dipole.

Smart Antenna Design   353

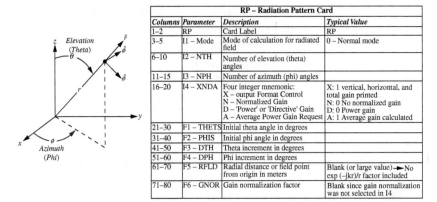

RP – Radiation Pattern Card			
Columns	Parameter	Description	Typical Value
1–2	RP	Card Label	RP
3–5	I1 – Mode	Mode of calculation for radiated field	0 – Normal mode
6–10	I2 – NTH	Number of elevation (theta) angles	
11–15	I3 – NPH	Number of azimuth (phi) angles	
16–20	I4 – XNDA	Four integer mnemonic: X – output Format Control N – Normalized Gain D – 'Power' or 'Directive' Gain A – Average Power Gain Request	X: 1 vertical, horizontal, and total gain printed N: 0 No normalized gain D: 0 Power gain A: 1 Average gain calculated
21–30	F1 – THETS	Initial theta angle in degrees	
31–40	F2 – PHIS	Initial phi angle in degrees	
41–50	F3 – DTH	Theta increment in degrees	
51–60	F4 – DPH	Phi increment in degrees	
61–70	F5 – RFLD	Radial distance or field point from origin in meters	Blank (or large value)→No exp (–jkr)/r factor included
71–80	F6 – GNOR	Gain normalization factor	Blank since gain normalization was not selected in I4

FIGURE 11.29 NEC2 radiation pattern card (RP) and spherical coordinate system definition.

would contain N EX cards to specify the complex-valued voltage source applied to each dipole.

Figure 11.29 defines the inputs to the RP card to evaluate the far-field radiation pattern in spherical coordinates along with typical values. The user specifies the total number of angles for the elevation and azimuth planes along with the respective starting values, and increments between angles. Columns 16 to 20 define a four-integer mnemonic value (XNDA), where each letter represents a mnemonic for the corresponding digit in I4. X controls the format printed in the output file, which can be in terms of the major/minor axis field components and total gain, or the vertical/horizontal components and total gain. N controls normalization of the gain data printed to file. Typically, realized gain is of importance, so no normalization is applied. D controls whether power or directive gain is computed and printed to the output file. A controls the computation of average gain over angle. The user can also specify the radial distance for calculation of the pattern. Leaving this value blank will omit the factor $e^{-jkr/r}$ from the radiated electric field calculated. Multiple RP cards can be specified within a given program. If the program contains a ground plane, no field points should be specified for values of θ greater than 90°.

11.5.3 Integrating NEC2 with MATLAB

The integration of NEC2 with MATLAB is accomplished using file input/output (I/O) and the MATLAB *dos* command for executing the NEC2 program. Here, we use the C++ version of NEC2 called `nec2++.exe`, which is executed on a Windows operating system using the typical command window (cmd.exe) and the following command line:

```
C:\>nec2++ -iInput_Filename.nec -oOutput_Filename.out
```

where the –i switch indicates the input .nec file to nec2++.exe and –o the corresponding output .out file. In MATLAB, this can be executed with either the *dos* or ! commands as

```
NEC_filename = 'Dipole_Example.nec';
Out_filename = 'Dipole_Example.out';

dos(['nec2++ -i',NEC_filename,' -o',Out_filename]);
```

or

```
!nec2++ -iDipole_Example.nec -oDipole_Example.out
```

Using the dos command allows the user to define the input and output filenames as variables.

The flow of executing a NEC2 program using nec2++ is accomplished by first using MATLAB to write the .nec input file containing the desired wire geometry, simulation frequency, excitation, radiation pattern requests, etc. Then the nec2++.exe command is executed from within MATLAB using the *dos* operating system command, which creates the .out output file. The data in the output file is then read into MATLAB where postprocessing of the results and visualization is performed.

11.5.4 Example: Simple Half-Wavelength Dipole Antenna

Consider the example NEC2 program `Dipole_Example.nec` shown in Fig. 11.25 for a half-wavelength dipole antenna. Here, the desired center frequency f_o is 900 MHz. This frequency corresponds to a wavelength $\lambda_o = \frac{299,792,458}{900,000,000} = 0.3331$ m and theoretical length $\lambda_o/2 = 0.1665$ m. The dipole wire is set up with the GW card such that the wire is z-directed and symmetric about the x-y plane as shown in Fig. 11.30. The wire radius is equal to 1 mm with 21 segments of

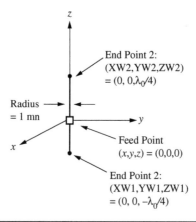

Figure 11.30 NEC2 geometry for simple half-wavelength dipole antenna example.

length 8 mm each, which satisfies the rule of thumb for the thin wire approximation. A 1-V voltage source excitation is added at segment 11 corresponding to the wire center and the coordinate system origin. A frequency sweep is performed from 800 to 1000 MHz in 2.5-MHz steps corresponding to 81 sample points. The radiation pattern is also computed in the x-z plane (i.e., $\phi = 0°$) with 73 points over 360° corresponding to a 5° spacing in elevation.

The results from the output file Dipole_Example.out are shown in Fig. 11.31. The first plot is the return loss (S_{11}) relative to the theoretical half-wavelength dipole antenna impedance $Z_o = 73.2\ \Omega$ computed as

$$S_{11}(\text{dB}) = 20\log_{10}\left(\left|\frac{Z_{ant} - Z_o}{Z_{ant} + Z_o}\right|\right) \qquad (11.24)$$

where $Z_{ant} = R_{ant} + jX_{ant}$ is the antenna impedance at a given frequency with resistive component R_{ant} and reactive component X_{ant} as shown in the upper right subplot of Fig. 11.31.

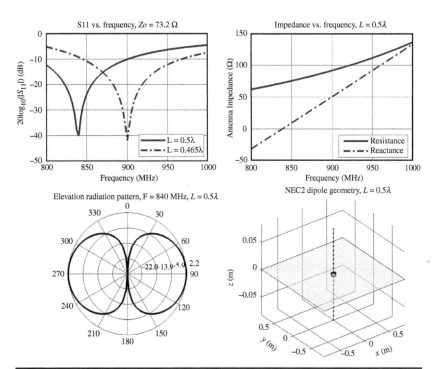

FIGURE 11.31 Results of example NEC2 dipole antenna program for dipole length $L = 0.5\lambda$ at 900 MHz. Note that in the top left subplot a dipole length $L = 0.465\lambda$ results in the minimum return loss (S_{11}) at the desired frequency.

Theoretically, the return loss should have a value of −inf dB at the resonant frequency of the dipole, which here was designed to be 900 MHz; however, we see that the theoretical dipole length of 0.5λ results in a minimum S_{11} value around 840 MHz. There, the impedance value is $72.3 + j1.1\ \Omega$, which is close to the theoretical value of $73.2\ \Omega$. This deviation from the theoretical frequency is attributed to the fact that the wire radius is not infinitesimally small. One would see that in the limit as the wire radius tends to zero, the S_{11} minimum will move closer to the theoretical resonant frequency. For a wire radius of 1 mm, the optimal length is close to 0.465λ or 0.155 m. The return loss versus frequency for the optimized length is shown in the top left subplot of Fig. 11.31.

The bottom left graphic is an elevation cut of the antenna power gain expressed in dBi for 840 MHz. We see that the peak value of the pattern is 2.2 dBi, which is close to the theoretical value of 2.15 dBi.

11.5.5 Monopole Array Example

Next, we consider an example of an 8-element uniform linear array of quarter-wavelength monopole antennas as shown in Fig. 11.32. Following the previous example, the optimal length of a monopole at the desired frequency of 900 MHz when the wire radius is 1 mm is equal to 0.0786 m, which is slightly less than the theoretical value of 0.0832 m. The spacing between elements is set to a half-wavelength or 0.1665 m to satisfy traditional Nyquist spatial sampling. Each element is excited by a voltage source of 1V on the first segment closest to an infinite perfectly conducting ground plane.

The corresponding NEC2 program for this setup is given in Fig. 11.33. There are eight separate GW cards for creating the monopole antennas as well as separate EX cards for each with 1V applied to each element. The radiation pattern at 900 MHz is computed in the x-y plane (elevation $\theta = 90°$) with an azimuth angle spacing of 1°.

The results of the NEC2 program are given in Fig. 11.34. The left-hand subplot illustrates a polar plot of the resulting array power gain

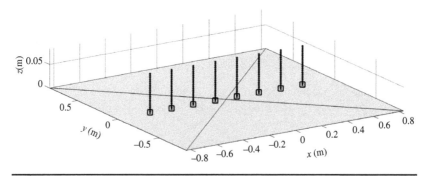

Figure 11.32 NEC geometry of 8-element $\lambda/4$ monopole array example.

Smart Antenna Design 357

```
CM Matlab NEC2 Monopole Array Example
CM [year month day hour minute seconds]
CM 2014 8 6 16 49 46.56
CE
GW 1 21 -0.58333 0.00000 0.00000 -0.58333 0.00000 0.083278 0.001
GW 2 21 -0.41667 0.00000 0.00000 -0.41667 0.00000 0.083278 0.001
GW 3 21 -0.25000 0.00000 0.00000 -0.25000 0.00000 0.083278 0.001
GW 4 21 -0.08333 0.00000 0.00000 -0.08333 0.00000 0.083278 0.001
GW 5 21  0.08333 0.00000 0.00000  0.08333 0.00000 0.083278 0.001
GW 6 21  0.25000 0.00000 0.00000  0.25000 0.00000 0.083278 0.001
GW 7 21  0.41667 0.00000 0.00000  0.41667 0.00000 0.083278 0.001
GW 8 21  0.58333 0.00000 0.00000  0.58333 0.00000 0.083278 0.001
GE 1
FR 0 1 0 0 900 1
EX 0 1 1 0 1 0
EX 0 2 1 0 1 0
EX 0 3 1 0 1 0
EX 0 4 1 0 1 0
EX 0 5 1 0 1 0
EX 0 6 1 0 1 0
EX 0 7 1 0 1 0
EX 0 8 1 0 1 0
GN 1
RP 0 1 360 1001 90 0 5 1
XQ
EN
```

FIGURE 11.33 NEC2 program for 8-element uniform monopole array. This program analyzes the radiation pattern in the azimuth plane (x-y) for $\lambda/4$ monopoles resonant at 900 MHz.

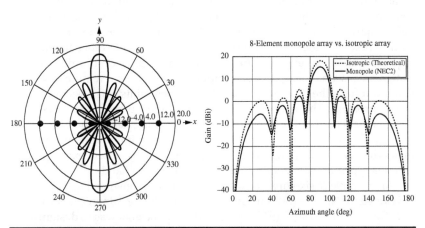

FIGURE 11.34 Array radiation pattern for 8-element $\lambda/4$ monopole array example. (Left) Polar plot of the NEC2-generated monopole array pattern showing the mainlobe broadside to the x axis. (Right) Comparison of NEC2 monopole array and ideal array of isotropic antennas.

in the azimuth (*x-y*) plane computed by NEC. The black circles are intended to show the vertical orientation of the monopoles coming out of the page (+z) as in Fig. 11.32, so the peak of the array pattern is broadside to the array as would be expected. The right-hand subplot is a comparison of the NEC2-generated monopole array pattern versus the ideal array of isotropic antennas. The mainlobe peak of the isotropic array is $20\log_{10}(8) = 18.06$ dB as opposed to the monopole array, which is equal to 15.34 dB. The peak sidelobe level for the monopole array pattern is 12.8 dB down from the peak, which is equal to the isotropic array; however, the rolloff of the sidelobe taper toward end fire is greater for the monopole array. The null locations between the two arrays are similar, except that the null depth for the monopole array is less due to the mutual coupling between the elements. This example illustrates the importance of including the element pattern and mutual coupling in the antenna array design process.

11.6 Evolutionary Antenna Design

An excellent example of a marriage between evolutionary optimization algorithms and computational electromagnetics code is the *crooked wire* antenna developed by Linden in [1]. The antenna design presented in that work was the first of its kind and inspired the field of evolutionary antenna design. Using evolutionary algorithms to design antennas was a departure from the traditional antenna design techniques that required exquisite knowledge of electromagnetic physics. These traditional designs were typically decoupled from computational electromagnetic techniques, which were used only to confirm or tweak the design in a man-in-the-loop iterative fashion before fabrication and final testing/validation of performance inside an anechoic chamber. An evolutionary antenna, on the other hand, requires little or no expertise in electromagnetic physics in order to design an antenna for a particular application nor does it require any real input from the designer other than the constraints and objectives of the problem. The outcome of the optimization process produces designs that could not be achieved with preexisting design methods. As a result, evolutionary antennas have become an important design methodology for creating unique antenna designs when existing antenna types are not adequate for the application at hand.

An example of six-wire crooked wire antenna is shown in Fig. 11.35 developed for NASA's (National Aeronautics and Space Administration) 2006 Space Technology 5 (ST5) spacecraft [48]. The goal was to design an antenna to fit onto miniaturized satellites called *microsats* and provide low-voltage standing wave ratio (VSWR) as well as a high peak gain and associated pattern smoothness.

This antenna is similar to the original seven-wire crooked wire antennas designed in Linden's dissertation [1], where a genetic algorithm

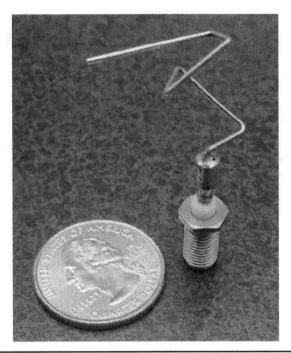

Figure 11.35 Example of six-wire crooked wire antenna for the 2006 NASA ST5 spacecraft.[2]

was used to optimize the geometry of a wire antenna to fit within a cubic volume of 0.5λ on each side at a design frequency of 1600 MHz as shown in Fig. 11.36.

The cubic volume constrains the search space for the end point coordinates (x_n, y_n, z_n) for wire n; thus there are 21 total variables to optimize. The starting point of the first wire is connected to an infinite PEC ground plane much like an ideal monopole and fed with a voltage source on the first segment.

The goal is to produce a constant gain hemispherical pattern over all angles 10° above the ground plane, which is expressed mathematically as

$$\text{Score} = \sum_{\text{over all }\theta,\phi} (\text{Gain}(\theta,\phi) - \text{Avg. Gain})^2 \qquad (11.25)$$

which is the sum-squared error between the average gain over all angles in the range $\theta = [0\ 80°]$, $\phi = [0\ 360°]$. As the values $\text{Gain}(\theta,\phi)$

---

[2] "St 5-xband-antenna." Licensed under Public domain via Wikimedia Commons; http://commons.wikimedia.org/wiki/File:St_5-xband-antenna.jpg#mediaviewer/File:St_5-xband-antenna.jpg.

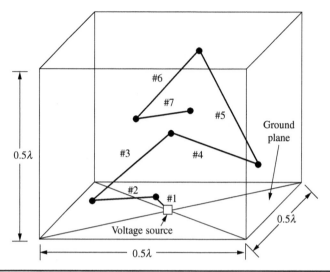

Figure 11.36 Evolutionary algorithm search space for seven-wire crooked wire antenna.

become closer to the average gain over all angles, the Score function tends toward 0.

The NEC program is set up as shown in Fig. 11.37. In the geometry, there are seven total GW cards. The starting point of wire 1 is connected to the coordinate system origin $(x, y, z) = (0, 0, 0)$. The end

```
CM Matlab GA Antenna Example
CM [year month day hour minute seconds]
CM 2014 8 6 9 28 22.776
CE
GW 1 11 0.0000 0.0000 0.0000 x1_end y1_end z1_end 0.0010
GW 2 11 x1_end y1_end z1_end x2_end y2_end z2_end 0.0010
GW 3 11 x2_end y2_end z2_end x3_end y3_end z3_end 0.0010
GW 4 11 x3_end y3_end z3_end x4_end y4_end z4_end 0.0010
GW 5 11 x4_end y4_end z4_end x5_end y5_end z5_end 0.0010
GW 6 11 x5_end y5_end z5_end x6_end y6_end z6_end 0.0010
GW 7 11 x6_end y6_end z6_end x7_end y7_end z7_end 0.0010
GE 1
FR 0 1 0 0 1600 1
EX 0 1 1 00 1 0
GN 1
RP 0 33 36 1001 -80 0 5 5
XQ
EN
```

Figure 11.37 Evolutionary antenna example NEC program template.

point of each wire connects to the starting point for the next wire. There are 11 segments per wire with a wire radius of 1 mm. Here, the GE card is set to 1 to inform the program that a ground plane is present. For the FR card, only 1 frequency value is analyzed at the center frequency of 1600 MHz. Next, an infinite perfectly conducting ground plane is added using the GN card and a value of 1. In the RP card, the angle ranges in elevation and azimuth are evaluated in 5° steps above the horizon as values below the ground plane will not be computed.

The CE method was used to perform the optimization with a population size of 100, smoothing coefficient $\alpha = 0.7$, and sample selection parameter $\rho = 0.1$. The results of the optimization are shown in Figs. 11.38 through 11.41. Figure 11.38 illustrates the final optimized wire antenna design as a function of the wavelength. The coordinates for the optimal antenna as a function of the wavelength λ are given in Table 11.9. This is one instance of the optimization procedure, and as shown in [1], there may exist multiple solutions to this problem.

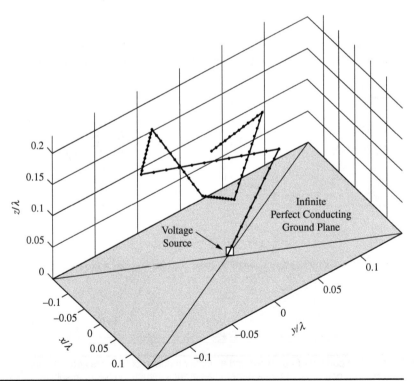

FIGURE 11.38 Optimized evolutionary "crooked wire antenna" for constant hemispherical gain.

362 Chapter Eleven

Element n	1	2	3	4	5	6	7
x	0.0101	−0.1275	−0.1355	0.0731	−0.0005	0.0368	−0.0192
y	0.0576	−0.0539	−0.0384	−0.0560	0.0080	0.0277	−0.0117
z	0.1339	0.1072	0.1637	0.1712	0.0843	0.2261	0.1659

TABLE 11.9 Optimal Crooked Wire Antenna Coordinates for Wire End Points as a Function of the Wavelength λ

Figure 11.39 is a plot of the radiation pattern for the optimized antenna in the x-z plane as well as an additional radiation pattern for a wire antenna drawn randomly for comparison. The optimization is observed to have been successful as the uniformity of the pattern is apparent. The random wire antenna geometry produces a pattern with a sharp null near the z axis as well as asymmetric lobing on each side of the x axis. For the optimized pattern, the peak gain is 2.9 dBi, which is great for most applications.

Figure 11.40 illustrates the progression of the score and position coordinates over the optimization. Only 80 iterations were necessary, equal to 8000 score function evaluations, in order to achieve a score depth of −21.6 at 1600 MHz.

Figure 11.41 depicts the effective bandwidth of the optimized antenna. The score function was evaluated over the frequency range

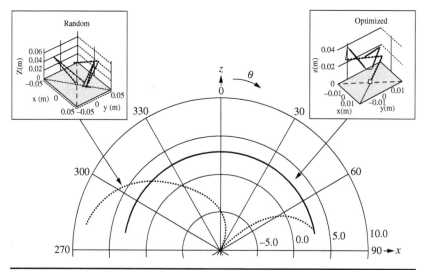

FIGURE 11.39 Comparison of optimized and a single random crooked wire geometry for f_o = 1600 MHz. The optimized antenna pattern has great symmetry in the hemisphere above the ground plane, whereas the random antenna geometry has asymmetry and a null near zero elevation.

Smart Antenna Design

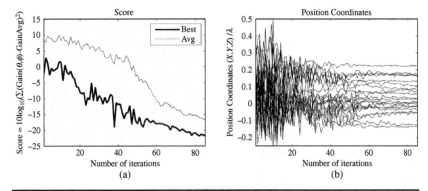

FIGURE 11.40 (Left) Progression of best and average score for crooked wire antenna. (Right) Progression of individual coordinates for each wire.

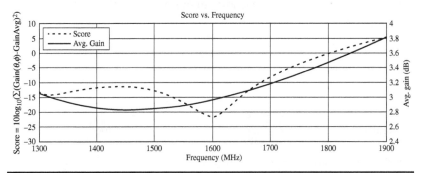

FIGURE 11.41 Score and average gain versus frequency. This reveals the effective bandwidth of the current point design at 1600 MHz.

1300 to 1900 MHz in 10-MHz steps and the average gain was saved for each frequency. The overall best pattern uniformity clearly occurs at the design frequency of 1600 MHz; however, the score function remains well below 1 dB over the range of 1300 to 1800 MHz, which is a very wide range for practical applications with very low ripple.

11.7 Current and Future Trends

11.7.1 Reconfigurable Antennas and Arrays

Traditionally, antennas and arrays were statically designed and mass-produced without much thought given to the performance in deployed environments. However, the surrounding environment has a significant impact on the performance of the antenna. Reconfigurable antennas have the characteristic whereby they alter their physical, chemical, and/or electrical properties of the structure in situ to effect a change in the radiation pattern, frequency, bandwidth, or polarization of the antenna. These changes can be effected to either optimize the antenna performance for the surrounding environment or to perform multifunction operation. Traditionally, the modern reconfigurable antenna uses RF switches such as PIN diode switches, or microelectromechanical switches (MEMS) to excite currents along different parts of the antenna structure to effect operation. Recently, switches based on tunable conductivity, permeability, and permittivity have been used. For any reconfigurable antenna, a large number of configuration options may exist to achieve a given performance. This requires efficient means for searching these potential configurations to find an optimal or acceptable solution. This again is a good application for the global optimization algorithms and computational electromagnetic techniques studied in this chapter.

11.7.2 Open-Source Computational Electromagnetics Software

The NEC, discussed in Sec. 11.5 is essentially an open-source code, and users can access the source code for version 2 and modify it and repackage without restriction. However, this is not representative of the computational EM community at large. Much of the software used today is developed commercially such as ANSYS HFSS, CST Microwave Studio, and FEKO. They are powerful software suites for solving complex and computationally intensive problems. They are generally very expensive (including perpetual maintenance and support fees) and not affordable for the average researcher, nor are they absolutely necessary for performing research in antenna and array applications. In addition, many application specific computational electromagnetics codes exist and are available on sites like MATLAB's Central File Exchange and other public domains; however, they may be difficult to adapt to other problems. A trend in the broader research community has been to collaboratively develop open-source or open-access software, which is freely available to the public. This trend has crept into the computational electromagnetics community over the past few years. Most of these codes are based on the finite-difference-time-domain (FDTD) method in computational electromagnetics. FDTD codes are useful in the analysis of microstrip antennas like those found in handheld commercial electronics such as cell phones and radios. Some examples include MEEP developed by MIT [49] and openEMS developed by Liebig [50, 51]. openEMS uses MATLAB as a flexible scripting interface for creating the geometry, performing analysis, and postprocessing the results. The openEMS wiki page has some interesting tutorials for analyzing arrays of microstrip line-fed patch antennas, which also includes the use of an equivalent circuit simulation coupled with the full-wave FDTD simulation to optimize the antenna performance.

11.8 References

1. Linden, D., "Automated design and optimization of wire antennas using genetic algorithms," Dissertation, Massachusetts Institute of Technology, 1997. [Online]. Available: http://hdl.handle.net/1721.1/10207.
2. Kennedy, J., and R. Eberhart, "Particle Swarm Optimization," *IEEE International Conference on Neural Networks*, Vol. 4, pp. 1942–1948, 1995.
3. Robinson, J., and Y. Rahmat-Sami, "Particle Swarm Optimization in Electromagnetics," *IEEE Trans. on Antennas and Propagation*, Vol. 52, No. 2, pp. 397–407, 2004.
4. Boeringer, D. W., and D. Werner, "Particle Swarm Optimization versus Genetic Algorithms for Phased Array Synthesis," *IEEE Trans. on Antennas and Propagation*, Vol. 52, No. 3, pp. 771–779, 2004.
5. Khodier, M., and C. Christodoulou, "Linear Array Synthesis with Mimimum Sidelobe Level and Null Control Using Particle Swarm Optimization," *IEEE Trans. on Antennas and Propagation*, Vol. 53, No. 8, pp. 2674–2679, 2005.
6. Dorigo, M., "Optimization, Learning and Natural Algorithms," Ph.D. Thesis, Politecnico di Milano, Italy, 1992.
7. Van Laarhoven, P., and E. Aarts, *Simulated Annealing*, Springer, The Netherlands, 1987.
8. Pham, D., A. Ghanbarzadeh, E. Koc, S. Otri, S. Rahim, and M. Zaidi, "The bees Algorithm—A Novel Tool for Complex Optimisation Problems," in *Proceedings of the 2nd Virtual Conference on Intelligent Production Machines and Systems*, 2006.
9. Yang, X.-S., "Firefly Algorithms for Multimodal Optimization," in *Stochastic Algorithms: Foundations and Applications*, Springer, Berlin Heidelberg, 2009, pp. 169–178.
10. Yang, X., and S. Deb, "Cuckoo Search via Levy Flights," *IEEE World Congress on Nature and Biological Inspired Computing (NaBIC 2009)*, pp. 210–214, 2009.
11. Yang, X., "A New Metaheuristic Bat-Inspired Algorithm," in *Nature Inspired Cooperative Strategies for Optimization (NICSO 2010)*, Springer, Berlin Hedelberg, 2010, pp. 65–74.
12. Guney, K., and S. Basbug, "Interference Suppression of Linear Antenna Arrays by Amplitude-Only Control Using a Bacterial Foraging Algorithm," *Progress in Electromagnetics Research*, No. PIER 79, pp. 475–497, 2008.
13. Haupt, R., and D. Werner, *Genetic Algorithms in Electromagnetics*, John Wiley & Sons, New York, 2007.
14. Sivanandam, S., and S. Deepa, *Introduction to Genetic Algorithms*, Springer, 2007.
15. Rubinstein, R., and D. Kroese, *The Cross Entropy Method: A Unified Approach to Combinatorial Optimization, Monte-Carlo Simulation, and Machine Learning*, Springer, New York, 2004.
16. Cover, T., and J. Thomas, *Elements of Information Theory*, Wiley, New York, 1991.
17. Kapur, J., and H. Kesavan, *Entropy Optimization Principles with Applications*, Academic Press, New York, 1992.
18. Rubinstein, R., "Optimization of Computer Simulation Models with Rare Events," *European Journal of Operational Research*, Vol. 99, pp. 89–112, 1997.
19. Rubinstein, R., "The Cross Entropy Method for Combinatorial and Continuous Optimization," *Methodology and Computing in Applied Probability*, Vol. 1, pp. 127–190, 1999.
20. Connor, J., "Antenna Array Synthesis Using the Cross Entropy Method," Ph.D. Dissertation, Florida State University, 2008.
21. Connor, J., S. Foo, and M. Weatherspoon, "Synthesizing Antenna Array Sidelobe Levels and Null Placements Using the Cross Entropy Method," in *Proceedings of the 2008 IEEE Industrial Electronics Conference*, Orlando, FL, 2008.
22. de Boer, P., "Analysis and Efficient Simulation of Queuing Models of Telecommunications Systems," Ph.D. Dissertation, University of Twente, 2000.
23. Keith, J., and D. Kroese, "Sequence Alignment by Rare Event Simulation," in *Proceedings of the 2002 Winter Simulation Conference*, San Diego, CA, 2002.

24. Chepuri, K., and T. Homem de Mello, "Solving the Vehicle Routing Problem with Stochastic Demands Using the Cross Entropy Method," in *Annals of Operations Research*, Kluwer Academic, 2004.
25. Szita, I., and A. Lorincz, "Learning Tetris Using the Cross Entropy Method," *Neural Computation*, Vol. 18, No. 12, pp. 2936–2941, 2006.
26. Chen, Y., and Y. Su, "Maximum Likelihood DOA Estimation Based on the Cross Entropy Method," *2006 IEEE Int. Symposium on Information Theory*, pp. 851–855, 2006.
27. Joost, M., and W. Schiffmann, "Speeding Up Backpropagation Algorithms by Using Cross-Entropy Combined with Pattern Normalization," *International Journal of Uncertainty, Fuzziness and Knowledge-Based Systems*, Vol. 6, No. 2, pp. 117–126, 1998.
28. Dorigo, M., M. Zlochin, N. Meuleau, and M. Birattari, "Updating ACO Pheromones Using Stochastic Gradient Ascent and Cross Entropy Methods," *Applications of Evolutionary Computing in Vol. 2279 of Lecture Notes in Computer Science*, pp. 21–30, 2002.
29. Liu, Z., A. Doucet, and S. Singh, "The Cross Entropy Method for Blind Multi-User Detection," in *Proceedings of the Int. Symp. on Information Theory (ISIT 2004)*, July 2004.
30. Zhang, Y., "Cross Entropy Optimization of Multiple-Input Multiple-Output Capacity by Transmit Antenna Selection," *IET Microwaves, Antennas and Propagation*, Vol. 1, No. 6, pp. 1131–1136, 2007.
31. Margolin, L., "On the Convergence of the Cross-Entropy Method," in *Annals of Operations Research*, Vol. 134, pp. 201–214, 2004.
32. Costa, A., O. Jones, and D. Kroese, "Convergence Properties of the Cross-Entropy Method for Discrete Optimization," *Operations Research Letters*, Vol. 35, No. 5, pp. 573–580, 2007.
33. Panduro, M., C. Brizuela, D. Covarrubias, and C. Lopez, "A Trade-off Curve Computation for Linear Antenna Arrays Using an Evolutionary Multi-Objective Approach," *Soft-computing—A Fusion of Foundations, Methodologies, and Applications*, Vol. 10, No. 2, pp. 125–131, 2006.
34. Tennant, A., M. Dawoud, and A. Anderson, "Array Pattern Nulling by Element Position Perturbations Using a Genetic Algorithm," *IEEE Electronic Letters*, Vol. 30, No. 3, pp. 174–176, 1994.
35. Murino, V., A. Trucco, and C. Regazzoni, "Synthesis of Unequally Spaced Arrays by Simulated Annealing," *IEEE Transactions on Antennas and Propagation*, Vol. 44, No. 1, pp. 119–123, 1996.
36. Rodriguez, J., L. Landesa, J. Rodriguez, F. Obelleiro, F. Ares, and A. Garcia-Pino, "Pattern Synthesis of Array Antennas with Arbitrary Elements by Simulated Annealing and Adaptive Array Theory," *Microwave and Optical Tech. Letts.*, Vol. 20, No. 1, pp. 48–50, 1999.
37. Jin, N., and Y. Rahmat-Samii, "Advances in Particle Swarm Optimization for Antenna Designs: Real-Number, Binary, Single-Objective and Multi-Objective Implementations," *IEEE Transactions on Antennas and Propagation*, Vol. 55, No. 3, pp. 556–567, 2007.
38. Lo, Y., "A Mathematical Theory of Antenna Arrays with Randomly Spaced Elements," *IEEE Transactions on Antennas and Propagation*, Vols. AP-12, pp. 257–268, 1964.
39. Steinberg, B., *Microwave Imaging with Large Antenna Arrays*, Wiley, New York, 1983.
40. Skolnik, M., "Statistically Designed Density-Tapered Arrays," *IEEE Transactions on Antennas and Propagations*, Vol. AP-12, pp. 408–417, 1964.
41. Haupt, R., "Thinned Arrays Using Genetic Algorithms," *IEEE Transactions on Antennas and Propagation*, Vol. 42, No. 7, pp. 993–999, 1994.
42. Mahanti, G., "Synthesis of Thinned Linear Antenna Arrays with Fixed Sidelobe Level Using Real-Coded Genetic Algorithms," *Progress in Electromagnetics Research*, Vol. PIER 75, pp. 319–328, 2007.
43. Meijer, C., "Simulated Annealing in the Design of Thinned Arrays Having Low Sidelobe Levels," *Proc. 1998 South African Symposium on Communications and Signal Processing*, pp. 361–366, 1998.
44. Quevedo-Teruel, O., and E. Rajo-Iglesias, "Ant Colony Optimization in Thinned Array Synthesis with Minimum Sidelobe Level," *IEEE Antennas and Propagation Letters*, Vol. 5, No. 1, pp. 349–352, 2006.

45. Shore, R., "A Proof of the Odd-Symmetry of the Phase for Minimum Weight Perturbation Phase-Only Null Synthesis," *IEEE Transactions on Antennas and Propagation*, Vol. AP-32, pp. 528–530, 1984.
46. Haupt, R., "Phase-Only Adaptive Nulling with a Genetic Algorithm," *IEEE Transactions on Antennas and Propagation*, Vol. 45, No. 6, pp. 1009–1015, 1997.
47. Burke, G., and A. Poggio, "Numerical Electromagnetics Code (NEC)—Method of Moments," *NOSC TD 116*, Jan. 1981.
48. Hornby, G., "Automated Antenna Design with Evolutionary Algorithms," in *Proceedings of 2006 A/AA Space Conference*, 2006.
49. "MEEP," [Online]. Available: http://ab-initio.mit.edu/wiki/index.php/Meep. [Accessed 27 08 2014].
50. Leibig, T., "openEMS—A Free and Open Source Equivalent Circuit (EC) FDTD Simulation Platform Supporting Cylindrical Coordinates Suitable for the Analysis of Traveling Wave MRI Applicaitons," *Int. Journal of Numerical Modeling: Electronic Networks, Devices, and Fields*, Vol. 26, No. 6, pp. 680–696, 2013.
51. Liebig, T., "openEMS," [Online]. Available: www.openems.de/. [Accessed 27 08 2014].

11.9 Problems

11.1 Optimization of the Rosenbrock function is a popular nonconvex performance test for optimization algorithms. The Rosenbrock function is defined as

$$f(x, y) = (a - x)^2 + b(y - x^2)^2$$

where the global minimum occurs at $(x, y) = (a, a^2)$, for $a = 1$ and $b = 100$. We would like to use this test function to evaluate the performance of the continuous variable genetic algorithm using roulette wheel selection and single-point crossover like the procedure in Fig. 11.4. Perform the following:

(a) Create a 2D image plot (using function `imagesc`) in MATLAB of the 3D Rosenbrock function $\log_{10}(f(x, y))$ for $a = 1$, $b = 100$, and $x \in [-1.5, 2]$, $y \in [-0.5, 3]$.

(b) For a minimum of 500 trials, use the continuous variable genetic algorithm with roulette wheel selection and single-point crossover (as in Fig. 11.5) to find the minimum of the Rosenbrock function. Plot a histogram of the final score for the best performer in the population for each of the trials. Use the following parameters for the genetic algorithm.

Parameter	Value
Population size	30
Mutation rate	0.15
Chromosomes kept	15
Chromosomes discarded	15
Number of chromosomes not mutated	1
Roulette wheel odds vector (from best performer to worst)	[0, 0.0083, 0.0250, 0.0500, 0.0833, 0.1250, 0.1750, 0.2333, 0.3000, 0.3750, 0.4583, 0.5500, 0.6500, 0.7583, 0.8750, 1.0000]
Maximum number of generations	3000
Minimum cost for termination	0

11.2 For the array thinning example in Sec. 11.3.1, we want to compare the performance versus a random search. Modify the MATLAB code of Fig. 11.14 to create a random thinned linear array drawn from a Bernoulli distribution with equal success probability ($p = 0.5$) for elements 2 through 25. Perform the following:

   (a) At each iteration, save the best performer in the population for both the CE method and the random search, and create a plot illustrating the peak sidelobe level (dB) versus iteration number for a single instance.

   (b) Repeat part (a) at least 500 times and, on separate figures, plot a histogram of the best performer's peak sidelobe level for the CE method and for random search. Plot with the same axis dimensions for ease in comparison.

11.3 In the example in Sec. 11.5.4, the optimal dipole length to achieve a minimum return loss value (S_{11}) at 900 MHz was equal to 0.465λ. We seek to use the continuous form of the CE method to validate the optimal value stated in the text. The score function for this problem is given by

$$\text{Score} = 20\log_{10}(S_{11})$$

where $Z_o = 73.2\,\Omega$ in Eq. (11.24) for computing return loss S_{11}. Use the following parameters for the CE method.

Parameter	Value
Population size	100
Smoothing coefficient, α	0.7
Sample selection parameter, ρ	0.1
Initial means of Gaussian distributions, μ	0.5λ
Initial variances of Gaussian distributions, σ^2	1
Terminate optimization when all σ^2 values fall below	1×10^{-6}
Lower bound on population sampling	0
Upper bound on population sampling	λ

   (a) Create a plot of the best and average population score in each iteration to illustrate the convergence of the optimization.

   (b) Create a plot of the dipole length for the best performer in the population and for the average length in the population for each iteration. Plot the length as a function of the wavelength λ.

11.4 For the dipole example in Fig. 11.31, it was observed that the thickness of the wire diameter shifts the resulting resonant frequency from its theoretical value. Show that this is true for a theoretical dipole resonant at 900 MHz by plotting a family of S_{11} in dB versus frequency curves from 800 to 1000 MHz for wire radii of 10^{-3}, 10^{-4}, 10^{-5}, and 10^{-6} meter.

11.5 Consider the $N = 8$ element monopole array example in Sec. 11.5.5. We wish to create a plot similar to Fig. 11.34 of the total gain in dBi versus azimuth angle over the range $[0, 180°]$ when the array is beamsteered toward $80°$. Recall from Chap. 4, that we can steer the beam toward an angle ϕ_o by introducing a progressive phase shift across the elements as

$$\delta_n = -kd(n-1)\cos\phi_o, \text{ for } n = 1 \text{ to } N,\ k = \tfrac{2\pi}{\lambda},\ d = 0.5\lambda$$

(a) What are the values δ_n in radians applied to each monopole in the array?

(b) The voltage source for the EX card in NEC is expressed in real (F1) and imaginary (F2) parts. What are the complex voltage values in volts applied to each monopole in the array expressed as $v = v_{real} + jv_{imag}$?

(c) Modify the code of Fig. 11.34 to produce the beamsteered plot in the linear axes of total gain in dBi versus azimuth angle in degrees. [Hint: You will need to change the inputs to the EX card to accommodate the new voltage values computed in (b).]

Index

A

Accelerated gradient approach (AG), 236
Active retrodirective arrays, 98–99
Adaptive arrays, 197, 198, 199
Adaptive beamforming, 198, 215–248
Adaptive nulling, 344–348
Adcock dipole antenna array, 267–272
Ampere's law, 10, 19
Analog-to-digital converters (ADC), 2–3
Angle-of-arrival (AOA), 4, 139, 163
   Bartlett estimation, 171–172
   Capon estimation, 172–174
   ESPRIT estimate, 189–193
   linear prediction of, 175–176
   maximum entropy estimate, 176–177
   Minimum-Norm estimate, 178–179
   MUSIC estimate, 179–183
   Pisarenko harmonic decomposition estimate, 177–178
   root-MUSIC estimate, 183–189
Angular spread, 140–145
Antenna:
   boresight, 45
   loop. See Loop antennas
   region, 38
   smart. See Smart antennas
   traditional fixed beam array, 1–2
Array correlation matrix, 169–171
Array covariance matrices, 205
Array elements:
   optimizing positions, 340–344
   thinning of, 335–340
Array factor (AF), 65, 67–68, 74, 77–79, 88, 89, 92, 95

Array vector, 94
Array weighting, 77–87
   beamsteering and, 86–87
   binomial, 79–81
   Blackman, 81
   Gaussian, 81–82
   Hamming, 81
   Kaiser-Bessel, 83–85
AS-145 antenna, 265
Attenuation constant, 11, 13–14
Autocorrelation, 114–115
Automatic repeat request (ARQ) codes, 153
Autoregressive estimator, 230
Autoregressive method, 175
Azimuth plane, 46

B

Bartlett AOA estimate, 171–172
Bartlett correlation, 270–272
Beam solid angle, 48
Beamsteered linear array, 73
   maximum directivity of, 76
Beamsteering:
   and circular arrays, 88–89
   and weighted arrays, 86–87
Beamwidth of smart antenna, 47
Bearing estimation, 163
Bellini-Tosi radio goniometer, 260–265
Bi-phase chipping, 245
Binomial weights, 79–81
Blackman weights, 81
Block-adaptive approach, 221
Block codes, 153

Index

Boundary(ies):
    conditions for electric field, 16–19
    conditions for magnetic field, 19–21
Broadside array, 71–72
    maximum directivity of, 75
Butler matrix, 91–93

C

Capon AOA estimate, 172–174
CART electromagnetic vector sensor antenna, 289
Cartesian basis vector, 167, 175, 203
Channel, definition of, 119
Channel coding, 153–154
Channel dispersion, 146–147
Channel equalization, 150–151
Channel impulse response, 136
Circular arrays, 87–89
Clarke flat fading model, 133
Coherence bandwidth of channel, 146–147
Column vector, 164
Computational electromagnetics, 314
Confidence error ellipse, 276–278
Conjugate gradient method (CGM), 234–238
Conjugate matched array, 96
Constant modulus algorithms (CMA), 227–234
Constant phasor current loop, 57–60
Convergence check step in genetic algorithm, 322
Convolutional codes, 153–154
Cooked wire antenna, 315, 358–363
Cost-based roulette wheel approach, 319–320
Cost function, 2, 209, 214, 216, 227–228, 230, 236
Covariance matrix, 116
Cramer-Rao lower bound for vector sensor, 304–308
Cross-entropy (CE) method, 314, 325–334
Cross-loop array, 260–265
    modern direct finding applied to, 270–272
Cross-product direction finding, 294–297

D

Damped sinusoidal signal, 295
Determinant of a matrix, 165–166
Diffraction, 124

Digital beamforming (DBF), 86–87, 197–198
Digital signal processing (DSP), 2–3, 197
Direct matrix inversion (DMI), 219–220
Direction finding:
    cross-product, 294–297
    radio, 256
    smart antenna for, 4
    super resolution, 297–300
Direction-of-arrival (DOA), 163
Directivity, 54
    of beamsteered linear array, maximum, 76
    of broadside array, maximum, 75
    of end-fire array, maximum, 76
    of N-element linear array, 73–77
    of smart antenna, 47–48
Disk of scatterers, 143–144
Dispersion function, 227
Diversity, 151–152
Doppler power density spectrum, 134–135
Doppler shift, 132
Dot product of row vector, 164
Doublets, 189
Dynamic LS-CMA algorithm, 233–234

E

E-plane pattern of smart antenna, 46
Effective aperture of antenna, 49
Eig function, 298
Eigenvalues of a matrix, 168–169
Eigenvectors of a matrix, 168–169
Electric field, boundary conditions for, 16–19
Electromagnetic fields:
    fundamentals of, 9–33
    Helmholtz wave equation, 11–12
    Maxwell's equations, 9–10
Electromagnetic vector sensors, 286–287
Elevation plane, 46
Encoding impact on genetic algorithm, 319
End-fire linear array, 72–73
    maximum directivity of, 76
Energy conservation, 16
Equal amplitude two-ray profile, 139
Erlang density, 110–111
Erlang distribution, 127
Error correction codes, 153
Error detection codes, 153

ESPRIT AOA estimate, 189–193, 200
Estimation of Signal Parameters via Rotational Invariance Techniques (ESPRIT). *See* ESPRIT AOA estimate
Evolutionary algorithms, 317–318
Excess delay, 137
Excitation card (EX), 352–353
Exponential density, 110–111
Ezmesh, 45

F

Fading signal:
  fast, 125, 126–135
  flat, 125
  frequency selective, 126, 149
  large-scale, 125
  lognormal, 125, 148
  Rayleigh, 125
  slow, 125, 147–149
  small-scale, 125
Faraday's law, 9
Fast fading, 125, 126–135
  MATLAB program for, 133–134
Field pattern, 44
Field regions of smart antenna, 37–39
Finite-difference-time-domain (FDTD) method, 364
Finite length dipole, 54–57
First arrival delay, 137
Fisher Information Matrix (FIM), 305–306
Fixed beam arrays, 91–93
Fixed weight beamforming, 201–215
Flat earth model, 120–122
Flat fading signal, 125
Forward error correcting (FEC) codes, 153
Fraunhofer region, 38–39
Frequency diversity, 152
Frequency selective fading, 126, 149
Fresnel coefficients, 24
Fresnel region, 38
Friis transmission formula, 50

G

Gauss method, 229
Gaussian density, 107–108
Gaussian profile, 139
Gaussian weights, 81–82
Gauss's law, 10, 51

Genetic algorithm, 314, 318–334
  convergence check step in, 322
  crossover in, 321
  encoding impact on, 319
  mating step in, 319–322
  trigonometric test function example, 322–325
Geolocation radio signals, 272–281
Giselle antenna, 287
Global optimization algorithms, 313–314, 315–334
Godard cost function, 227

H

H-plane pattern of smart antenna, 46
Hamming weights, 81
Helmholtz wave equation, 11–12
Hermitian transpose, 164, 167
High-frequency direction finding (HFDF), 262, 268
Howells-Applebaum algorithm, 199
Hybrid-ARQ code, 153

I

Identity matrix, 166
Improving signal quality, 149–157
Impulse response of radio channel, 136
Incident fields of planewave, 21
Indoor distribution of scatterers, 144–145
Infinitesimal dipole, 52–53
Intersymbol interference (ISI), 149
Inverse of a matrix, 167–168

K

Kaiser-Bessel weights, 83–85
Knife-edge diffraction of planewave, 31–33
Kolster, Dr. Frederick, 256

L

Laplace density, 112–113
Large-scale fading signal, 125
Least mean squares (LMS) algorithm, 216–218, 222–223
Least squares constant modulus algorithm. *See* LS-CMA
Linear antennas:
  finite length dipole of, 54–57
  infinitesimal dipole of, 52–53

Index

Linear array of antenna, 63–77
  beamsteered, 73
  beamwidth, 69–70
  broadside, 71–72
  end-fire, 72–73
  maxima, 69
  N-element array, 65–77
  nulls, 69
  two-element array, 63–65
Linear prediction AOA estimate, 175–176
Log-likelihood function, 211
Lognormal fading signal, 125, 148
Loop antennas, 57–60
  calibration, 265–267
  early direction finding with, 255–256
  fundamentals of, 256–257
  matched polarization, 258
  vertical, 257–260
Loss tangent, 11
LS-CMA, 229–234
  dynamic, 233–234
  static, 232–233

M

Magnetic field, boundary conditions, 19–21
Magnetic vector potential, 51–52
Magnitude of the error squared, 2
Mahalanobis statistics, 279–281
Mating step in genetic algorithm, 319–322
MATLAB:
  commands for three-dimensional field pattern plot, 45
  integrating with NEC2, 353–354
  plotting pseudospectrum, 172, 174, 175, 176, 177, 179, 181–183
  program for fast fading with velocity, 133–134
Matrix:
  addition, 166
  algebra, 163–169
  array correlation, 169–171
  basics, 165–169
  determinant, 165–166
  eigenvalues and eigenvectors of, 168–169
  Hermitian transpose, 167
  identity, 166
  inverse of, 167–168
  multiplication, 166
  trace of, 167
  transpose, 167

Maximum arrival angle (θ_M), 140
Maximum entropy AOA estimate, 176–177, 200
Maximum excess delay, 137
Maximum likelihood (ML) method, 200, 211–212
Maximum signal-to-interference ratio, 201–207
Maxwell's equations, 9–10
Mean arrival angle (θ_0), 140
Mean excess delay, 137
Minimum mean-square error, 207–210
Minimum-Norm AOA estimate, 178–179, 200
Minimum variance distortionless response (MVDR), 172–174, 200, 212, 298–299, 307
Minimum variance solution, 212–215
Moment, 105–107
Monopole array, 356–358
Multipath propagation mechanisms, 122–124
Multiple-in-multiple-out (MIMO) communications, 154–157
Multiple-transmit multiple-receive (MTMR) communications, 154
MUSIC AOA estimate, 179–183, 200

N

N-element linear array, 65–77
Narrowband array, traditional, 203–204
nec2+ +.exe, 353–354
Normal incidence of planewave, 21–24
Numerical Electromagnetics Code (NEC):
  integrating with MATLAB, 353–354
  overview, 314, 348
  program execution order, 350
  resources of, 348–349
  setting up NEC2 stimulation, 349–353

O

Oblique incidence of planewave, 24–25
One-sided exponential profile, 139
Open-source computational electromagnetics software, 364
Optimizing array element positions, 340–344
Outrage probability, 128

Index

P

Parallel polarization of planewave, 25–27
Parameter vector, 305
Pascal's triangle, 79–80
Passive retrodirective arrays, 96–98
Path gain factor (F), 30
Pattern multiplication, 65
Performance surface function, 209
Periodic antenna arrays, 334
Perpendicular polarization of planewave, 27–28
Phase-locked loop (PLL) systems, 199
Phased arrays, 197, 300
Pincushion beams, 91
Pisarenko harmonic decomposition (PHD) AOA estimate, 177–178
Planewave:
  incident fields of, 21
  knife-edge diffraction of, 31–33
  normal incidence of, 21–24
  oblique incidence of, 24–25
  parallel polarization of, 25–27
  perpendicular polarization of, 27–28
  propagation over flat earth, 28–30
  reflected fields of, 21
  reflection and transmission coefficients of, 21–28
  transmitted fields of, 22
Poincaré sphere representation, 290–291, 300
Polarization diversity, 152
Power angular profile, 139–142
Power delay–angular profile, 145–146
Power delay profile (PDP), 137–139
Power density, 39–42, 54
Power pattern, 44
Power spectral density, 114–115
Poynting vector, 39–40, 256, 285
Probability density function (pdf), 104–105, 107–113, 211
  Bernoulli, 333
  Rayleigh, 127
Propagation model:
  basics, 124–149
  flat earth model, 120–122
  multipath propagation mechanisms, 122–124
Pseudospectrum $P(\theta)$, 171

Q

Quadratic convergence, 237

R

Radiating near-field region, 38
Radiation intensity, 42–43, 54
Radiation pattern (RP) card, 353
Radio direction finding, 256
  in World War I, 262
Radio goniometer, 260–265
RAKE receiver, 152–153
Random variable, 103–104
Rank-based roulette wheel approach, 320–321
Rayleigh density function, 108–109
Rayleigh fading signal, 125
Rayleigh probability density function, 127
Rayleigh scattering, 124
Reactive near-field region, 38
Received signal vector, 243
Reciprocity, 49
Reconfigurable antennas and arrays, 363–364
Rectangular coordinates, propagation in, 12–14
Rectangular planar arrays, 89–90
Recursive least squares algorithm, 223–226
Reflected fields of planewave, 21
Reflection, 124
Reflection coefficient of planewave, 21–28
Refraction, 124
Retrodirective arrays, 96–99, 199
Riccati equation, 225
Rician density/distribution, 111–112, 130–131
Rician factor, 130
Ring of scatterers, 142–143
RMS angular spread (σ_θ), 140
RMS delay spread, 138–139
Root-MUSIC AOA estimate, 183–189
Roulette wheel technique, 319–321
Row vector, 164

S

Sample matrix inversion (SMI) technique, 200, 218–226
Scalloped beams, 92
Scattering, 124
Self-phased array, 199
Sherman Morrison-Woodbury (SMW) theorem, 225
Sidelobe cancellation (SLC), 93–95, 199
Signal subspace, 171

Signal-to-interference-plus-noise ratio (SINR), 303–304
Signal-to-interference ratio (SIR), 172, 201–207
Simple half-wavelength dipole antenna, 354–356
Single-input single-output (SISO) systems, 139, 154–155
Single vector sensor:
  instantaneous response of, 293
  steering vector derivation, 289–293
Slow fading signal, 125, 147–149
Small-scale fading signal, 125
Smart antennas:
  applications of, 3
  array weighting. See Array weighting
  basic nomenclature, 43–50
  beam solid angle, 48
  beamwidth of, 47
  benefits of, 3–4
  boresight of, 45
  circular arrays, 87–89
  current and future trends of, 363–364
  definition of, 197
  demand growth of, 2–3
  designer disciples, 37
  designing with NEC, 348–358
  as direction-finding techniques, 4
  directivity of, 47–48
  disciplines of, 5–6
  effective aperture of, 49
  field regions, 37–39
  fixed beam arrays of, 91–93
  gain of, 48–49
  historical development of, 199–200
  linear arrays of. See Linear array of antenna
  optimizing arrays, 334–344
  overview of, 1–2
  pattern of, 44–45
  power density of, 39–42
  principal plane patterns of, 45–47
  radiation intensity of, 42–43
  rectangular planar arrays of, 89–90
  retrodirective arrays of, 96–99
  role in MIMO communications, 4
Snell's law of reflection, 25
Space diversity, 152
Space division multiple access (SDMA), 3–5
  receiver, 240–241

Spectral estimation, 163
Specular reflection, 25
Spherical coordinates, propagation in, 14–15
Spherical spreading, 15
Spot beams, 91
Spreading sequence array weights, 238–240
Stansfield algorithm, 272–275, 278
Static LS-CMA, 232–233
Statistical average, 105–106
Steepest descent method, 217
Steering vector for an array of vector sensors, 289–294
Steradians, 48
Stochastic optimization algorithms, 316–317
Strict-sense stationary process, 113
Subspace method, 180
Super resolution direction finding, 297–300

T

Thinning antenna arrays, 335
Thinning array elements, 335–340
Time diversity, 152
Time-reversal mirror, 96
Total least-squares (TLS) criterion, 191–192
Tournament selection technique, 319
Trace of a matrix, 167
Transmission coefficient of planewave, 21–28
Transmitted fields of planewave, 22
Transverse electromagnetic wave, vector components of, 289–291
Turbo codes, 154
Two-element array, 63–65

U

Underdamped case, 218
Uniform density/distribution, 109–110

V

V-BLAST algorithm, 156–157
Vandermonde vector, 165
Vector dot product, 164
Vector Helmholtz wave equation, 11
Vector Hermitian transpose, 164
Vector potential:
  of finite length dipole, 58
  of infinitesimal dipole, 52–53

Vector sensors:
    advantages of, 288
    array signal model and steering vector, 293–294
    beamforming, 300–304
    challenges associated with, 288
    Cramer-Rao lower bound, 304–308
    direction finding, 294–300
    electromagnetic, 286–287
    introduction of, 285–288
    steering vector for an array of, 289–294
    3-dB beamwidth of, 302–304
    types of, 286–287

Vector transpose, 164
Vertical loop antennas, 257–260
Volume-to-volume communications, 154

W

Watson-Watt direction-finding algorithm, 268–270
Weighted least-square solution, 275–276
Wide-sense stationary process, 113
Widrow algorithms, 199
Wiener-Hopf equation, 209
Wiener-Khinchin pair, 115